Compendium of Plant Genomes

Series editor

Chittaranjan Kole, Raja Ramanna Fellow, Department of Atomic Energy, Government of India, Kalyani, Nadia, West Bengal, India

Whole-genome sequencing is at the cutting edge of life sciences in the new millennium. Since the first genome sequencing of the model plant Arabidopsis thaliana in 2000, whole genomes of about 70 plant species have been sequenced and genome sequences of several other plants are in the pipeline. Research publications on these genome initiatives are scattered on dedicated web sites and in journals with all too brief descriptions. The individual volumes elucidate the background history of the national and international genome initiatives; public and private partners involved; strategies and genomic resources and tools utilized; enumeration on the sequences and their assembly; repetitive sequences; gene annotation and genome duplication. In addition, synteny with other sequences, comparison of gene families and most importantly potential of the genome sequence information for gene pool characterization and genetic improvement of crop plants are described.

Interested in editing a volume on a crop or model plant? Please contact Dr. Kole, Series Editor, at ckole2012@gmail.com

More information about this series at http://www.springer.com/series/11805

Masayoshi Shigyo · Anil Khar
Mostafa Abdelrahman
Editors

The Allium Genomes

Springer

Editors
Masayoshi Shigyo
College of Agriculture, Graduate School of Sciences and Technology for Innovation
Yamaguchi University
Yamaguchi, Japan

Anil Khar
Division of Vegetable Science
ICAR-Indian Agricultural Research Institute
New Delhi, India

Mostafa Abdelrahman
Laboratory of Molecular Breeding
Arid Land Research Center
Tottori University
Tottori, Japan

ISSN 2199-4781 ISSN 2199-479X (electronic)
Compendium of Plant Genomes
ISBN 978-3-030-07103-5 ISBN 978-3-319-95825-5 (eBook)
https://doi.org/10.1007/978-3-319-95825-5

This Springer imprint is published by the registered company Springer Nature Switzerland AG
The registered company address is: Gewerbestrasse 11, 6330 Cham, Switzerland

This book series is dedicated to my wife Phullara,
and our children Sourav, and Devleena

Chittaranjan Kole

We respectfully dedicate this volume to Prof. Michael Havey and Dr. Chris Kik for their work on Allium genome construction. Without their pioneer spirit the Allium research community could not have launched DNA marker breeding and whole genome sequencing.

With respect to the photo:

From left to right: Dr. E.R.J. Keller, Dr. G. Galvan, Prof. M. Shigyo, Dr. O.E. Scholten, Prof. M. Havey, Dr. A.W. van Heusden, Dr. C. Kik.

The photo was taken in Dronten, the Netherlands on November 9 2007.

Preface to the Series

Genome sequencing has emerged as the leading discipline in the plant sciences coinciding with the start of the new century. For much of the twentieth century, plant geneticists were only successful in delineating putative chromosomal location, function, and changes in genes indirectly through the use of a number of 'markers' physically linked to them. These included visible or morphological, cytological, protein, and molecular or DNA markers. Among them, the first DNA marker, the RFLPs, introduced a revolutionary change in plant genetics and breeding in the mid-1980s, mainly because of their infinite number and thus potential to cover maximum chromosomal regions, phenotypic neutrality, absence of epistasis, and codominant nature. An array of other hybridization-based markers, PCR-based markers, and markers based on both facilitated construction of genetic linkage maps, mapping of genes controlling simply inherited traits, and even gene clusters (QTLs) controlling polygenic traits in a large number of model and crop plants. During this period, a number of new mapping populations beyond F_2 were utilized and a number of computer programs were developed for map construction, mapping of genes, and for mapping of polygenic clusters or QTLs. Molecular markers were also used in studies of evolution and phylogenetic relationship, genetic diversity, DNA-fingerprinting, and map-based cloning. Markers tightly linked to the genes were used in crop improvement employing the so-called marker-assisted selection. These strategies of molecular genetic mapping and molecular breeding made a spectacular impact during the last one and a half decades of the twentieth century. But still they remained "indirect" approaches for elucidation and utilization of plant genomes since much of the chromosomes remained unknown and the complete chemical depiction of them was yet to be unraveled.

Physical mapping of genomes was the obvious consequence that facilitated the development of the "genomic resources" including BAC and YAC libraries to develop physical maps in some plant genomes. Subsequently, integrated genetic–physical maps were also developed in many plants. This led to the concept of structural genomics. Later on, emphasis was laid on EST and transcriptome analysis to decipher the function of the active gene sequences leading to another concept defined as functional genomics. The advent of techniques of bacteriophage gene and DNA sequencing in the 1970s was extended to facilitate sequencing of these genomic resources in the last decade of the twentieth century.

As expected, sequencing of chromosomal regions would have led to too much data to store, characterize, and utilize with the-then available computer software could handle. But development of information technology made the life of biologists easier by leading to a swift and sweet marriage of biology and informatics, and a new subject was born—bioinformatics.

Thus, the evolution of the concepts, strategies, and tools of sequencing and bioinformatics reinforced the subject of genomics—structural and functional. Today, genome sequencing has traveled much beyond biology and involves biophysics, biochemistry, and bioinformatics!

Thanks to the efforts of both public and private agencies, genome sequencing strategies are evolving very fast, leading to cheaper, quicker, and automated techniques right from clone-by-clone and whole-genome shotgun approaches to a succession of second-generation sequencing methods. Development of software of different generations facilitated this genome sequencing. At the same time, newer concepts and strategies were emerging to handle sequencing of the complex genomes, particularly the polyploids.

It became a reality to chemically—and so directly—define plant genomes, popularly called whole-genome sequencing or simply genome sequencing.

The history of plant genome sequencing will always cite the sequencing of the genome of the model plant Arabidopsis thaliana in 2000 that was followed by sequencing the genome of the crop and model plant rice in 2002. Since then, the number of sequenced genomes of higher plants has been increasing exponentially, mainly due to the development of cheaper and quicker genomic techniques and, most importantly, development of collaborative platforms such as national and international consortia involving partners from public and/or private agencies.

As I write this preface for the first volume of the new series "Compendium of Plant Genomes", a net search tells me that complete or nearly complete whole-genome sequencing of 45 crop plants, eight crop and model plants, eight model plants, 15 crop progenitors and relatives, and 3 basal plants is accomplished, the majority of which are in the public domain. This means that we nowadays know many of our model and crop plants chemically, i.e., directly, and we may depict them and utilize them precisely better than ever. Genome sequencing has covered all groups of crop plants. Hence, information on the precise depiction of plant genomes and the scope of their utilization is growing rapidly every day. However, the information is scattered in research articles and review papers in journals and dedicated Web pages of the consortia and databases. There is no compilation of plant genomes and the opportunity of using the information in sequence-assisted breeding or further genomic studies. This is the underlying rationale for starting this book series, with each volume dedicated to a particular plant.

Plant genome science has emerged as an important subject in academia, and the present compendium of plant genomes will be highly useful both to students and teaching faculties. Most importantly, research scientists involved in genomics research will have access to systematic deliberations on the plant genomes of their interest. Elucidation of plant genomes is of interest not only for the geneticists and breeders, but also for practitioners of an array of plant science disciplines, such as taxonomy, evolution, cytology, physiology, pathology, entomology, nematology, crop production, biochemistry, and obviously bioinformatics. It must be mentioned that information regarding each plant genome is ever-growing. The contents of the volumes of this compendium are, therefore, focusing on the basic aspects of the genomes and their utility. They include information on the academic and/ or economic importance of the plants, description of their genomes from a molecular genetic and cytogenetic point of view, and the genomic resources developed. Detailed deliberations focus on the background history of the national and international genome initiatives, public and private partners involved, strategies and genomic resources and tools utilized, enumeration on the sequences and their assembly, repetitive sequences, gene annotation, and genome duplication. In addition, synteny with other sequences, comparison of gene families, and, most importantly, the potential of the genome sequence information for gene pool characterization through genotyping by sequencing (GBS) and genetic improvement of crop plants have been described. As expected, there is a lot of variation of these topics in the volumes based on the information available on the crop, model, or reference plants.

I must confess that as the series editor, it has been a daunting task for me to work on such a huge and broad knowledge base that spans so many diverse plant species. However, pioneering scientists with lifetime experience and expertise on the particular crops did excellent jobs editing the respective volumes. I myself have been a small science worker on plant genomes since the mid-1980s and that provided me the opportunity to personally know several stalwarts of plant genomics from all over the globe. Most, if not all, of the volume editors are my longtime friends and colleagues. It has been highly comfortable and enriching for me to work with them on this book series. To be honest, while working on this series I have been and will remain a student first, a science worker second, and a series editor last. And I must express my gratitude to the volume editors and the chapter authors for providing me the opportunity to work with them on this compendium.

I also wish to mention here my thanks and gratitude to the Springer staff, Dr. Christina Eckey and Dr. Jutta Lindenborn in particular, for all their constant and cordial support right from the inception of the idea.

I always had to set aside additional hours to edit books beside my professional and personal commitments—hours I could and should have given to my wife, Phullara, and our kids, Sourav, and Devleena. I must mention that they not only allowed me the freedom to take away those hours from them but also offered their support in the editing job itself. I am really not sure whether my dedication of this compendium to them will suffice to do justice to their sacrifices for the interest of science and the science community.

Kalyani, India Chittaranjan Kole

Preface to the Volume

The *Allium* vegetable crops including onion, garlic, leek and Japanese bunching onion are one of the earliest domesticated crops, that has been broadly cultivated and prized for its flavor, medicinal, and nutritional properties. The word "*Allium*" is derived from the Greek word aleo "ἀλόη" which means "to avoid", because of its offensive smell. References to these plants in the Bible and the Quran reflect their importance to ancient civilizations both as flavorful foods and as healing herbs. Currently, bulb onion annual production was estimated to be approximately 93.16 million tons in 2016 according to FAOSTAT (accessed 2018), which ranked onion third after tomato and watermelon in the global vegetable crops. The rigorous estimation of the economic effect of these cash crops seems to be outside the bounds of possibility due to the traditional utilization of some minorities as an indigenous plant in the remote and backward village. Thus, *Allium* species have made a significant contribution toward our dietary life.

During domestication, the genomes of many crop species have become more complex, with increasing numbers of repetitive DNA sequences. For instance, the paleontological species such as bulb onion, bunching onion, garlic, etc., the transposon would be responsible for the enormousness of genome size attributively together with a minor DNA restructuring by point mutation. Such a complicated genome reorganization is estimated to cause the speciation of *Allium*, which is the primary factor for reproductive isolation followed by the enlargement of habitat range and the creation of noble utilization forms in the species of this genus. Next-generation sequencing (NGS) technologies allow hundreds of thousands to billions of nucleotide reads, which help in understanding the genetics of complex traits and to provide insights into genomic variation, (single-nucleotide polymorphisms (SNPs), insertions/deletions (InDels) and other structure of variances (SVs)), quantitative trait locus (QTL) introgressions, and have facilitated the development of gene expression atlases and increased our understanding of the signaling pathways involved in the responses of plants to biotic and abiotic stressors. To proceed genome analysis of *A. cepa* and its functional study, *Allium* international research community have developed several types of artificially manipulated genetic stocks and applied these stocks to the latest modern analysis technology.

The present book entitled "*The Allium genomes*" aims to present a full picture of the state-of-the-art research and development of *Allium* vegetable crops, including species and genetic resources, genetics and gene mapping, breeding and cytogenetics as well as taxonomical and ethnobotanical aspects. The 14 chapters represented in this book will provide the readers with the recent advances in *Allium* crop genomics, including (1) the economic and academic importance of *Allium* species such as onion, leek, garlic and Japanese bunching onion, (2) taxonomical and ethnobotanical aspects of *Allium* species from Middle Asia with particular reference to subgenus *Allium*, which could be of interest for innovative breeding concerning pharmaceutical, ornamental, and edible traits, (3) an update of *Allium* genetic resources, (4) the classical genetics of gene mapping for perspective onion breeding, (5) the molecular cytogenetic progress in the study of the *Allium* chromosomes and its applications in onion breeding, (6) the novel knowledge of cytoplasmic genomes, (7) cytogenetic aspects of repetitive sequences, (8) molecular genetics of simple sequence repeat, (9) theoretical description of gene annotation, (10) gene family evolution in *Allium* species, (11) strategies and tools for sequencing in structural and functional genomic resources, (12) the targeted and untargeted metabolomic profiling of bioactive metabolites in *Allium* crops, (13) Progress to date and development of new population resources for molecular mapping by NGS technology, (14) the clarification of the impact on *Allium* plant breeding along with the background history of the national and international genome initiatives. This book attempted to maximize the involvement and collaboration between *Allium* breeders and researchers all over the world to enhance the transfer of laboratory-based innovations to end-user practice for the improvement of *Allium* crops and food sustainability.

We are grateful to all our colleagues for their contribution. We wish to record our thanks and appreciations for Prof. Chittaranjan Kole, the Series Editor, for his assistance and guidance right from the inception till the publication of this book.

For our international research community on *Allium* genomics, the leadership of two scientific authorities in the last three decades will never be forgotten in everyone's mind. Dr. Michael Havey has been a USDA Research Geneticist and Professor in the Department of Horticulture at the University of Wisconsin–Madison since 1988. He had spent 30 years for his research on the breeding, genetics, and genomics of the Alliums (onion and garlic). Since 1988 at Wageningen, Dr. Chris Kik had been a senior researcher for 17 years in the Plant Research International (PRI) formerly known by the Centre for Plant Breeding and Reproduction Research (CPRO-DLO). The activity of his research group was intense along with a number of the EU projects including "Garlic & Health" during that time. Dr. Kik also organized several collecting missions for plant genetic resources as a head curator for crop plants in Centre for Genetic Resources, the Netherlands since 2005. Both of them are scientifically advanced and have high standards of practical research on pre-breeding or DNA marker breeding in *Allium*. Furthermore, both of global leaders, Dr. Havey and Dr. Kik have spent a lot of time for training of young scientists from different parts of the world. In this book, most of the contributors consist of their favorite disciples in different countries. As shown in a previous page, we would like to dedicate this book to Prof. Michael Havey and Dr. Chris Kik with our great compliments.

Yamaguchi, Japan Masayoshi Shigyo
New Delhi, India Anil Khar
Tottori, Japan Mostafa Abdelrahman

Contents

Abbreviations

2-CPGTH	S-2-Carboxypropylglutathione
ABS	Access and Benefit Sharing
ACSOs	S-alk(en)yl-$_L$-Cysteine Sulfoxides
AFLP	Amplified Fragment Length Polymorphism
AGL6	AGAMOUS LIKE6
AIRDC	Association of International Research and Development Centers
ALL1	Alliinase
AMOVA	Molecular Variance
ANS	Anthocyanidin Synthase
AOX	Alternative Oxidase
AP2	APETELLA 2
ATC	*Arabidopsis thaliana* Relatives of Centroradialis
ATP	Adenosine Triphosphate
AVRDC	Asian Vegetable Research and Development Center
BAC	Bacterial Artificial Chromosome
C	Color Factor
CAN-ABS	Competent National Authority on ABS
CBD	Convention on Biodiversity
CGIAR	Consultative Group on International Agricultural Research
CgMS	Cytoplasmic-Genic Male Sterility
CGN	Centre for Genetic Resources, the Netherlands
CHI	Chalcone Isomerase
ChIP	Chromatin Immuno Precipitation
CHS	Chalcone Synthase
cM	Centi-Morgans
CMS	Cytoplasmic Male Sterility
CO	CONSTANS
DDA	Data-Dependent Analysis

DFR	Dihydroflavonol 4-Reductase
DH	Doubled-Haploid
DSC	Differential Scanning Calorimetry
EALLDB	European Allium Database
ECPGR	European Cooperative Programme for Plant Genetic Resources
ESF	European Science Foundation
EST	Expressed Sequence Tags
FAOSTAT	Food and Agriculture Organization Statistics
FISH	Fluorescence In Situ Hybridization
FKF1	Flavin-Binding, Kelch Repeat, F-Box1
FT	Flowering Locus T
GBIF	Global Biodiversity Information Facility
GBIS	Genebank Information System
GCLV	Garlic Common Latent Virus
GG-PRENCS	Gamma-Glutamyl-S-1-Propenyl-$_L$-Cysteine Sulfoxide
GG-PRENCSO	Gamma-Glutamyl-S-1-Propenyl-$_L$-Cysteine Sulfoxide
GI	GIGANTEA
GISH	Genomic In Situ Hybridization
GPAT2	Glycerol-3-Phosphate Acyltransferase 2
GRIN	Germplasm Resource Information Network
GSCC	Garlic and Shallot Core Collection
HMW	High Molecular Weight
HRP	Horse Radish Peroxidase
IBPGP	International Board of Plant Genetic Resources
IGS	Intergenic Spacer
InDel	Insertion/Deletion
IPGRI	International Plant Genetic Resources Institute
IPK	Institute für Pflanzengenetik und Kulturpflanzenforschung
IPNI	International Plant Names Index
IT-PGRFA	International Treaty on Plant Genetic Resources for Food and Agriculture
ITS	Internal Transcribed Spacer
LC-MS	Liquid Chromatography–Mass Spectrometry
LC-QqQ-MS	Liquid Chromatography–Tandem Quadrupole–Mass Spectrometry
LD	Long Day
LFS	Lachrymatory Factor Synthase
LFY	Leafy
LIMS	Laboratory Information Management Systems
LSC	Large Single-Copy

LYSV	Leek Yellow Stripe Virus
MALs	Alien Monosomic Addition Lines
MAT	Mutually Agreed Terms
MRR	Mitochondrial Retrograde Regulation
MS	Mass Spectrometry
MS/MS	Tandem Mass Spectrometry
MT	Mitochondrial
MTA	Material Transfer Agreement
N	Normal
NEP	Nucleus-Encoded RNA Polymerase
NGS	Next-Generation Sequencing
N-msms	Maintainer Line
NOR	Nucleolus Organizer Regions
OYDV	Onion Yellow Dwarf Virus
PCR	Polymerase Chain Reaction
PEBP	Phosphatidylethanolamine Binding Domain Proteins
PEP	Plastid-Encoded RNA Polymerase
PGR	Plant Genetic Resources
PI	PISTILLATA
PIC	Prior Informed Consent
PPR	Pentatricopeptide Repeat
PRENCSO	*trans*-S-1-Propenyl-L-Cysteine Sulfoxide
QTLs	Quantitative Trait Loci
QTOF-MS	Liquid Chromatography–Quadrupole Time of Flight–Mass Spectrometry
RAPD	Random Amplified Polymorphic DNA
Rf	Fertility Restoration
RFLP	Restriction Fragment Length Polymorphism
ROS	Reactive Oxygen Species
S	Sterile
SD	Short Day
SDR	Second Division Restitution
SGR	Structural Genomic Resources
SiR	Sulphite Reductase
SLV	Shallot Latent Virus
SNP	Single Nucleotide Polymorphisms
SPE3	SEPALLATA3
SRAP	Sequence-Related Amplified Polymorphism
SRM	Selected Reaction Monitoring
SSC	Small Single-Copy

SSR	Simple Sequence Repeat
TCA	Tricarboxylic Acid
TFL1	Terminal Flower Like 1
TSA	Transcriptome Shotgun Assembly
WIEWS	World Information and Early Warning System

Economic and Academic Importance

1

Claudio R. Galmarini

Abstract

Allium species have been cultivated for thousands of years for its therapeutic properties, religious significance, taste and aroma. Most of the cultivated crops evolved from wild relatives that grow in central Asia. Nowadays, *Alliums* are major vegetable crops cultivated worldwide. The most important crops of this family are onion, garlic, shallot, leek and Japanese bunching onions. Worldwide the consumption of these vegetable crops is increasing, one of the reasons is consumer awareness of the potential of these vegetables to enhance health, to improve welfare and reduce the risk of diseases. In this chapter, the economic importance of the principal edible *Alliums* is described, information is given about the academic contribution to the knowledge of *Allium* crops in recent years, and also future challenges are mentioned, especially those related to plant breeding, genetics and genomics, that will contribute to increase yields, quality and to have a more sustainable production.

1.1 *Allium* Economic Importance

Allium vegetables are the fourth most abundant group of commercially produced nonleguminous vegetables after potatoes, cassava and tomatoes according to FAO (2017) statistics. Onion, garlic, Japanese bunching onion, shallot and leek are the most crucial edible *Allium* crops. Nevertheless, over 20 other *Allium* species have been consumed by humans (van der Meer 1997). There are also *Allium* species used for ornamental purposes, such as *Allium aflatunense*. Seed production of *Allium* crops is also an important economic activity. *Allium* vegetables gross production value for 2014 was US$61,348 million (FAO 2017); about 70% of this value is for dry onion bulbs, 25% for garlic, 4% for green onions and shallots and 1% for leeks. The production of *Allium* crops, especially garlic and onion shows an important increase since 2000, in great part, this fact is explained by an increase in *per capita* consumption, one of the reasons may be that consumers are aware of the potential of these vegetables to enhance health and prevent chronic diseases.

C. R. Galmarini (✉)
Instituto Nacional de Tecnología Agropecuaria (INTA), Estación Experimental Agropecuaria La Consulta, CC8 (5567), La Consulta, Mendoza, Argentina
e-mail: galmarini.claudio@inta.gob.ar

C. R. Galmarini
Consejo Nacional de Investigaciones Científicas y Técnicas (CONICET), Buenos Aires, Argentina

C. R. Galmarini
Facultad de Ciencias Agrarias, Universidad Nacional de Cuyo, Almirante Brown 500, M5528 AHB Chacras de Coria, Luján, Mendoza, Argentina

M. Shigyo et al. (eds.), *The Allium Genomes*, Compendium of Plant Genomes,
https://doi.org/10.1007/978-3-319-95825-5_1

1.2 Onion (*Allium cepa* L.)

Onion is the most important *Allium* vegetable crop. Nearly 88 million tons per annum are worldwide produced (Table 1.1). Bulb onions are produced in different latitudes, from subarctic regions to regions close to the equator, although they are best adapted to production in temperate and subtropical areas (Brewster 2008). Onion gross production value for 2014 was US $42,743 million (FAO 2017). Due to the storage and transport facility of the crop, there is a significant international trade; about 20 million tons are traded annually (FAO 2017). Due to this global trade, and because of the many techniques for growing and storing onions for sale year-round, bulb onions are available throughout the year in most countries. The use of onion F_1 hybrids is increasing worldwide; this tendency reaches different kinds of onion germplasm, such as Granex, long-day storage and Spanish onions.

The leading producers are China (20,507,759 tons), followed by India (around 13,000,000 tons), United States of America, Egypt, Iran, Turkey, Pakistan, Brazil, Russia and Republic of Korea (Data for 2014, FAO 2017).

The leading exporters include The Netherlands, China, Mexico, India, USA, Egypt, Spain, New Zealand and Argentina. The main importers are USA, United Kingdom, Malaysia, Germany, Saudi Arabia, Japan, Canada, Czech Republic, Republic of Korea and Brazil. As an example, The Netherlands produces large quantities of long-storing, spring-sown, pungent onions which are exported between September and April, mainly to Germany and UK. In a Mediterranean country, like Spain, crops are planted in autumn and harvested during spring and summer. In large countries, like the USA, the market is supplied year-round with onions from different regions within the country and from bulbs that come from Mexico or countries from the Southern Hemisphere. The main producing areas are Idaho-eastern Oregon, California, Washington, Georgia and Texas. Countries of the Southern Hemisphere like Australia, New Zealand, Argentina and Chile have an essential export market in Europe, especially Germany and UK, and also their production is sold in Asian countries, like Japan and Malaysia, and even in USA and Canada. In the case of South America, Argentina is a vital onion exporter to Brazil (Galmarini 2001).

Onion prices tend to fluctuate widely from year to year making onion production a risky business. Since there is a global market in onion bulbs, there is little that producers in any one region can do to control the market and stabilize prices. One strategy is to add value to the production or produce particular products. An example of the latter is the sweet onion market. The production of this type of onions is very well developed in the United States, especially in Georgia and Texas. Moreover, in countries like Peru and Chile, sweet onions are produced to export to the United States market from January to May.

About 1% of the world onion cropped area is produced for the dehydration industry; nevertheless, this is an essential activity for

Table 1.1 World onion cultivated area, production, yield and consumption from 1970 to 2014

Year	Area (ha)	Production (tons)	Yield (kg/ha)	Consumption (kg/capita/year)
1970	1,334,693	16,748,643	12,549	4.15
1980	1,616,802	22,400,015	13,854	4.63
1990	1,885,700	30,609,791	16,228	5.29
2000	2,838,022	49,966,784	17,606	7.46
2010	4,174,459	78,984,889	18,921	10.53
2014	5,298,873	88,475,089	16,679	10.97

Source FAOSTAT (2017)

diversification. The production is concentrated in China, USA, India and Mexico in the Northern Hemisphere and in Argentina in the Southern Hemisphere. The industry uses cultivars that have high solids and pungent white bulbs.

Onion dry bulb production has increased through the years, mainly due to an increase in yields and the cropped area. The world average yield is about 16 t ha^{-1}. Nevertheless, there are growers that harvest more than 100 tons per hectare, due to good growing conditions, irrigation and excellent agronomic practices.

Since 1970, onion consumption has increased almost three times (Table 1.1). The average consumption is around 10 kg/capita/year, nevertheless in countries like Greece, Albania, Iran, Kuwait, Uzbekistan, and yearly onion consumption per capita is over 20 kg (FAO 2017). One reason for consumption increase may be the health-benefits that this vegetable brings to humans. The beneficial health-effects attributed to onions are essential for consumers and breeders (Galmarini 2010). Onion consumption has been associated with decreased cardiovascular events, because of their hypocholesterolemic, hypolipidemic, anti-hypertensive, antidiabetic, antithrombotic and anti-hyperhomocysteinemia effects, as well as with many other biological activities including antimicrobial, antioxidant, anticarcinogenic, antimutagenic, antiasthmatic, immunomodulatory and prebiotic activity (Corzo-Martinez et al. 2007). Garlic and onion have phytochemicals categorized as functional foods; some of them are fructans, flavonoids and organo-sulphur compounds (Galmarini et al. 2001).

Onion seed market is also important, around 50,000 tons are yearly produced (FAO 2017). Onion represents an important percentage of the total vegetable seed economic value commercialized in the world. Onion seed production is performed both in the Northern and Southern hemispheres. The main producing countries are the USA, The Netherlands, Japan, Turkey, China, Spain, Italy, France, Australia, Chile, Argentina and South Africa.

1.3 Garlic (*Allium sativum* L.)

Garlic is the second most widely cultivated *Allium* with a current worldwide production of 24.9 million tons per annum, cultivated in 1.5 million hectares (FAO 2017). Garlic gross production value for 2014 was US$15,129 million (FAO 2017). Since 1970, world garlic production has increased more than 10 times, while cultivated area increased around four times, indicating an increment in yield (Table 1.2). In addition to fresh consumption, the production of dried and processed garlic products for use in the food industry and as dietary health food supplements is an important activity.

Garlic is mainly grown in temperate areas. The garlic bulb is well adapted to storage and transportation, and there is an important international trade. The main producing and exporting country is China, which produces around 80% of the total world production. Garlic output in China exceeded the figures recorded by the world's second-largest producer, India, more than tenfold. Other important producers are South Korea, Egypt, Russia, Myanmar, Spain, USA, Uzbekistan, Argentina and Brazil. The leading exporters are China, Spain, and Argentina, while the leading importers are Indonesia, Brazil, Vietnam, Syria, USA, Pakistan, Russia, Germany, Italy and France (FAO 2017). China dominates the fresh and dehydrated garlic international market, being the leading exporter.

According to FAO (2017), the world average yield is about 16 t ha^{-1} of dry garlic; although there are countries with an average yield over 30 t ha^{-1}, due to good growing conditions, irrigation and excellent agronomic practices. Although garlic production history shows a steady increase during the past 45 years, the most dramatic increase in total production and yield has been observed in the past 15 years; the increase in yields of garlic produced in China is in great part responsible of this change.

Regarding consumption, in some countries like the Republic of Korea 10 kg/capita/year

Table 1.2 World garlic cultivated area, production and yield from 1970 to 2014

Year	Area (ha)	Production (tons)	Yield (kg/ha)
1970	416,385	2,854,876	6856
1980	619,410	4,251,797	6864
1990	811,519	6,463,996	7965
2000	1,082,515	11,086,692	10,241
2010	1,336,544	22,557,355	16,877
2014	1,547,381	24,939,965	16,117

Source FAOSTAT (2017)

have been reported. In the USA, the consumption is around one kg/capita/year. Garlic consumption rose from 1.6 kg/capita/year to almost 3 kg/capita/year, in the last decades (Cavagnaro and Galmarini 2007); one of the reasons, as was mentioned for onion, is the consumer awareness of health benefits associated with garlic intake (Gonzalez et al. 2009). Different types of garlic, hard neck, soft neck, Asian, Mediterranean or Russian types are grown and consumed around the world (Burba 2013a, b). Growers and consumers' preferences vary with geographic region, cultural background and end-use. Although fresh intact bulbs are the usual form of garlic traded in commerce, fresh garlic preparations (e.g. chopped, sliced or minced), as well as dehydrated forms of condiments, are common in several countries. In parts of Asia and North Africa, garlic leaves are marketed. Garlic preparations for the nutraceutical industry are also necessary (Cavagnaro and Galmarini 2007).

1.4 Shallot (*Allium cepa* L. Aggregatum Group)

Shallots are of much less economic importance than onion and garlic. Shallot gross production value for 2014 was US$2505 million (FAO 2017). The worldwide production is about 4 million tons in about 200 thousand hectares (Table 1.3). This production can be overestimated since FAO statistics add to shallots other *Alliums*. Shallots are mainly produced by small-scale growers. There are noticeable differences between shallots in many countries. Some differences are linked to traditions or local customs. In Asia, shallots are predominantly small and round with a deep red colour, while in France, Europe's major shallot producer prefers shallots that are more elongated and reddish brown in colour. In the Netherlands and Belgium, yellow-skinned shallots used to be very common, but they are almost only grown by hobby gardeners and on allotments (Brewster 2008).

The major production regions are France, The Netherlands, the United States, New Zealand and Great Britain for the so-called "long day varieties". These varieties grow slower and firmer, and therefore can be stored for longer. They are available all year-round.

Other important production regions can be found in South East Asia and Africa. China, Taiwan, Korea, Indonesia, Thailand, Turkey, Tunisia and Nigeria are important shallot producers. In America, the production is important in Mexico and Ecuador. These are tropical cultivars that grow with fewer hours of daylight and humid environments, and also have pest and disease resistance to grow in those environments (Brewster 2008).

1.5 Leek (*Allium ampeloprasum* L.)

This vegetable, unlike an onion, is indifferent to day-length, and the same genotype can be grown and produce economic yield over a wide range of latitudes (De Clercq and Van Bockstaele 2002). Leeks are well adapted to cool conditions and are harvested through the winter in Western Europe. Leek gross production value for 2014 was US$971 million (FAO 2017). Worldwide

Table 1.3 World shallots cultivated area, production and yield from 1970 to 2014

Year	Area (ha)	Production (tons)	Yield (kg/ha)
1970	103,783	1,329,878	12,814
1980	125,642	1,924,840	15,320
1990	162,870	2,509,379	15,407
2000	193,808	3,359,098	17,332
2010	230,052	4,129,546	17,950
2014	219,367	4,165,600	18,989

Source FAOSTAT (2017)

Table 1.4 World leek and other *Alliums* cultivated area, production and yield from 1970 to 2014

Year	Area (ha)	Production (tons)	Yield (kg/ha)
1970	61,205	835,974	13,658
1980	57,425	741,939	12,920
1990	81,708	1,499,235	18,349
2000	99,748	1,625,155	16,293
2010	130,202	2,122,492	16,302
2014	133,433	2,236,771	16,763

Source FAOSTAT (2017)

production is around 2 million tons in about 130,000 ha (Table 1.4); nevertheless, this production can be overestimated because FAO statistics add to leek other alliums.

The main producers are Indonesia, Turkey, Belgium, France, China, Republic of Korea, Poland and Germany. It is an important crop in Europe were about 30,000 ha are grown. France is an important European Community (EC) producer with nearly 9 thousand ha yielding about 200,000 tons. The average yield is 16.7 t ha^{-1} (Table 1.4). The monetary value of the crop is quite high due to the high price per unit weight.

1.6 Japanese Bunching Onion (*Allium fistulosum* L.)

The production of this vegetable is important in East Asia. The annual production of Japanese bunching onion in Japan is about 500,000 tons in a cultivated area of 23,000 ha, in South Korea, there is around 27,000 ha that produced around 723,000 tons, and in China, 545,000 ha produce 20,754,000 tons (Park 2012). The main production areas are distributed in the southern part of Japan, South Korea and China. In South America, Colombia has a considerable production of Japanese bunching onion for domestic consumption.

Data on annual consumption per person shows a marked difference in the consumption of Japanese bunching onion; in Korea 6.6 kg/capita/year are consumed, in Japan 1.7 kg/capita/year and in China 5.1 kg/capita/year (Park 2012).

1.7 Other Minor Edible Alliums

Rakkyo (*Allium chinense* G. Don) and Chinese chives (*Allium tuberosum* Rottl.) are also crops of commercial importance in East Asia. Annual Japanese production of rakkyo is about 30,000 tons; a high proportion of this production is used for pickles. The annual Chinese chive production is about 66,000 tons (Brewster 2008).

Chives (*Allium schoenoprasum* L.) are widely grown by small farmers for use as a flavouring herb. The total world area of commercial production is about 1000 ha with large areas in Denmark, New Zealand and Germany (Brewster 2008).

1.8 *Allium* Academics

Since the publication of the book Onions and their Allies in 1963 by Jones and Mann (1963), there has been great advances in scientific and agronomical knowledge of *Alliums*. The book of Jones and Mann contributed to agronomical, physiological and breeding of Alliums. In 1985, Fenwick and Hanley (1985), published a review focused on food science aspects and uses of *Allium* crops. Five years later Rabinowitch and Brewster (1990a, b, c) edited a three-volume book, Onions and Allied Crops. This book provided a very good review of aspects such as: genetic resources, anatomy, pollination biology, sulfur biochemistry, medicinal values and physiology, among others, of Allium crops. A great contribution to summarize *Allium* knowledge was James Brewster′s book Onions and other Vegetable Alliums published in 1994 and updated in 2008. Dr. Brewster, passed away in 2015, the *Allium* community should recognize his contribution to *Allium* science, not only for his book, which is a reference book for students and agronomist all over the world and has been translated to several languages, but also for his studies about onion physiology. Although Dr. Brewster is usually thought of like an onion physiologist, his passion for knowledge extended into the genetics and breeding of the onion crops. His book relates the production and utilization of *Allium* crops to many aspects of plant science underpinning their production and storage technologies. It covers species and crop types, plant structure, genetics and breeding, physiology of growth and development as well as pests and diseases, production agronomy, storage after harvest and the biochemistry of flavour, storage carbohydrates and colour and how this relates to nutritional and health benefits. It provides many examples where scientific knowledge helps to explain and improve agronomic practice. This book is still the standard reference work for many people. In 2002, the book *Allium* Crop Science: Recent Advances, edited by Rabinowitch and Currah (2002) made a great contribution to review the advances occurred since the publication of Onions and Allied Crops in 1990. Aspects such as genome organization, exploitation of wild and cultivated relatives for breeding purposes, fertility and seed production of garlic; genetic transformation of onions; molecular markers in *Alliums*; detection of garlic viruses and the propagation of virus-free crops; bacterial and fungal diseases of the Alliums; strategies of integrated pest management, among other aspects were covered.

Another important contribution was the review regarding genome mapping and molecular breeding made by Cavagnaro and Galmarini for garlic and McCallum for onion in 2007 in the book Genome Mapping and Molecular Breeding in Plants edited by Kole.

For garlic research advances, there is a comprehensive review in Spanish, edited by Burba (2013a, b) in a five-volume book where aspects of garlic crop situation, cultivars and seed production, agronomy and post harvest are summarized.

Important contributions to divulge academic and agronomic advances of *Allium* Science are the Symposia organized by Edible Alliaceae Working Group, established in 1994 by the International Society for Horticultural Sciences (ISHS). The Proceedings of these Symposia, published in Acta Horticulturae, provide a good coverage not only of scientific and agronomic advances but also economic and marketing information about edible *Allium* crops (Amstrong 2001; Burba and Galmarini 1997; Gokce 2016; Guangshu 2005; Wako and Shigyo, 2012). The first Symposium was hosted in Mendoza (Argentina) in 1994, and then were hosted at Adelaide (Australia) in 1997, Athens (United States) in 2000, Beijing (China) 2004, Dronten (The Netherlands) in 2007, Fukuoka (Japan) in 2012, and the last one took place at Nigde (Turkey) in 2015. The Symposia have allowed opportunities for international scientists, growers, seed companies, among others, to meet and discuss the present and future developments of Edible *Allium* crops. Important items like Molecular Breeding and Genomics, Growth Physiology, Cultivation Techniques, Storage and Processing, Pest and Diseases, Secondary Metabolites and Phytochemicals and Genetic Resources are discussed in these Symposia.

A brief analysis of the contributions made in the last four Symposia indicates that 55% deal with onion, 30% with garlic, 3% are related to leek and 12% to other *Alliums*. These results suggested that there are few groups working with leek, shallot and other minor *Allium* crops and that most of the research is concentrated in onion and garlic. A disciplinary analysis indicates that 24% of the contributions are related to genetic and plant breeding, 22% to plant physiology, 21% to agronomy aspects of *Allium* crops, 15% to plant protection, 12% to genetic resources and 6% of the contributions are related to crop situation and economical value. This disciplinary analysis indicates a relatively balanced situation among the different approaches.

In recent years, there has been an outstanding contribution to *Allium* knowledge. Perhaps the field where more advances has been produced is Genetics, Genomics and Breeding. Important research groups like Dr. Mike Havey's group at the University of Wisconsin-Madison, Dr. Chris Kik's group at Wageningen, Dr. John McCallum in New Zealand, Dr. Masayoshi Shigyo at Yamaguchi University in Japan, Dr. Ludmila Khrustaleva, from the University of Moscow, Dr. Colin Eady in New Zealand, Dr. Borut Bohanec in Slovenia, among others, have made relevant contributions.

In onion physiology it was already mentioned the contribution of Dr. J. Brewster. Dr. Brian Thomas group at the University of Warwick (UK) has contributed to understand the physiology and genetic control bulbing in *Allium* species in response to day-length. Regarding garlic physiology, genetics and agronomy outstanding contributions were made by Dr. C. Messiaen from France, Dr. José Luis Burba, from Argentina, Dr. Takeomi Etoh from Japan, Dr. P. Simon from USA, and Dr. Rina Kamenetsky from the Volcani Center of Israel, among others.

Dr. Olga Scholten from Wageningen has contributed to study genetic aspects of *Allium* diseases. Also, the work of Dr. Astley in the UK and Dr. Keller in Germany has been very important to the conservation and characterization of *Allium* genetic resources.

All the recent advances have contributed to an increase not only the knowledge, but also yields and to have a more sustainable production.

1.9 Lessons Learnt and Challenges for the Future

Although recent advances, especially in plant breeding and genetics, have contributed to increase yields and also to have a more sustainable production; the demand for high-quality *Allium* vegetables will continue to increase in the future. Vegetable production will be done using less agrochemical products, so a great effort should be made to introduce resistance to pests and diseases into *Allium* crops, especially important are resistance against soil-borne pathogens and nematodes. Water will be used less wastefully, not only its quantity but also its quality (e.g. salinity) will be important; so the introduction of resistances to abiotic stress should be a priority. The production of *Allium* cultivars with better health-benefits will also be required by consumers. Perhaps, higher garlic productivity and quality will come from work on fertile garlics (Etoh and Simon 2002; Simon and Jenderek 2003). In addition, the market for oriental vegetables, such as Chinese chives, blanched Japanese bunching onions is likely to increase. More efforts should be made to study the so-called "minor edible *Allium* crops".

Breeding strategies will need a multidisciplinary approach to solve *Allium* production problems in the coming years. A team where breeders work together with agronomists, geneticists, molecular biologists, physicians, plant pathologists, biochemists, nutritionists and food science experts will be required. This strategy will contribute to increasing competitiveness of *Allium* agro-industrial chain, to diversify varietal offer to access new markets, increase consumption and also to adopt production strategies that allow better yields, quality and sustainability. In the near future, genetics will have a very important role in the economic value of *Allium* crops incorporating biotechnology tools to plant breeding, such as: haploidy using gynogenesis, molecular markers and *Allium* transformation (Eady 2001). The generated knowledge will help for example to establish integrate control systems toward the management of soil-borne diseases. This teamwork will require a very good communication with growers and industry.

A big challenge is to incorporate in a practical way basic genetic and genomic knowledge to breeding programs. *Allium* species are notable for their very large genomes, typically in the range 10–20 Gbp (Ricroch et al. 2005), which have complicated genomic studies and precluded genome sequencing to date. Genetic map development in onion and other *Allium* has been limited by difficulty in developing, maintaining and exchanging genetic stocks, high degrees of heterozygosity and a dearth of sequence data (McCallum 2007).

Comparative genomic approaches have been widely used and proven in crop genetics, and are of growing interest as improved sequencing technologies enable ever broader and more detailed surveys of germplasm. Online databases integrating genetic map, marker, sequence and germplasm are now key tools for exploiting such data. Due to *Allium* economic significance, there is a clear and pressing need for such resources.

Despite the rapid advances in sequencing technologies, the enormous size of *Allium* nuclear genomes will preclude full sequencing in the short term. As large-scale DNA sequencing technologies become more efficient and less costly, the genomic DNAs of more and more plants are being sequenced, assembled and annotated. These complete sequences are extremely valuable for the identification of specific genes associated with important phenotypes. In the case of onion, I certainly agree with Havey (2016), who proposed an efficient strategy for an international effort to sequence the onion nuclear DNA and provide the sequences and annotations on a freely accessible website. Deep transcriptome sequencing is already underway. Reduced representation approaches can be used to select against repetitive DNAs and enrich for more unique regions of the onion genome. These genomic reads can be aligned against expressed sequences of the transcriptome to identify promoter and intronic regions. Dr. Havey suggests that the international community should select and focus on one doubled haploid line as the common reference material for collaborative sequencing efforts. A publically accessible website should provide unrestricted access to the sequence and annotations. These resources will enable translational genomics of onion by providing researchers with tools to more efficiently select for important traits in onion improvement.

References

Amstrong J (ed) (2001) Proceedings of the second international symposium on edible *Alliaceae*, Adelaide, South Australia, 10–13 November 1997. Acta Horticulturae 555, 304 p

Brewster JL (2008) Onions and other vegetable Alliums, 2nd edn. CAB International, Wallingford, UK, p 454

Burba JL (2013a) Grupos Ecofisiológicos (GE), de ajos en Argentina y su equivalencia internacional. In 100 temas sobre producción de ajo. Mejoramiento genético y producción de semillas de ajo. Ediciones INTA, La Consulta, Mendoza, Argentina, vol 2, pp 8–16

Burba JL (ed) (2013b) 100 temas sobre producción de ajo. Ediciones INTA, La Consulta, Mendoza, Argentina, vol 1–5, 500 p

Burba JL, Galmarini CR (eds) (1997) Proceedings of the first international symposium on edible Alliaceae, Mendoza, South Argentina, 14–18 March 1994. Acta Horticulturae 433, 652 p

Cavagnaro PF, Galmarini CR (2007) Garlic. In: Kole, C (ed) Genome mapping and molecular breeding in plants, vegetables, vol 5. Springer, Berlin, 375, pp 349–365

Corzo-Martinez M, Corzo N, Villamiel M (2007) Biological properties of onions and garlic. Trends Food Sci Technol 18:609–625

De Clercq H, Van Bockstaele E (2002) Leek: advances in agronomy and breeding. In: Rabinowitch HD, Currah L (eds) *Allium* crop science: recent advances. CABI Publishing, New York, USA, pp 431–458

Eady CC (2001) *Allium* transformation. In: Learmonth R, Khachatourians GG (eds) The handbook of transgenic plants. Marcel Dekker, New York, pp 655–671

Etoh T, Simon P (2002) Diversity, fertility and seed production of garlic. In: Rabinowitch HD, Currah L (eds) *Allium* crop science: recent advances. CABI Publishing, New York, USA, pp 101–107

FAO. 2017. FAOSTAT. http://www.fao.org/faostat/en/#data

Fenwick GR, Hanley AB (1985) The genus *Allium*, part 1. CRC Crit Rev Food Sci Nutr 22:199–271

Galmarini CR (2001) Onion Crop Situation in the Mercosur. Acta Hort 555:269–274

Galmarini CR, Havey MJ, Goldman IL (2001) Genetic analyses of correlated solids, flavor, and health related traits in onion (*Allium cepa* L.). Mol Genet Genomics 265:543–551

Galmarini CR (2010) Mejoramiento de hortalizas para incrementar sus propiedades funcionales. Horticultura Brasileira 28(2):S32–S38

Gokce AF (ed) (2016) Proceedings of the VII international symposium on edible Alliaceae, Nidge, Turkey, May 2015. Acta Horticulturae 1143, 352 p

Gonzalez R, Soto V, Sance M, Camargo A, Galmarini CR (2009) Variability of solids, organosulfur compounds, pungency and health-enhancing traits in garlic (*Allium sativum* L.) cultivars belonging to different ecophysiological groups. J Agric Food Chem 57:10282–10288

Guangshu L (ed) (2005) Proceedings of the IV international symposium on edible Alliaceae, Beijing, China, April 2004. Acta Horticulturae 688, 370 p

Havey MJ (2016) Lessons from 25 years of genetic mapping in onion, where next? Acta Hort 1143:1–6

Jones HA, Mann LK (1963) Onions and their allies. InterScience, New York, p 286

McCallum J (2007) Onion. In: Kole C (ed) Genome mapping and molecular breeding in plants, vegetables, vol 5. Springer, Berlin, 375 pp 331–347

Park H (2012) Food system of *Allium* vegetables in east Asia. Acta Hort 969:23–40

Rabinowitch HD, Brewster JL (eds) (1990a) Onions and allied crops, I: botany, physiology and genetics. CRC Press, Boca Raton, Florida, p 273

Rabinowitch HD, Brewster JL (eds) (1990b) Onions and allied crops, II: agronomy, biotic interactions, pathology and crop protection. CRC Press, Boca Raton, Florida, p 320

Rabinowitch HD, Brewster JL (eds) (1990c) Onions and allied crops, III: biochemistry, food science and minor crops. CRC Press, Boca Raton, Florida, p 265

Rabinowitch HD, Currah L (eds) (2002) *Allium* crop science: recent advances. CABI Publishing, New York, USA, p 515

Ricroch A, Yockteng R, Brown S, Nadot S (2005) Evolution of genome size across some cultivated *Allium* species. Genome 48:511–520

Simon PW, Jenderek MM (2003) Flowering seed production and the genesis of garlic breeding. In: Janick J (ed) Plant breeding reviews, vol 23. Wiley, New York, pp 211–244

Van Der Meer QP (1997) Old and new crops within edible *Alliums*. Acta Hort 433:17–31

Wako T, Shigyo M (eds) (2012) Proceedings of the VI international symposium on edible Alliaceae, Fukuoka, Japan, May 2012. Acta Horticulturae 969, 327 p

Taxonomical and Ethnobotanical Aspects of *Allium* Species from Middle Asia with Particular Reference to Subgenus *Allium*

2

Furkat O. Khassanov

Abstract

The genus *Allium* L. is a complicated polymorphous group of vascular plants of which a number of species are used since ancient times as vegetable crops. The number of wild *Allium* species growing naturally in the Old and New World is increasing every year due to new discoveries and is currently estimated around 1000 species. The region encompassing the Mediterranean, Middle Asia, Iran and Afghanistan is the main centre of diversity of the largest subgenus of the genus *Allium*, namely *Allium* with circa 400 species and subspecies. Nearly all ancestors of garlic and leek originate from this region and some of their closest relatives are still present. Furthermore, there are numerous wild *Allium* species present in this region, which could be of interest for innovative breeding concerning pharmaceutical, ornamental and edible traits. A revised and updated list of the subgenus is presented in this paper.

F. O. Khassanov (✉)
Institute of Botany, Academy of Sciences of Uzbekistan, Tashkent, Uzbekistan
e-mail: fkhasanov1@mail.ru

2.1 Introduction

Allium L. (*Amaryllidaceae* J. St.-Hil.) is the largest genus of the petaloid monocotyledons, excluding orchids and is one of the most studied groups of vascular plants because of its economic importance. Every year a dozen new species (mainly from the Eastern Mediterranean and Middle Asian regions) are described and the current total number of wild species described exceeds 1000 (with about 50 subspecies). One of the first descriptions of *Allium* species was given by Halleir (1745) as "Moly foliis caulinis lanceolatis floribus umbellatis, ex ala bulbiferum". A few years later Linneus (1753) described plant species (including *Allium*) via binomial names (*Allium sativum, Allium schoenoprasum, Allium angulosum, Allium nutans, Allium senescens* and *Allium porrum*). Afterwards, several taxonomic classifications of the genus were proposed, starting from Don (1832), who divided *Allium* species into seven groups without taxonomical distinctive generative characters. Then, Koch (1837) authorized a classification in sections, and he distinguished the following sections: *Anguinum* G. Don f. ex Koch, *Rhizirideum* G. Don f. ex Koch, *Molium* G. Don f. ex Koch, *Porrum* G. Don f. ex Koch, *Schoenoprasum* G. Don f. ex Koch and *Codonoprasum* G. Don f. ex Koch. Fifty years

M. Shigyo et al. (eds.), *The Allium Genomes*, Compendium of Plant Genomes,
https://doi.org/10.1007/978-3-319-95825-5_2

later Regel (1875) wrote the first comprehensive monograph of the genus in which more than new 100 species were described. Vvedensky (1935), published a revision of the genus *Allium* occurring in the Old World in the Flora of the Union of Soviet Socialist Republics (USSR) and in 1971 he described 17 new species in the Conspectus Florae Asiae Mediae (Vvedensky and Kovalevskaya 1971). A number of European and American botanists also made important contributions to *Allium* taxonomy (Hermann 1939; Stearn 1944; Traub 1968, 1972; Wendelbo 1969, 1971). It was Kamelin (1973) who presented the first subdivision of the genus into 6 subgenera and 20 sections. His classification was based on the idea that primitive *Allium* species had 12 stamens. According to him, these 12 stamens evolved into two circles of stamens. The outer (=lateral) circle of three separate stamens remained the same whereas the inner circle stamens evolved into three cusps each bearing three connate filaments of which the central one carries an anther. The stamens of the outer circle alternate with the cusps of the inner circle. The form of the inner circle cusps is the main character for taxonomic research in several sections within this subgenus.

Hanelt and co-workers undertook a new approach to elucidate the taxonomy of the Alliums using molecular methods involving 450 Old World *Allium* species (Hanelt et al. 1992). One of the outcomes of this research became also the monography "Nomenclator Alliorum: *Allium* names and synonyms—a world guide" (Gregory et al. 1998). Four subgenera, 39 sections and 682 species in the Old World and 84 American species were identified, as well as 1170 species names were included, over half of which were generally considered to be synonyms. Two years before, Mathew (1996) published a full revision of subgenus *Allium* with multidisciplinary data. Finally, a new classification of the genus based on nuclear ribosomal DNA internal transcribed spacer (ITS) sequences was published (Friesen et al. 2006), and 15 subgenera and 72 sections were recognized, namely (*Nectaroscordum* (Lindl.) Asch. & Graebn., *Microscordum* (Maxim.) N. Friesen, *Amerallium* Traub., *Caloscordum* (Herb.) R.M. Fritsch, *Anguinum* (G. Don ex Koch) N. Friesen, *Porphyroprason* (Ekberg) R.M. Fritsch, *Vvedenskya* (Kamelin) R.M. Fritsch, *Melanocrommyum* (Webb. & Berth.) Rouy., *Butomissa* (Salisb.) N. Friesen, *Cyathophora* (R.M. Fritsch) R.M. Fritsch, *Rhizirideum* (G. Don ex Koch) Wendelbo, *Allium*, *Reticulatobulbosa* (Kamelin) N. Friesen, *Polyprason* Radic and *Cepa* (Mill.) Radic). Authors revealed three evolutionary lines: the first line consists of first three above-mentioned subgenera; the second line includes five next ones and third line includes seven retained ones including subgenus *Allium*.

A new *Allium* list with descriptions and results of lectotypification already made by Allium specialists in the past decade became necessary and appropriate especially for scientists dealing with *Allium* genetic resources. One of the best revisions of subgenus *Melanocrommyum* was made by Fritsch and co-workers (Fritsch et al. 2010; Fritsch and Abbasi 2013; Fritsch 2016) based on nearly 30 years of studies on living material originating from the whole area of distribution. His new classification included 160 species, affiliated to the 20 sections and 22 subsections was based on molecular and morphological characters and marks the end-point of the Gatersleben *Allium* classification approach.

2.2 *Allium* Subgenus *Allium*

The largest group within the genus is subgenus *Allium* with more than 375 species and 35 subspecies divided into 18 sections, however, a detailed phylogenetic classification of this taxonomically complex subgenus is still missing. Altogether only about 40 species of this subgenus have been studied via DNA analysis. Future research will probably detect more infrasubgeneric groups, which occurs in temperate Europe, the Mediterranean region, Near and Middle East countries and North-west China, with only a few species occurring outside this area (Fig. 2.1). For example, *Allium synnotii* G. Don (=*Allium dregeanum* Kunth) occurs in the Southern hemisphere in South Africa, which perhaps resulted from an early introduction of

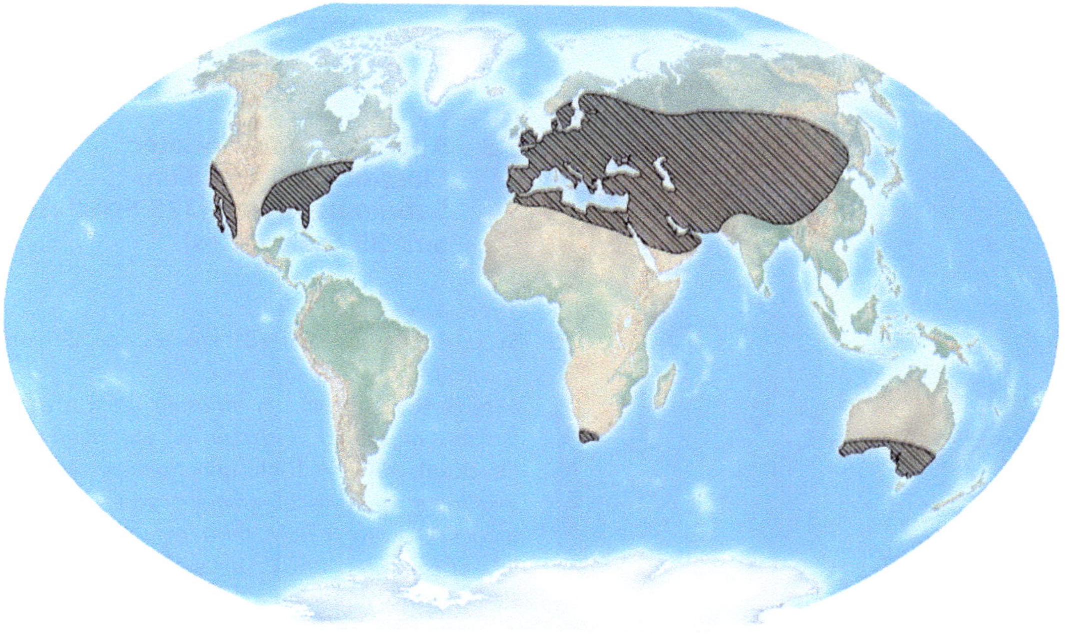

Fig. 2.1 Distribution of *Allium* subgenus *Allium*

European settlers of a related taxon (de Wilde-Duyfjes 1976). Furthermore, species like *Allium vineale* L. and *Allium ampeloprasum* L. are naturalized in North America (McNeal and Jacobsen 2007) and Australia.

Taxonomical studies based on herbarium material only is often inadequate, even on the species level, because during the drying process several diagnostic characters can change. For example, important traits in *Allium* taxonomy, like pedicels and perianth form and colour, do change during drying. Thus, the large number of newly discovered *Allium* taxa from the end of the twentieth century could be explained because of this phenomena. Subgenus *Allium* consists of two large groups; one having simple cuspidated inner filaments while another have three-cuspidated inner filaments. Moreover, recently described sections show different types of filaments (e.g. teeth-like, lateral cusps shorter than anther-bearing cusps or sometimes numerous), as well as a keeled or tubercular-pitted form of bulblets (Fig. 2.2) which never occur in other subgenera.

This might point to the presence of centres of recent speciation in the region of Middle Asia, Pakistan, Iran, Afghanistan and the Middle East. The majority of the newly described sections are widely supported by molecular data, whereas the informal groups proposed by Mathew (1996), were not supported (Friesen et al. 2006). Species from subgenus *Allium* are ecologically restricted to dry, open habitats mostly with a sparse vegetation cover, and are rarely found in open forests or forest margins. Some species like *Allium rubellum* and *Allium griffithianum* can occur in monotypic associations in deserts and low mountains in Middle Asia. In years with sufficient precipitation and high spring temperatures, they can form flowering carpets. The largest section within the subgenus, namely *Codonoprasum*, together with section *Allium,* is distributed mainly in the Mediterranean. In Middle Asia, only several species of tall garlic (*Allium ampeloprasum, Allium atroviolaceum*) are growing in the pre-mountainous cultivated zone. Section *Codonoprasum* can be characterized as a heterogeneous complex of small size *Allium* species occurring in desert and mountainous regions having a broad spectrum of vegetative and generative periods. Sections *Longivaginata* and *Brevispatha* are

Fig. 2.2 Bulblets of *Allium haneltii* F.O. Khass. & R.M. Fritsch

taxonomically close showing a large overlap of asymmetric spathe traits as well as unusual late flowering (in September). Furthermore, there are several monotypic (*Haneltia, Rechingeria, Kopetdagia*) and oligotypic (*Mediasia, Spathulata*) sections which are endemic in Middle Asia. Moreover, *Allium kujukense* Vved. was earlier included in subgenus *Allium*, but Kamelin (1973) surprisingly placed it into subgenus *Melanocrommyum* as a separate section, namely *Vvedenskya* Kamelin. Molecular analysis indicated its unique origin and therefore this section was later raised to the subgenus level (Friesen et al. 2006). However, it will be clear that more research on this interesting subgenus is warranted to obtain an in-depth insight into the taxonomic and evolutionary relationships between the sections and species occurring in the various sections.

The taxonomic classification below of subgenus *Allium* presents the current state of the art.

2.3 A Revised and Updated Classification of *Allium* Subgenus *Allium* (375 Species and 35 Subspecies)

Type: *A. sativum* L.

Section 1. ***Allium*** (ca 125 species and 20 subspecies). Typus: *A. sativum* L.

abbasii R.M. Fritsch, *acutiflorum* Loisel., *affine* Ledeb., *aksekiense* Özhatay & al., *albiflorum* Omelczuk, *alibile* A. Rich., *altoatlanticum* Seregin, *amethystinum* Tausch, *ampeloprasum* L., *anatolicum* Özhatay & B. Mathew, *antalyense* Eren & al., *armerioides* Boiss., *artemisietorum* Eig & Feinbrun, *ascalonicum* L., *asirense* B. Mathew, *asperiflorum* Miscz. ex Grossh., *atroviolaceum* Boiss., *aucheri* Boiss., *aznavense* R.M. Fritsch, *baeticum* Boiss., *barthianum* Asch. & Schweinf., *baytopiorum* Kollmann & Özhatay, *bourgeaui* Rech.f. (ssp. *creticum* Bothmer, ssp. *cycladicum* Bothmer), *brevipes* Ledeb., *calyptratum* Boiss., *cappadocicum* Boiss. & Balansa, *chamaespathum* Boiss., *clivorum* R.M. Fritsch, *commutatum* Guss., *curtum* Boiss. & Gaill. (ssp. *aegyptiacum* Täckh. & Drar, ssp. *palaestinum* Feinbrun), *damascenum* Feinbrun, *deserti-syriaci* Feinbrun, *dictyoprasum* C.A. Mey. ex Kunth, *dictyoscordum* Vved., *dilatatum* Zahar., *drusorum* Feinbrun, *ebusitanum* Font Quer, *ekimianum* Ekşi & al., *eldivanense* Özhatay, *enginii* Özhatay & B. Mathew, *erubescens* K. Koch, *erzincanicum* Özhatay & Kandemir, *esfandiarii* Matin, *fethiyense* Özhatay & B. Mathew, *filidens* Regel (ssp. *filidens*, ssp. *mogianense* R.M. Fritsch & F.O. Khass.), *filidentiforme* Vved., *gabardaghense* Fırat, *fominianum* Miscz. ex Grossh., *fritschii* F.O. Khass. & Yengal., *fuscoviolaceum* Fomin, *goekyigitii* Ekim & al., *gomphrenoides* Boiss. & Heldr. ex Boiss., *gorumsense* (Regel) Boiss., *gramineum* K. Koch, *guttatum* Steven (ssp. *guttatum*, ssp. *dalmaticum* (A. Kern. ex Janch.) Stearn, ssp. *sardoum* (Moris) Stearn, ssp. *tenorei* (Parl.) Soldano), *heldreichii* Boiss., *hemisphaericum* (Sommier) Brullo, *gypsodictyum* Vved., *hamrinense* Hand.-Mazz., *ilgazense* Özhatay, *integerrimum* Zahar., *iranicum* (Wendelbo) Wendelbo, *jajlae* Vved., *jubatum* J.F. Macbr., *junceum* Sm.(ssp. *junceum*, ssp. *tridentatum* Kollmann, Koyuncu & Özhatay), *karakense* Regel, *karyeteini* Post, *kayae* Özhatay & Koyuncu, *koyuncui* H. Duman & Özhatay, *leucanthum* K. Koch, *longicollum* Wendelbo, *longipapillatum* R.M. Fritsch & Matin, *machmelianum* Post, *macrochaetum* Boiss. & Hausskn., *mareoticum* Bornm. & Gauba, *margaritiferum* Vved., *melananthum* Coincy, *melitense* (Sommier & Caruana ex Borg) Cif. & Giacom., *multiflorum* Desf., *nathaliae* Seregin, *nevsehirense* Koyuncu & Kollmann, *notabile* Feinbrun, *oltense* Grossh., *pardoi* Loscos, *pervariensis* Firat & al., *pervestitum* Klokov, *phanerantherum* Boiss. & Hausskn. (ssp. *deciduum* Kollmann & Koyuncu, ssp. *involucratum* Ekşi & al.), *pruinatum* Link ex Spreng., *pustulosum* Boiss. & Hausskn., *polyanthum* Schult. & Schult.f., *ponticum* Miscz. ex Grossh. & Grossh., *proponticum* Stearn & Özhatay (ssp. *proponticum*, ssp. *parviflorum* (Kollmann) Koyuncu), *pseudoampeloprasum* Miscz. ex Grossh., *pseudocalyptratum* Mouterde, *pseudophanerantherum* Rech.f., *pyrenaicum* Costa & Vayr., *regelianum* A.K. Becker ex Iljin, *reuterianum* Boiss., *robertianum* Kollmann, *rollovii* Grossh., *rotundum* L., *rubrovittatum* Boiss. & Heldr., *sandrasicum* Kollmann, Özhatay & Bothmer, *sannineum* Gomb., *sativum* L., *scaberrimum* M. Serres, *scabriflorum* Boiss., *scorodoprasum* L., *sinaiticum* Boiss., *sintenisii* Freyn, *synnotii* G. Don, *sphaerocephalon* L.(ssp. *sphaerocephalon*, ssp. *arvense* (Guss.) Arcang., ssp. *laxiflorum* (Guss.) Giardina & Raimondo, ssp. *trachypus* (Boiss. & Spruner) K. Richt.), *stearnianum* Koyuncu, Özhatay & Kollmann (ssp. *stearnianum*, ssp. *vanense* Kollmann & Koyuncu), *stylosum* O. Schwarz, *subnotabile* Wendelbo, *talijevii* Klokov, *talyschense* Miscz. ex Grossh., *trachycoleum* Wendelbo, *truncatum* (Feinbrun) Kollmann & D. Zohary, *tuncelianum* (Kollmann) Özhatay, B. Mathew & Siraneci, *ugami* Vved., *valentinae* Pavlov, *vineale* L., *waldsteinii* G. Don f., *wendelboi* Matin, *willeanum* Holmboe.

Section 2. ***Brivedentia*** F.O. Khass. & Yengal. in Öztürk, Seçmen & Gork (eds.), Plant Life in South-West and Central Asia: 147 (1996) (12 species, 2 subsp.). Typus: *Allium brevidens* Vved.

brevidens Vved.(ssp. *brevidens*, ssp. *pshikharvicum* R.M. Fritsch & F.O. Khass.), *chirchikense* ined., *circumflexum* Wendelbo, *freitagii* ined., *hedgei* Wendelbo, *ionandrum* Wendelbo, *michaelis* F.O. Khass. & Tojibaev, *micranthum* Wendelbo, *miserabile* Wendelbo, *nikolai* F.O. Khass. & Achilova, *ophiophyllum* Vved., *turcomanicum* Regel.

Section 3. ***Crystallina*** F.O. Khass. & Yengal. in Öztürk, Seçmen & Gork (eds.), Plant Life in South-West and Central Asia: 147 (1996). (2 species). Typus: *Allium crystallinum* Vved.

crystallinum Vved., *brevidentiforme* Vved.

Section 4. ***Multicaulia*** F.O. Khass. & Yengal. in Öztürk, Seçmen & Gork (eds.), Plant Life in South-West and Central Asia: 148 (1996). (6 species). Typus: *Allium borszczowii* Regel.

boissieri Regel, *borszczowii* Regel, *ferganicum* Vved., *lehmannianum* Merckl. ex Bunge, *rinae* F.O. Khass., Shomur. & Tojibaev, *oxianum* F.O. Khass. & Tojibaev.

Section 5. ***Unicaulia*** F.O. Khass. in Stapfia 95: 174 (2011). (3 species).Typus: *Allium kotschyi* Boiss.

kotschyi Boiss., *ravenii* F.O. Khass., Shomur. & Kadyrov, *sosnovskyanum* Miscz. ex Grossh.

Section 6. ***Spathulata*** F.O. Khass. & R.M. Fritsch in O.A. Ashurmetov & al. (eds.), Plant. Life in South-West. and Central Asia: Vth Internat. Symp. Tashkent 1998 109 (2000) (2 species). Typus: *Allium spathulatum* F.O. Khass. & R.M. Fritsch.

spathulatum F.O. Khass. & R.M. Fritsch, *formosum* Sennikov & Lazkov.

Section 7. ***Mediasia*** F.O. Khass., Yengal. & N. Friesen in Aliso 22: 389 (2006) (2 species). Typus: *Allium turkestanicum* Regel.

turkestanicum Regel, *yuchuanense* Y.Z. Zhao & J.Y. Chao.

Section 8. ***Avulsa*** F.O. Khass. in Öztürk, Seçmen & Gork (eds.), Plant Life in South-West and Central Asia: 149 (1996) (15 species). Typus: *Allium griffithianum* Boiss.

bracteolatum Wendelbo,? *bungei* Boiss.,? *chloroneurum* Boiss., *choriotepalum* Wendelbo, *fibrosum* Regel, *griffithiamum* Boiss., *jacquemontii*

Kunth, *lamondiae* Wendelbo, *leucosphaerum* Aitch. & Baker, *oshense* ined., *pamiricum* Wendelbo, *panjaoense* Wendelbo, *rubellum* M. Bieb., *sabzakense* Wendelbo, *umbilicatum* Boiss.

Section 9. ***Brevispatha*** Valsecchi in Giorn. Bot. Ital. 108(1–2): 92 (1974) (15 species and 2 subspecies). Typus: *Allium parciflorum* Viv.

antonii-bolosii P. Palau (ssp.*antonii-bolosii* ssp. eivissanum (Garbari & Miceli) N. Torres & Rosselló), *azaurenum* Gomb., *balansae* Boiss., *callidyction* C.A. Mey. ex Kunth, *callimischon* Link (ssp. *callimischon,* ssp. *haemostictum* Stearn), *cupani* Raf., *inops* Vved., *kirindicum* Bornm., *kollmannianum* Brullo, Pavone & Salmeri, *lojaconoi* Brullo, Lanfr. & Pavone, *margaritae* B. Fedtsch., *pentadactyli* Brullo, Pavone & Spamp, *peroninianum* Azn., *ritsi* Iatroú & Tzanoud., *sindjarense* Boiss. & Hausskn. ex Regel.

Section 10. ***Coerulea*** (Tzagolova) F.O. Khass. in Öztürk, Seçmen & Gork (eds.), Plant Life in South-West and Central Asia: 150 (1996) (19 species and 2 subspecies). Typus: *Allium coeruleum* Pall.

aemulans Pavlov, *coeruleum* Pall., *caesioides* Wendelbo, *caesium* Schrenk, *delicatulum* Siev. ex Schult. & Schult., *elegans* Drobow, *eremoprasum* Vved., *gillii* Wendelbo, *glomeratum* Prokh., *lasiophyllum* Vved., *montelburzense* R.M. Fritsch, Salmaki & Zarre, *najafdaricum* R.M. Fritsch, *oreophiloides* Regel (ssp. *oreophiloides,* ssp. *salangense* Wendelbo), *renardii* Regel, *scharobitdinii* F.O. Khass. & Tojibaev, *svetlanae* Vved. ex Filim., *taciturnum* Vved., *tatyanae* F.O. Khass. & F. Karimov, *zaissanicum* Kotukhov.

Section 11. ***Eremoprasum*** (Kamelin) F.O. Khass., R.M. Fritsch & N. Friesen in Aliso 22: 389 (2006) (8 species). Typus: *Allium sabulosum* Stev. ex Bunge.

confragosum Vved., *incrustatum* Vved., *jaxarticum* Vved., *joharchii* F.O. Khass. & Memariani, *popovii* Vved., *sabulosum* Steven ex Bunge, *scrobiculatum* Vved., *transvestiens* Vved.

Section 12. ***Kopetdagia*** F.O. Khass. in M.A. Öztürk, Ö. Seçmen & G. Görk (eds.), Plant Life in South-West and Central Asia 1: 150 (1996) (3 species). Typus: *Allium kopetdagense* Vved.

kopetdagense Vved. in B. Fedtsch. & al., *spirophyllum* Wendelbo, *stocksianum* Boiss.

Section 13. ***Longivaginata*** (Kamelin) F.O. Khass., R.M. Fritsch & N. Friesen in Aliso 22: 389 (2006) (6 species). Typus: *Allium longivaginatum* Wendelbo.

arlgirdense Blakelock, *autumniflorum* F.O. Khass. & Akhani, *dolichovaginatum* R.M. Fritsch, *jaegeri* R.M. Fritsch, *longivaginatum* Wendelbo, *petri* F.O. Khass. & R.M. Fritsch.

Section 14. ***Minuta*** F.O. Khass. in Öztürk, Seçmen & Gork (eds.), Plant Life in South-West and Central Asia: 150 (1996) (7 species). Type: *Allium minutum* Vved.

aktauense F.O. Khass. & Esankulov, *anisotepalum* Vved., *minutum* Vved., *namanganicum* ined., *orunbaii* F.O. Khass. & R.M. Fritsch, *parvulum* Vved., *registanicum* Wendelbo.

Section 2.15. ***Pallasia*** (Tzagolova) F.O. Khass. in Sennikov: Flora of Uzbekistan. Vol. 1 *Amaryllidaceae*: 83 (2016). (11 species). Typus: *Allium pallasii* Murr.

anacoleum Hand.-Mazz., *capitellatum* Boiss., *eusperma* Airy Shaw, *grisellum* J.M. Xu, *lalesaricum* Freyn & Bornm., *maowenense* J.M. Xu, *pallasii* Murray, *sairamense* Regel, *schoenoprasoides* Regel, *songpanicum* J.M. Xu, *tuchalense* F.O. Khass. & Noroozi.

Section 16. ***Rechingeria*** F.O. Khass. & Tirkash. in Stapfia 99: 214 (2013) (monotypic). Typus: *Allium rechingerii* Wendelbo.

Section 17. ***Codonoprasum*** Rchb., Flora Germanica Excursoria. Carolum Cnobloch, Lipsiae: 878 (1830) (137 species and nine subspecies). Typus: *Allium oleraceum* L.

achaium Boiss. & Orph. ex Boiss., *aegilicum* Tzanoud., *aeginiense* Brullo & al., *aetnense* Brullo & al., *agrigentinum* Brullo & Pavone, *albotunicatum* O. Schwarz (ssp. *albotunicatum,* ssp. *hermaneum* Kollmann & Shmida), *antiatlanticum* Emb. & Maire, *anzalonei* Brullo & al., *apergii* Trigas & al., *apolloniensis* Biel, Kit Tan & Tzanoud., *apulum* Brullo & al., *archeotrichon* Brullo & al., ? *autumnale* P.H. Davis, *balcanicum* Brullo & al., *bassitense* J. Thiébaut, *bingoelense* Yild. & Kılıç, *birkinshawii* Mouterde, *brachyspathum* Brullo & al., *brevicaule* Boiss. & Balansa, *brulloi* Salmeri, *brussalisii* Tzanoud. & Kypr., *calabrum*

(N. Terracc.) Brullo & al., *candargyi* Karavok. & Tzanoud., *carinatum* L., *castellanense* (Garbari & al.) Brullo & al., *cephalonicum* Brullo & al., *chloranthum* Boiss., *convallarioides* Grossh., *corsicum* Jauzein & al., *croaticum* Bogdanovic & al., *cyprium* Brullo & al.(ssp. *cyprium*, ssp. *lefkarense* (Brullo & al.) Christodoulou & Hand), *daninianum* Brullo & al., *decaisnei* C. Presl, *deciduum* Özhatay & Kollmann (ssp. *deciduum*, ssp. *retrorsum* Özhatay & Kollmann), *desertorum* Forssk., *diomedeum* Brullo & al., *dirphianum* Brullo & al., *djimilense* Boiss. ex Regel, *dodecanesi* Karavok. & Tzanoud., *ekeri* E. Kaya & Koçyiğit, *euboicum* Rech.f., *exaltatum* (Meikle) Brullo & al., *exile* Boiss. & Orph., *favosum* Zahar., *flavum* L. (ssp.*flavum*, ssp. *ionochlorum* Maire, ssp. *tauricum* (Besser ex Rchb.) K. Richt.), *flexuosum* d'Urv., *fuscum* Waldst. & Kit., *galileum* Brullo & al., *garbarii* Peruzzi, *garganicum* Brullo & al., *glumaceum* Boiss. & Hausskn. ex Boiss., *greuteri* Brullo & Pavone, *guicciardii* Heldr., *hirtovaginatum* Kunth, *hirtovaginum* Candargy, *huber-morathii* Kollmann, Özhatay & Koyuncu, *hymettium* Boiss. & Heldr., *ionicum* Brullo & Tzanoud., *julianum* Brullo & al., *karistanum* Brullo & al., *kastambulense* Kollmann, *kossoricum* Fomin, *kunthianum* Vved., *kurtzianum* Asch. & Sint. ex Kollmann, *kyrenium* Giusso & al., *lazikkiyense* Koçyiğit & al., *lehmannii* Lojac., *lenkoranicum* Miscz. ex Grossh., *liliputianum* Koçyiğit & al., *litardierei* J.-M. Tison, *longispathum* F. Delaroche, *lopadusanum* Bartolo & al., *luteolum* Halácsy, *macedonicum* Zahar., *maghrebinum* Brullo & al., *makrianum*Brullo & al., *maraschicum* Koçyiğit & Özhatay, *marathasicum* Brullo & al., *marginatum* Janka, *mauritanicum* Brullo & al., *meikleanum* Brullo & al., *melanantherum* Pancic, *microspathum* Ekberg, *myrianthum* Boiss., *nebrodense* Guss., *occultum* Tzanoud. & Trigas, *oleraceum* L., *olympicum* Boiss., *opacum* Rech.f., *oporinanthum* Brullo & al., *optimae* Greuter, *orestis* Kalpoutz. & al., *paczoskianum* Tuzs., *pallens* L., *paniculatum* L., *panormitanum* Brullo & al., *pelagicum* Brullo & al., *perpendiculum* Koçyiğit & al., *phalereum* Heldr. & Sart., *phitosianum* Brullo & al., *phrygium* Boiss. & Balansa, *pictistamineum* O. Schwarz, *pilosum* Sm., *platakisii* Tzanoud. & Kypr., *podolicum* Blocki ex Racib. & Szafer, *pseudoflavum* Vved., *pseudostamineum* Kollmann & Shmida, *rausii* Brullo & al., *retrorsum* (Özhatay & Kollmann) Brullo & al., *rhodopeum* Velen., *rumelicum* Koçyiğit & Özhatay, *rupestre* Steven, *rupicola* Boiss. ex Mouterde, *samniticum* Brullo & al., *samothracicum* Tzanoud. & al., *savii* Parl., *schergianum* Boiss., *sibthorpianum* Schult. & Schult.f., *sipyleum* Boiss., *stamineum* Boiss., *staticiforme* Sm., *tardans* Greuter & Zahar., *tardiflorum* Kollmann & Shmida, *tauricola* Boiss., *tauricum* Pall. ex G. Don f., *telmatum* Bogdanovic & al., *tenuiflorum* Ten., *therinanthum* Brullo & al., *tingitanum* Brullo & al., *trichocnemis* J. Gay, *turcicum* Özhatay & Cowley, *tzanoudakisanum* Brullo & al., *valdecallosum* Maire & Weiller, *valdesianum* Brullo & al, *variegatum* Boiss., *vescum* Wendelbo, *weissii* Boiss., *wiedemannianum* Regel, *yildirimlii* Dural.

Section 2.18. ***Haneltia*** F.O. Khass. in Stapfia 95: 174 (2011) (monotypic). Typus: *Allium haneltii* F.O. Khass. & R.M. Fritsch.

2.4 Ethnobotanical Aspects of Wild *Allium* Species from Middle Asia

More than 250 different *Allium* species are reported for the mountainous regions of Middle and South-West Asia (Khassanov 1997). The use of especially tasteful and curative wild *Alliums* has a long tradition in several Middle Asian populations with apparently deep historical roots. This might be illustrated by the fact that people living in the countryside (especially mountainous villages), as well as urban areas, do have a long standing tradition how to prepare special dishes from particular *Allium* plants. The economically most important *Allium* crop species (common onion and garlic) are worldwide used as spices, vegetables and medicinal plants. Traditionally, they play a very important role in the daily diet. However, there are also more than 80 endemic species in the mountainous regions in Middle Asia and therefore information about the use of these species by local population could be very important. During the PharmAl Project (2002–2006) funded by Volkswagen Stiftung, about 860 *Allium* samples were collected, around 20 new species were identified, and 36 wild *Allium* species were reported as edible ones in Uzbekistan (Khassanov 2007). First of all, local people are gathering and cultivating wild relative onion species like *Allium oschaninii* O. Fedtsch., *Allium praemixtum* Vved. and *Allium pskemense* B. Fedtsch. Therefore, populations of these species became rare and are currently included in the Red Data Book of Uzbekistan (Khassanov 2016a). The ancestral species of garlic *Allium longicuspis* Regel is naturally growing in the mountainous zone nearby the settlements but is normally not used, but the domesticated species *A. sativum* is cultivated in every private garden. Section *Allium* s. l. is also widely occurring in Uzbekistan and 21 species are known from this section growing in Uzbekistan (Khassanov 2016b). These species are used by the local population as a garlic substitute. Another group of important edible wild Alliums belong to subgenus *Melanocrommyum* and are represented by several species. One of the most interesting wild species from a pharmaceutical point of view is *Allium tschimganicum* B. Fedtsch (Fig. 2.3) = *Allium motor* Kamelin & Levichev (motor, moy-modor—in Tadzhik language means "health"). The natural distribution area of this rather tall species is rather small, comprising only the Chatkal mountain range in Uzbekistan and adjacent Kyrgyz and Kazakh territories.

Here it grows on shady slopes among bushes and is easily recognized by a smooth, thin, and quickly decaying sheath leaf. The leaves are much collected in April–May. They are freshly used as stuffing for a special variant of the national pie dish "somsa", which is highly esteemed for a specific activity as a stimulant. However, for persons having high blood pressure, the consumption is less advisable (Umarov 1992). The use of *A. tschimganicum* became known among people living in the Kirgiz territory of Chatkal mountain range during the 1970s, first in the Parkent area of Uzbekistan (Umarov 1992). Today this species is also much collected for consumption and has become rare in some areas. One of the main characters of the species is a reddish liquid coming out of the damaged tissue of freshly plucked leaves. Another species of pharmaceutical interest is *Allium rosenorum* R. M. Fritsch (Fig. 2.4), which is naturally growing in the shade under shrubs and trees in the Pamir-Alay mountainous system (mainly Hissar range) in Uzbekistan and Tajikistan.

Fresh and dried young leaves are also used for the national soup dish "'atolla", which is

Fig. 2.3 Umbel of *Allium tschimganicum* B. Fedtsch

Fig. 2.4 Umbel of *Allium rosenorum* R.M. Fritsch

much esteemed as an appetizer and general stimulant in Central Tajikistan (Keusgen et al. 2004, 2006).

2.5 Ornamental Wild *Allium* Species from Middle Asia

This list of ornamental *Allium* species and cultivars consists of around 250 species, 20 hybrids/intraspecific taxa and 240 cultivars. Most of the species of these genetic resources occur mainly in two subgenera namely *Allium* and *Melanocrommyum*. The list was compiled using recently published catalogues (Fritsch 2015; http://www.ipk-gatersleben.de/gbisipk-gaterslebendegbis-i/spezialsammlungen/allium-review), which were analysed for ornamental value. Species and cultivars solely offered as a vegetable, spice or medical plant were ignored, but those with a dual purpose including ornamental characters were accepted. Many members of subgenus *Melanocrommyum* are cultivated as ornamentals in European and North American gardens (Fritsch 2015). Most of these species occur naturally in South-west and Middle Asia where they are collected in nature and used as vegetable, spice and medicinal plants by the local people. The most cultivated species are *Allium cristophii* Trautv., *Allium stipitatum* Regel, *Allium altissimum* Regel, *Allium aflatunense*, *Allium giganteum* Regel and *Allium hollandicum* R.M. Fritsch. The latter species was described as a new species formed in the process of many years of cultivation. Furthermore new ornamental *Allium* species could be identified in Middle Asia, for example, *Allium backhousianum* Regel (Fig. 2.5) = *Allium gulczense* B. Fedtsch. ex O. Fedtsch. growing naturally in the mountains of Kyrgyzstan and Uzbekistan (Alay and Fergan ranges).

The plants can grow up to 2 m tall, have very beautiful multi-flowered umbels (10–15 cm in diameter) and could be successfully cultivated in regions with a humid climate. Another two rare endemic species which have ornamental value, *Allium majus* Vved., *Allium isakulii* R.M. Fritsch & F.O. Khass (Fig. 2.6), and can be cultivated in the countries with an arid climate in the pre-mountainous zone. The latter *Allium* species has the largest umbel within all wild Alliums (20–40 cm in diameter) and its pedicels and tepals remain straight-up after drying and do not drop or twist after many years.

Fig. 2.5 Umbel of *Allium backhousianum* Regel

Fig. 2.6 Umbel of *Allium isakulii* R.M. Fritsch & F.O. Khass

Acknowledgements The author is greatly indebted to Dr. R.M. Fritsch (IPK Gatersleben, Germany), Dr. N. Friesen (Botanischer Garten der Universität Osnabrück, Germany), Prof. Dr. M. Keusgen (Marburg University, Germany) for many years of cooperation and Dr. C. Kik (Centre for Genetic Resources, the Netherlands (CGN), Wageningen, the Netherlands) for critical reading of the manuscript and joint collecting missions.

References

Don G (1832) A monograph of the genus *Allium*. Mem Wernerian Nat Hist Soc 6:1–102

de Wilde-Duyfjes BEE (1976) A revision of the genus *Allium* L. (*Liliaceae*) in Africa, Proefschrift. Wageningen. Also published as Meded. Landbouwhogesch. Wageningen 76: no.ll

Friesen N, Fritsch RM, Blattner FR (2006) Phylogeny and new intrageneric classification of *Allium* L. (*Alliaceae*) based on nuclear rDNA ITS sequences. Aliso 22: 372–395

Fritsch RM (2015) Checklist of ornamental *Allium* species and cultivars currently offered in the trade, 64 p. http://www.ipk-gatersleben.de/gbisipk-gaterslebendegbis-i/spezialsammlungen/allium-review/

Fritsch RM (2016) A preliminary review of *Allium* subg. *Melanocrommyum* in Central Asia. IPK Gatersleben. https://doi.org/10.5447/ipk/2016/60

Fritsch RM, Abbasi M (2013) A taxonomic review of *Allium* subg. Melanocrommyum in Iran. IPK, Gatersleben

Fritsch RM, Blattner FR, Gurushidze M (2010) New classification of *Allium* L. subg. *Melanocrommyum* (Webb & Berthel.) Rouy (*Alliaceae*) based on molecular and morphological characters. Phyton 49:145–220

Gregory M, Fritsch RM, Friesen N, Khassanov FO, McNeal DW (1998) Nomenclator Alliorum: *Allium* names and synonyms—a world guide. Royal Botanic Gardens, Kew

Halleir A (1745) Allii genere natural. Gottingae

Hanelt P, Schulze-Motel J, Fritsch RM, Kruse J, Maass H, Ohle H, Pistrick K (1992) Infrageneric grouping of *Allium*—the Gatersleben approach. The Genus *Allium* —taxonomic problems and genetic resources. Gatersleben, Germany, pp 107–123

Hermann F (1939) Sectiones et subsections nonnullae europea generic *Allium*. Feddes Repert. Berlin Bd 46:57–58

Kamelin RV (1973) Florogeneticheskij analiz estestvennoj flory gornoj Srednej Azii. Nauka, Leningrad, p 133

Keusgen M, Fritsch RM, Hisoriev H, Kurbonova PA, Khassanov FO (2004) Wildwachsende zentralasiatische Allium-Arten, die als Gewürz oder Arzneimittel verwendet werden. Programm Abstracts und Postertexte, Fachtagung für Arznei-und Gewürzpflanzen, 7. bis 9. September 2004 in Jena, p 101

Keusgen M, Fritsch RM, Hisoriev H, Kurbonova PA, Khassanov FO (2006) Wild *Allium* species (*Alliaceae*) used in folk medicine of Tajikistan and Uzbekistan. J Ethnobiol Ethnomed 2:18. http://www.ethnobiomed.com/content/pdf/1746-4269-2-18.pdf

Khassanov FO (1997) Conspectus of the wild growing *Allium* species of Middle Asia. In: Ozturk M, Secmen O, Gork G (eds) Plant life in South West Asia. Ege University Press, Izmir, pp 141–159

Khassanov FO (2007) Edible Alliums of Uzbekistan. In: Keusgen M, Fritsch RM (eds) Proceedings, First Kazbegi workshop on „Botany, taxonomy and phytochemistry of wild *Allium* L. species of the Caucasus and Central Asia“, June 4–8, 2007, Kazbegi, Caucasus, Georgia. Marburg & Gatersleben, pp 131–133

Khassanov FO (2016a) *Amaryllidaceae* in Sennikov. Flora Uzbekistan 1:83

Khassanov FO (2016b) Red data book of Uzbekistan, 5th edn, vol 1. Plants, p 1–356

Linnaeus C (1753) Species plantarum. Holmiae, Stockholm, pp 294–302

Mathew B (1996) A review of *Allium* section *Allium*. Royal Botantic Gardens, Kew, pp 1–76

McNeal DW, Jacobsen TD (2007) Genus *Allium* in Flora of North America. New York and Oxford 26:53–336

Regel E (1875) Alliorum adhuc cognitorum monographia. Trudy Imp. S.-Peterb. Bot. Sada 3(2):1–266

Stearn W (1944) Notes on the genus *Allium* in the Old World. Herbertia. London, Kew Publisher, No. 11, pp 11–34

Traub HP (1968) The subgenera, sections and subsections of *Allium* L. Pl. Life. University Press, Stanford, No 24, pp 147–163

Traub HP (1972) Genus *Allium*—subgenera, sections and subsections. Ibid. 28:132–137

Umarov TA (1992) Dikorastushchie pishchevye rastenija gornoj chasti Tashkentskoj Oblasti Avtoreferat [Summary] of thesis. Tashkent, pp 1–22

Vvedensky AI (1935) Rod 267. Luk—*Allium* L. In: Komarov, V. L. (ed.) Flora SSSR; 4. tom Izd. Akad. Nauk SSSR, Leningrad, pp 112–280

Vvedensky AI, Kovalevskaya SS (1971) Rod 151, (7) *Allium* L. - Luk zhua (kaz.) piez (tadzh.). In: Vvedensky AI, Kovalevskaya SS (eds) Opredelitel rastenij Srednej Azii. Kriticheskij konspekt flory 2. Izd. "FAN" Uzb. SSR, Tashkent, pp 39–89, incl. Appendix. Descriptiones plantarum novarum in tomo II Conspectus Florae Asiae Mediae commematorum, pp 311–328

Wendelbo P (1969) New subgenera, sections and species of *Allium*. Bot. Notizer.- Goeteborg 122(1):25–37

Wendelbo P (1971) Flora Iranica, vol 71. Akadesche Druck, p 1–123

Allium Genetic Resources

3

E. R. Joachim Keller and Chris Kik

Abstract

An overview of the developments in *Allium* genetic resources during the past 25 years is presented in this chapter. A first important development has been the introduction and further development of web-based genebanking information systems (e.g. GENESYS, PLANT-SEARCH), which facilitated the exchange of data to a large extent between *Allium* collection holders worldwide. These information systems made it possible to obtain an overview of the *Allium* genetic resources managed worldwide and identify the gaps in collections which still need to be filled, especially in the face of the ongoing genetic erosion. A second important area of progress has been the development of new methods for the maintenance of *Allium* germplasm, especially cryopreservation. This method has made it possible to maintain *Allium* accessions in a cheap and effective way. The method is especially important for the conservation of vegetatively maintained germplasm. Other developments in *Allium* genebanking are the improvement of the health status of the germplasm kept in the collections and the continuing characterization and evaluation of germplasm, which stimulates the utilization of the *Allium* genetic resources held in genebanks. Significant changes could also be observed with respect to acquisition and exchange of plant genetic resources due to many and complex new regulations on the legal and organizational levels due to the adoption of the CBD and IT-PGRFA by many countries. It makes the handling of the plant accessions safer and more consistent but also more circumstantial. Finally, we need to underline that in an increasingly changing world with all the threats of genetic erosion and extinction due to disappearance of traditional cultivation methods, devastation of our environment and climatic change, the conservation of genetic resources is of prime importance for agriculture. Especially for breeders, a highly diverse genepool of a crop plant is an invaluable treasure. The importance to keep this treasure will no doubt become even more important in the future.

3.1 Introduction

This chapter is an update of the chapter on *Allium* genetic resources written around 25 years ago by Astley in the book entitled Onions and Allied

E. R. J. Keller
Leibniz Institute of Plant Genetics and Crop Plant Research (IPK), Corrensstrasse 3, 06466 Seeland, OT Gatersleben, Germany

C. Kik (✉)
Centre for Genetic Resources, the Netherlands (CGN), Wageningen University and Research, Droevendaalsesteeg 1, 6708 PB Wageningen, The Netherlands
e-mail: chris.kik@wur.nl

M. Shigyo et al. (eds.), *The Allium Genomes*, Compendium of Plant Genomes,
https://doi.org/10.1007/978-3-319-95825-5_3

Crops (Rabinowitch and Brewster 1990). In his chapter, Astley wrote in much detail the different genebanking procedures as regards to *Allium* crops (Astley 1990). The present chapter can be seen as an update of this chapter and will focus on new genebanking activities that emerged since 1990. These are first the improvements on genebank information systems, which currently allow for the first time a broad overview on global *Allium* genetic resources. In this context, also the *Allium* genetic resources present in botanical gardens around the world can be taken into account as a global portal is available for botanical gardens. Genebank information systems also allow for the storage of a large quantity of data, their efficient processing and easy presentation, such as images, scans of herbaria and molecular marker management. Furthermore, the standardization of characterization data using descriptor lists has improved the utilization of these information systems largely. Second, the development of the newly introduced cryopreservation of *Allium* and its further improvement will change the conservation of vegetative *Allium* genetic resources significantly. These advancements may also have an impact on breeding and storage of generative germplasm as well since pollen cryopreservation was developed too. Third, virus infections in vegetative germplasm are of major influence on agronomic performance. Therefore, increasing consideration has been given in the past two decades to this field. Last but not least, all these activities would not have been possible without cooperation among the various institutions and persons worldwide for mutual benefit using international organizations, concerted actions and projects.

3.2 Exploring and Collecting *Allium* Genetic Resources

3.2.1 International Arrangements on the Exchange of Plant Genetic Resources

Since the beginning of the previous century, researchers have warned about the detrimental effects of genetic erosion on the crop level (see for an overview: Frankel and Bennett 1970) and since the 1980s also to crop wild relatives growing in nature. Collecting germplasm of landraces and wild relatives of cultivated crops is of evident importance for a number of activities such as the breeding of new cultivars and the performance of innovative research. Genes are conserved which could become important for new resistances to (a)biotic stressors, yield or even novel and previously not observed traits of interest. In the past, expeditions were predominantly focussed on collecting landraces of many crops (multi-crop expeditions). Currently, the focus is more on collecting single crops and their crossable wild relatives (single-crop expeditions). The latter category has become increasingly important as these crops' wild relatives prove to be interesting gene reservoirs (Hajjar and Hodgkin 2007; Kik 2002).

Collecting of germplasm has become more difficult in the past three decades as a paradigm shift has taken place concerning the ownership of plant genetic resources (PGR). Before the early 1990s, PGR were seen as a heritage of mankind whereas after this time PGR were seen as national sovereignty. This paradigm shift resulted in the development of two international arrangements, namely the Convention on Biodiversity (CBD[1];) which entered into force in 1993 and the International Treaty on Plant Genetic Resources for Food and Agriculture (IT-PGRFA or IT[2]) which entered into force in 2004. The CBD was negotiated by the Ministries of Environment and focuses on all the biodiversity in a country. As it was felt that the CBD did not fully address PGR used in Food and Agriculture, a novel agreement (IT) was negotiated by the Ministries of Agriculture. The IT has a more restricted focus compared to the CBD as it concerns only the biodiversity that is under the control of the government, so the biodiversity that is present in state-owned nature reserves, genebanks, etc. Furthermore, the IT focuses on a very restricted part of the biodiversity, namely

[1]https://www.cbd.int/.

[2]http://www.fao.org/plant-treaty/en/.

the so-called Annex 1 species, referring to a species list in the annex of the IT.

A crucial component of both arrangements is the fair and equitable sharing between donor and recipient of the benefits arising from the use of the PGR collected, which is part of the so-called access and benefit sharing (ABS) component in both arrangements. Recently, this ABS component of the CBD was made more explicit and resulted in the Nagoya protocol of ABS which entered into force in 2014. As *Allium* species are not Annex 1 species, collecting will take place under the CBD. This implies that a bilateral agreement between parties has to be negotiated in which prior informed consent (PIC) and mutually agreed terms (MAT, mostly in the form of a material transfer agreement [MTA]) are of crucial importance. This has to be negotiated with the competent national authority on ABS (CNA-ABS) in the country in which the collecting takes place. However, as the CNA-ABS is often very difficult to contact, does not respond to any request, is reluctant to sign PIC and MAT due to failed negotiations, or other reasons, international collecting expeditions do not often take place. This would not be a real problem as the biodiversity in countries where *Allium* species are growing would be protected via in situ and ex situ measures. However, this is often not the case, and combined with the fact that genetic erosion is worldwide a continuing threat to the existing biodiversity, it results, therefore, indirectly in the loss of biodiversity. Hopefully, these problems can be solved in the nearby future when it is realized that access to and utilization of PGR is of primary importance in the development of improved cultivars (e.g. more climate-resilient cultivars).

3.2.2 Collecting *Allium* Genetic Resources

Several genebanks were active in the past collecting *Allium* biodiversity. However, the major activities took place at the Institute für Pflanzengenetik und Kulturpflanzenforschung (IPK) at Gatersleben, Germany, collecting local *Allium* plant genetic resources already since the 70s of the past century (Hammer et al. 1977). Intense relations developed with countries of the former Soviet Union, especially to Georgia, which had a climax between 1987 and 2003 (Keller et al. 2012a). Also in subsequent years important collecting missions took place especially focussing on subgenus *Melanocrommyum* (Fritsch 2016). An overview of the various collecting missions performed is presented in Table 3.1.

This resulted in the build-up of the largest and most diverse *Allium* collections worldwide: the Taxonomic Reference Collection and the *Allium* Crop Collection of IPK Gatersleben, covering, at present 2772 accessions of 363 taxa and 80 hybrids (Table 3.2).

Ad hoc collecting missions involving *Allium* species also took place at other organizations than IPK. For example, a large collecting mission in Greece focussing on *Allium ampeloprasum* and crop wild relatives was carried out in 2009 coordinated by Kik (CGN, Wageningen, the Netherlands) and Tzanoudakis (University of Patras, Patras, Greece) collecting over 100 accessions (Fig. 3.1).[3]

Other large collections also acquired their accessions by collecting and material exchange actions. Some of the largest *Allium* collections are in the USA (PGR Geneva 1135 seed-propagated [J. Labate]; WRPIS Pullman 1280 vegetatively propagated [B. Hellier]), 4 centres in South Korea (1173 [H.-H. Kim]), World Vegetable Center Taiwan (1129),[4] InHort Poland (895 [M. Olas-Sochacka)], CRI Czech Republic (817 [H. Stavělíková; Petrzelova et al. 2016]), CIAF Spain (ca. 500 [M. De Los Mozos Pascual]), and INRA France (439 [F. Esnault]).[5]

Currently, the Research Institute of Horticulture InHort at Skierniewice (Poland) performs each year 4–6 multi-crop collecting missions mostly in their own country but also in Latvia and Estonia, which involve *Allium* (Olas-Sochacka, pers. comm.). Most probably much more

[3] http://missions.cgn.wur.nl/TPSK/TPSK.htm.

[4] http://203.64.245.49/AVGRIS/#.

[5] Data were provided by their collection curators [names in brackets] in March 2017.

Table 3.1 Survey on missions to collect wild *Allium* germplasm for the Taxonomic Reference Collection of IPK Gatersleben, Germany, their main collectors and the number of accessions (from Keller et al. 2012a, updated by RM Fritsch pers. comm.)

Country and years	Number of accessions
Afghanistan 2013, 2015, 2016 (M. Keusgen)	16
Altaj (Russia) 1991 (N. Friesen, K. Pistrick)	64
Armenia 2010 (R. Fritsch)	22
Bulgaria 1987 (J. Kruse, J. Schultze-Motel)	35
China 1988 (P. Hanelt)	15
Korea (DPR) 1986 (P. Hanelt)	20
Georgia 2002–2006, 2011 (M. Akhalkatsi, R. Fritsch, M. Keusgen, K. Pistrick)	152
Iran 1994, 2004–2008, 2010–2012 (M. Abbasi, R. Fritsch, M. Keusgen)	413
Kazakhstan 1990, 1992–1994, 1997 (R. Fritsch, F. O. Khassanov, K. Pistrick, J. Ruksans)	163
Kyrgyzstan 1992, 1994, 1997, 1998, 2013 (R. Fritsch, F. O. Khassanov, K. Pistrick, M. Keusgen)	117
Mongolia 1985, 1987 (P. Hanelt, J. Kruse, K. Pistrick, Ch. Sanchir)	76
Rhodos (Greece) 2012–2014 (K. Pistrick, R. Kummer)	98
Tajikistan 1983, 1984, 1986–1988, 1990, 1991, 1994, 1997, 1998, 2003–2006, 2015–2016 (P. Hanelt, R. Fritsch, H. Hisoriev, M. Keusgen, I. Kudratov, P. Kurbonova, K. Pistrick)	490
Turkey 1995 (N. Friesen, R. Fritsch)	105
Turkmenistan 1995, 2002 (R. Fritsch, M. Keusgen, F. O. Khassanov)	62
Uzbekistan 1988, 1990, 1992–1995, 1997, 1998, 2003, 2005, 2006 (R. Fritsch, F. O. Khassanov, I.I. Maltzev, I. Kudratov, K. Pistrick)	267

collecting missions involving *Allium* species have been carried out in the past two decades. However, the extent of these activities is, unfortunately unknown, because there is no global information system on the acquisition of germplasm.

3.3 Management

The debate on the relative merits of in situ and ex situ management of PGR to which Astley (1990) was referring to, has ended in the view that both methods of management are complementary. In the meantime, in situ management has been split into two directions namely on farm management (the management of landraces on the farm) and nature management (in situ management) as it was felt that both methods involve different ways of management of plant genetic resources.

3.3.1 In Situ and on Farm Management of *Allium* PGR

Dealing with this issue Astley (1990) wrote that '*There has been little direct effort to conserve wild* Allium *taxa* in situ *although undoubtedly various species occur in existing nature reserves*'. Also he wrote that '*A further aspect of* in situ *conservation is its value in the maintenance of domesticates*'. This text was followed by some general remarks on this issue what was later on called on farm management of PGR. Since then very few studies/data have been presented on in situ and on farm management of *Allium* PGR. As currently both methods (ex situ and in situ/on farm) are recognized to be complementary, it would be important to obtain more data on these ways of management as regards to *Allium* crops and species.

Table 3.2 Survey on the living accessions of the Gatersleben *Allium* collections in 2016 (from Keller et al. 2012a, updated by RM Fritsch pers. comm.)

Subgenus	Crop *Allium* Collection	Taxonomic Reference Collection
subg. *Allium*	630 (10 species, 11 cultivar groups)	453 (85 species and subspp.)
subg. *Amerallium* Traub	2 (2 species)	149 (38 species)
subg. *Anguinum* (G. Don ex Koch) N. Friesen		24 (4 species)
subg. *Butomissa* (Salisb.) N. Friesen	30 (2 species)	44 (3 species)
subg. *Caloscordum* (Herb.) R. M. Fritsch		3 (1 species)
subg. *Cepa* (Mill.) Radić	556 (11 species, 2 cultivar groups)	129 (19 species and subspp.)
subg. *Cyathophora* (R. M. Fritsch) R. M. Fritsch	2 (1 species)	6 (1 species)
subg. *Melanocrommyum* (Webb & Berthel.) Rouy		298 (74 species and subspp.)
subg. *Nectaroscordum* (Lindl.) Asch. & Graebn.		3 (3 species)
subg. *Polyprason* Radić	15 (7 species)	68 (19 species)
subg. *Porphyroprason* (Ekberg) R. M. Fritsch		8 (1 species)
subg. *Reticulatobulbosa* (Kamelin) N. Friesen	1 (1 species)	65 (22 species)
subg. *Rhizirideum* (G. Don ex Koch) Wendelbo s. strictiss.	20 (6 species)	141 (19 species)
subg. *Vvedenskya* (Kamelin) R. M. Fritsch		2 (1 species)
Related genera of *Allioideae* or formerly affiliated there		37 (19 species)
Intersectional or species hybrids	80 (in 15 cross combinations)	13 hybrids with unknown parents
Sum of determined accessions	1336 (40 species, 13 cultivar groups, 80 named hybrids)	1436 (310 species, subspecies and hybrids)
Not yet determined living accessions	131	191

3.3.2 Ex Situ Management of *Allium* PGR

Standards for ex situ management have been further developed and updated in the last years for many species. Basic documents for all types of management are provided in manuals from Bioversity International (Cromarty et al. 1982; Ellis et al. 1985a, b; Sackville Hamilton and Chorlton 1997; Hong et al. 1998; Rao et al. 2006; Singh et al. 2012) and by FAO (2014).

3.3.2.1 Seeds

Current procedures concerning the various aspects of *Allium* seed management do not differ to a large extent to what Astley (1990) wrote on this subject (see also Walters 2004). What can be added are the effects of storage on seed germination. Mostly data obtained from rapid ageing experiments are used to model seed survival of a particular crop, but rapid ageing conditions do not necessarily mimic real storage conditions. However, data from real storage conditions are

Fig. 3.1 Collecting *Allium ampeloprasum* and wild relatives in Greece by Chris Kik in 2009

rare given the long time periods involved. In this context, the records over the past 25 years on storage and germination under real genebank conditions involving *Allium* accessions maintained at the Centre for Genetic Resources, the Netherlands (CGN) can be used as an example (see for a general crop overview: Van Hintum and van Treuren 2012; Van Treuren et al. 2013).

The storage method used at CGN has been described, in general, by Astley (1990) and involves the drying of the harvested seeds during minimally one month in a conditioned storage room (15 °C and 15% relative humidity). Before storage, the viability of the seed is checked and minimum germination thresholds of 60% (wild material) and 80% (cultivated material) are applied. After this check, the seed is vacuum sealed in three-layered Al bags and stored at −20 °C. On a regular basis, accessions are checked for their seed viability.[6]

It turned out that this way of storage can be considered as a safe method for the long-term storage of *Allium* seeds at least for cultivated material as there is a significant positive correlation over a period of 25 years between regeneration and germination testing (Fig. 3.2). For wild *Allium* material, only a limited number of data points are present; therefore, no conclusions can yet be drawn in case of this material.

Another example is the seed storage at IPK Gatersleben. It differs in some aspects from that in the CGN. After harvest, the drying process is performed in two steps. The first step is the preliminary drying in gauze sacs or large paper bags at 18 ± 2 °C and 20 ± 3% relative humidity, the second step is final drying at 18 ± 2 °C and 10 ± 3% relative humidity for at least 3 weeks. Prior to storage, measurement of seed moisture is performed targeting a value of 5–7%. For storage, the sample is split into three parts, active sample, basic sample (backup), and safety duplicates. Backup sample and safety duplicates are stored in aluminium bags like in CGN but at −18 °C. Targeted germination is 80%. The active sample is stored in tightly closed glass jars at −18 °C. To keep the seed moisture content at the targeted value of 5–7%, a silica gel bag is added to the sample. Germinability should be there at least 50% (K. Krusch and U. Lohwasser, pers. comm.). For the active *Allium* samples, the germination percentage over the years is presented in Fig. 3.3.

Both figures show that storage at −18 or −20 °C is a feasible method to keep the germplasm over some decennia without a decline of germinability. Concluding from these different storage strategies, one can say that the genebanks

[6]Tests performed according to the rules of the International Seed Testing Association ISTA.

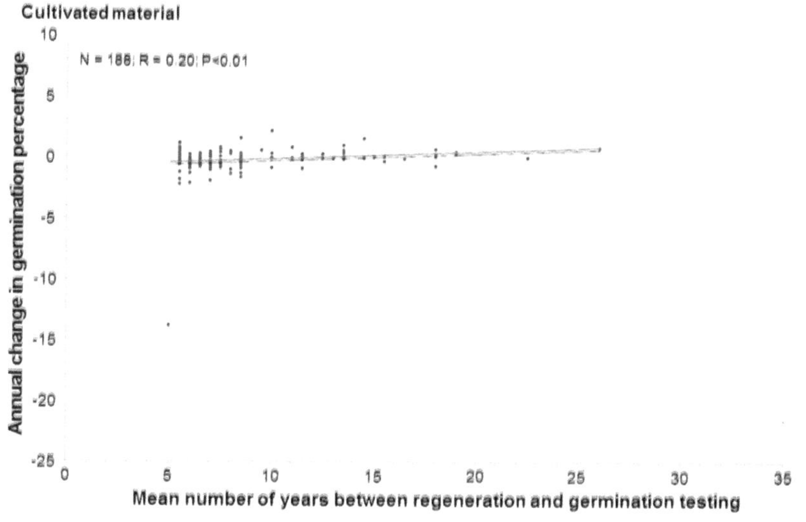

Fig. 3.2 *Allium* germination testing at CGN: the correlation between the number of years in between regeneration and germination testing and the annual change in germination percentage (modified from Van Treuren et al. 2013)

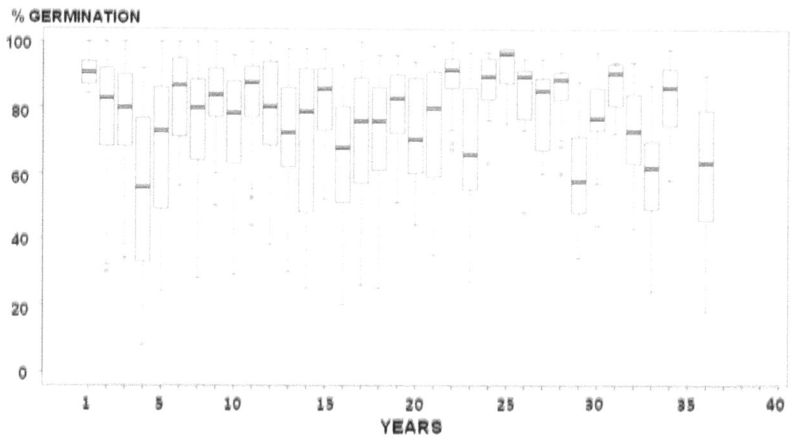

Fig. 3.3 Survey on 1473 germination tests on *A. cepa* seed samples in the year 2012 in relation to their storage duration (box plot provided by A Börner, IPK)

may vary their methods to some extent as long as the recommendations of low moisture content and cold temperature are followed.

3.3.2.2 Vegetative Material

A considerable part of the *Allium* genetic resources cannot be managed nor stored as seed either because of the general inability to form seeds (garlic, great-headed garlic, haploids, sterile hybrid material) or because of the traditional breeding strategy as clonal varieties (shallot, some ornamental *Allium* hybrids). This germplasm must be maintained in the vegetative state. This is also relevant for collection material growing in inappropriate conditions and, therefore, exhibiting only low seed set. Since the maintenance of vegetative material requires much higher efforts in terms of agricultural measures and phytosanitary treatments, it is always a special problem in plant collections.

Field culture is the most traditional method to keep such collections. Perennial *Allium* may be kept for several years in the same place. In most plant collections, garlic and shallot are replanted every year, shallot and a part of the garlic germplasm are planted in spring and harvested in summer. Another part of the garlic genepool can be planted already in autumn and sprouts in early spring. Garlic accessions forming bulbils can also be kept over winter using these propagules.

Reasons for this annual maintenance strategy are low winter hardiness and accumulation of soil-borne diseases (e.g. nematodes) which latter requires disinfection (e.g. by solarization). Since no seed phase is included which would be an obstacle for virus transmission, in most vegetative collections a considerable load of viruses can be observed.

In Vitro Culture

In vitro culture is the first step to keep the germplasm apart from the risks of infestation and possible endangerments by bad weather conditions. Taking shoot tips containing the meristems from bulb bases or, if available, from bulbils, disinfecting them, trimming the explants, and cultivating them on artificial media usually starts the culture, which then consists of a cycle of growth and multiplication. Multiple shoots develop from the original plantlets, which can be stimulated by additional cutting and/or hormones (Bhojwani 1980; Moriconi et al. 1990; Kahane et al. 1992). The growth phase may be delayed by cold storage between −1 and 10 °C, depending on the material, allowing maintenance of the material in slow-growth cycles (Keller et al. 1997; Kästner et al. 2001; Keller and Senula 2013). For mass propagation, the hormone concentration may be augmented. The upper limit of such high concentrations depends on the occurrence of induced alterations of the metabolism (hyperhydricity) and possible mutagenic effects of the hormones once the concentration is too high. Alternatively to the shoot tips from bulbs or bulbils, pieces of the whole basal plate or from young inflorescence bases may be a good source of material. This had been described for micropropagation of onion already earlier (Dunstan and Short 1979). Inflorescence bases are usable especially in species that form voluminous inflorescences such as garlic, great-headed garlic, onion and shallot. There are also reports on regeneration from callus culture (Havranek and Novak 1973; Barandiaran et al. 1999) and somatic embryogenesis as being usable for in vitro culture (Bockish et al. 1997; Féréol et al. 2002). No successes have been reported to cultivate protoplasts for conservation. This was mainly due to microbial contamination (Fellner 1993; Fellner et al. 1996).

The characteristic feature of in vitro culture, namely the need to transfer the material from one subculture to the next one and risks of infection by fungi, latent bacteria or active vectors like mites and thrips, are drawbacks of in vitro storage. This weakness becomes evident after perpetuation of culture cycles for many years, which does not only delimitate the duration of in vitro storage but can also influence the success of cryopreservation when using in vitro cultures as source material (Keller 2005). For *Allium*, in vitro storage is mainly used as a background for micropropagation, haploid production in onion, and cryopreservation, not as a basis for independent in vitro collections.

Cryopreservation

Cryopreservation, the storage of living material at between −196 and −140 °C in or over liquid nitrogen, is able to circumvent these drawbacks (Fig. 3.4). For *Allium*, first success was described by Niwata (1995), Hannan and Garoutte (1996) and Makowska et al. (1999). Overviews were given by Keller (2002), Keller et al. (2011a, 2012c), and Keller and Senula (2013). The main crop for cryopreservation in *Allium* has been until present garlic.

But other germplasms, such as shallot and top onion (Kim et al. 2007a; Senula et al. 2010, Keller and Senula 2016), and some wild species (Keller et al. 2011a; Keller and Senula 2016) are also considered, so far usually with lower success. When the material is properly introduced into cryopreservation, it is no longer endangered by metabolic or genetic changes or infections. Cryopreservation is a procedure requiring relatively high labour input by well-skilled personnel, but this is counterbalanced by the very low needs for the maintenance once the material is in storage. Several critical factors need to be considered:

(1) The size of the explant: usually explants need to have a diameter of 0.5–1 mm and a length of maximally 1–2 mm. If the explant is too large, thermal tensions may destroy it

Fig. 3.4 Transferring shoot tip explants excised from garlic bulbils into cryotubes for storing them in liquid nitrogen by E.R. Joachim Keller and Jiří Zámečník

during cooling or warming and/or cryoprotective substances may not enter all its parts. If it is too small, the capacity to regenerate may be lost (Baek et al. 2003; Keller 2002).

(2) The water status in the tissue: water exists in manifold forms in the cell. It is the matrix of cellular solutions (free water, e.g. in the vacuole), and it forms hydration films around ions and polar compounds (bound water). During cooling, the free water molecules form ice crystals at temperatures between 0 and −40 °C, depending on the concentration of the solutes. Ice crystals may disrupt the cellular structures. Then, on the way down to the temperature of liquid nitrogen, several structural changes of this ice may happen. The same is true when the object is rewarmed, where expansion of ice may be a critical factor blowing and breaking the cells open. These effects may be diminished by removing a major part of the free water by dehydrating treatments (pre-drying by osmotically active solutions, cryoprotectants; even by air drying).

Destruction of the tissue may follow two ways: depending on the cooling speed, ice crystals form first in the cell wall space and the cell content is dried out because of the osmotic water fluxes (slow freezing) or ice is formed inside the cell (fast freezing), disrupting its content (Mazur 1965; Mazur and Schmidt 1968). In an equilibrated stage, both tendencies counteract and this equilibration was reached in slow-freezing programmes. These were the first cryopreservation programmes usable for plants. Recently a more efficient method can be used for a part of the objects, mainly for shoot tips. When the cooling speed and the concentration of cryoprotective solutions are high enough, the object may be transferred into vitrification. This is an amorphous ('glassy'—'*vitrum*': lat. glass) state of water in solutions, not disrupting cellular structures. However, breaks by thermal tension at a very low temperature and during warming as well as structural changes into ice ('recrystallization') in the rewarming phase cannot be excluded completely. Since the state of the tissue water is one of the crucial factors, investigations are running also for garlic (Zámečník et al. 2001, 2012; Zámecník and Faltus 2014; Lynch et al. 2016).

The influences of these factors depend on the tissue type and its structural and biochemical composition. It may vary rather individually in the diverse explants, their conditions before excision and the preparation method. Pre-treatments and the cryo-procedure itself exhibit stress to the object and may hurt it. Not all explants may survive. Therefore, sufficiently high numbers of individual explants need to be collected for the given germplasm and tests on the usability of the material and suitability of the method are needed in every case. Usually establishing optimum curves is needed in which the reduction of survival caused by the treatments in a control sample without liquid nitrogen is recorded against the cryoprotective effects of the treatments in liquid nitrogen. Furthermore, since the regeneration of viable plants is the ultimate criterion of cryopreservation, a regeneration control sample needs to be taken with every charge of the cryoprotected material. The sum of these calculations needs to give a result satisfying the expectation that at least one (or better more than one) living plant can be derived from a charge of cryoprotected explants. Statistical considerations about that were done by Dussert et al. (2003).

The most detailed and comprehensive investigations on garlic cryopreservation were performed by Kim in the South Korean Genebank at Suwon. In course of his dissertation (Kim 2004), he worked on all aspects of the cryopreservation protocol (Kim et al. 2004a), beginning with the explant preparation over cryoprotectant comparisons, details of the cryoprotectant uptake (Kim et al. 2004b), features of the tissue water (Kim et al. 2005b) and contents of cryoprotectant in the cell (Kim et al. 2005a) until regeneration efficiency. Furthermore, he invented alternative loading and cryoprotectant solutions (Kim et al. 2009a, b), and he used alternative sources for explants such as inflorescence bases using also new protocols (Kim et al. 2006, 2007b). He also worked on the use of cryopreservation for virus elimination (cryotherapy—Kim et al. 2007a). Finally, on the base of his experience, he was able to include the major part of the South Korean garlic field collection and some other *Allium* species to build up a large cryobank of *Allium* (Kim et al. 2007b), actually the biggest one in the world (Kim et al. 2012).

Other laboratories also contributed with methodical investigations on comparing cryoprotectants (Volk et al. 2004b; Volk and Walters 2006), features of the tissue water measured by thermal analyses (Zámečník and Šesták 2011; Zámečník et al. 2001, 2012), age of usable explants (Keller 2005), alternative explants such as inflorescence bases (Keller and Senula 2016), bulb bases (Lynch et al. 2011, 2012; Souch 2006; Lynch et al. 2016), embryogenic callus (Sudarmonowati 2001) and embryogenic cell suspensions (Féréol et al. 2005) and management of the collections including implementation of linked research (Ellis et al. 2006; Keller et al. 2014, 2016; Zámečník 2002; Zámečník et al. 2007). The largest garlic cryobank is that of the National Agrobiodiversity Center at Suwon, South Korea amounting to 1158 accessions (Kim et al. 2012). For further cryo-collections of vegetative *Allium* see Table 3.4.

International actions and projects considerably contributed to the consolidation of this technique (COST, EURALLIVEG, AEGIS—Keller et al. 2011b, c, 2012a, b; Zanke et al. 2011).

It will not come as a surprise that the need of skilled technical personnel and the observation of such sophisticated factors is one of the main reasons to have some reservations to use cryopreservation in the conservation of vegetatively maintained PGR despite the fact that it is, indeed, the only safe way to keep vegetative material in the long time (Keller et al. 2008). The counterbalance of the relatively high introduction costs against the very low maintenance costs was, therefore, already from early on a matter of investigation (Dulloo et al. 2009; Keller 2006; Schäfer-Menuhr et al. 1996). For garlic, it could be demonstrated that in dependence on the type of material and the specific maintenance methods a break-even point is reached after 6–10 years, after which cryopreservation is more cost-effective than field culture. Since garlic is relatively easy to maintain in the field, in contrast to, e.g. potato, this critical point is reached later in garlic than for potato (Keller et al. 2013b). Thus, the maintenance strategy widely depends on the germplasm considered. Very valuable material (rare genotypes, virus-free clones, etc.) may be more acceptable for cryopreservation than ordinary varieties. Since recovering of the plants from cryopreservation causes specific costs, rarely requested material may come more

Table 3.4 Survey on some cryobanks of vegetative *Allium*

Country	Total in cryo	*Allium* in cryo	Data provided by	Info date
Poland	167	167	Marta Olas-Sochacka	16.1.17
Czech Republic	413	127	Jiří Zamečník	7.2.17
South Korea	1675	1158[a]	Haeng-Hoon Kim	5.4.17
USA	4362 (3903[b])	135 (100[b])	Maria Jenderek	1.3.17
Germany	1491	160	Angelika Senula	6.3.17

[a]Including some accessions of *A. cepa Aggregatum* group, *A.* x *proliferum*, and *A. macrostemon*; [b]accessions with >40% post-cryo-viability

into focus to be 'put away' into cryotanks than frequently required germplasm. Increasing labour costs for field workers may be an effect in favour of cryopreservation moving the break-even point to earlier times. In any case, cryopreservation may serve as a safety backup for all kind of material.

Pollen as natural propagules of genetic information is also well suitable for storage of *Allium* germplasm. Representing insect-pollinated species, *Allium* has highly dehydrated pollen which can already be stored at ambient temperature for some time. Mean storability of Amaryllidaceae pollen in air dry state was reported to be 38 days and at favourable humidity even 75 days (Johri and Vasil 1961). All the more, longer storage is possible at low temperatures. Cryopreservation is a suitable method for a long-term preservation of *Allium* pollen. First protocols for *Allium* were developed by Ganeshan and co-workers (Ganeshan 1986; Ganeshan et al. 2008). It is easy to keep the pollen in a dry state in capsules or cryotubes. Senula and Keller (2014) used strips of sandpaper for harvesting the pollen grains. The strips were subsequently placed into dry empty cryotubes. In principle, pollen storage is a good complement to storage of seeds and vegetative plant material. If, like in some wild species, low seed set or poorly germinating seeds are available only, pollen represents an additional source of genetic material. Due to the small size of pollen grains, a more diverse material can be stored than it may be possible by seeds. It has, however, also some complications, because for recovering a sample, always a crossing partner must be present to regenerate the plants, and only the genome of the germplasm can be kept, not the plastome. Insofar, pollen storage is a complementary strategy, not a method by its own.

In the collection of IPK Gatersleben, pollen storage of *Allium* was started in 2010. To check the suitability of the method, microscopical comparisons of viability staining with germination in a suitable medium (Brewbaker and Kwack 1963) were done on the pollen. Usually staining results suggested better viability than germination tests did, due to possibly still functioning enzymes reacting with the dyes when the grains were already damaged. The ultimate criterion is the seed set. It was demonstrated in model cases using onion and the wild species *A. obliquum* L. In the years 2010–2012, a pollen harvest campaign was performed in 176 accessions of the Taxonomic Reference Collection of IPK. The mean pollen germination rate was 21%. Direct comparisons of pollen germination were made in 159 accessions of *Allium* and some other species of Amaryllidaceae. The germination rate of cryopreserved pollen was 78% of the rate of fresh pollen. In the species *A. obliquum*, seed set with cryo-pollen was 60% of that by using fresh pollen. In onion, this rate was only 25% (Senula and Keller 2014). It was demonstrated that differential scanning calorimetry (DSC) and cryo-microscopy are suitable tools to investigate water content before and demonstrate the intactness of pollen after cryopreservation (Senula and Keller 2014; Keller and Senula 2016).

Recent surveys on cryopreservation, including *Allium*, are published in the comprehensive manual edited by Reed (2008) and further book chapters and articles (Engelmann 2012; Nirmal Babu et al. 2012; Keller et al. 2013a; Keller and Senula 2013; Popova et al. 2015).

Fertility Restoration in Garlic

Significant progress has been made in the past three decades concerning the ex situ maintenance of garlic as in the past this was only possible via vegetative maintenance. However, Etoh (1986) collected potentially fertile garlic from the centre of biodiversity of this species namely Central Asia. Later on, it was proven that garlic having the power to fertility restoration can be found in many locations in Central Asia, which also opened up the way to maintain at least part of the garlic collections held worldwide via seeds. Furthermore, fertile garlic also opened the way to develop improved garlic cultivars by means of sexual hybridization (Jenderek and Hannan 2000; Etoh et al. 2001; Kamenetsky and Rabinowitch 2001; Kamenetsky et al. 2005; Shemesh-Mayer et al. 2015).

3.4 Documentation

The utilization of collections depends for a considerable part on their documentation, which is needed to select accessions on the basis of their value compared to other accessions. For this passport, characterization and evaluation data need to be arranged and administered. Whereas in former times, data were collected on cards and in notebooks, digitalized data management is now the rule. Passport data covering the origin of the germplasm are stored in databases, often together with characterization data. One of the best measures to document an accession is its deposit in herbaria and fruit/seed collections. Since some characters such as perianth form or colour may change in the drying process, collections of photographic pictures are valuable complements, which can be presented in databases. Management data are handled in genebank information systems and laboratory information management systems (LIMS). The material becomes increasingly valuable when evaluation data are recorded which may be the subject of permanent observations over several growth periods. Close cooperation with breeders and other users can support these activities (see Sect. 3.5).

There are several databases which can be consulted on various information levels. It begins with the correct botanical names. There is still a considerable confusion concerning exact botanical terms. For *Allium*, the Nomenclator Alliorum should be used as the botanical standard (Gregory et al. 1998). Additional information on this topic can be gained by consulting databases implemented by the Royal Botanic Garden Kew (The Plant List[7]; WCSP[8]), The International Plant Names Index IPNI,[9] the Tropicos database of the Missouri Botanical Garden[10] and the Germplasm Resource Information Network (GRIN) taxonomy database.[11] Here, the user can find whether the name he/she has available is an accepted one, a synonym or simply an error. Further information can be derived from the Global Biodiversity Information Facility GBIF.[12] A survey over the number of accessions present in a selected genebank and other information is available amongst others via the GENESYS and World Information and Early Warning System (WIEWS)[13] databases; harvest data of the main crops including *Allium* are documented in the FAOSTAT database,[14] both implemented by the Food and Agriculture Organization of the United Nations FAO. In Europe, the European Cooperative Programme for Plant Genetic Resources provides a survey of the *Allium* accessions within the EURISCO[15] database (Weise et al. 2017). Furthermore, there is the European *Allium* Database (EALLDB),[16] recently running at IPK

[7]http://www.theplantlist.org/.

[8]http://apps.kew.org/wcsp/home.do.

[9]http://www.ipni.org/ipni/plantnamesearchpage.do.

[10]http://www.tropicos.org/.

[11]https://npgsweb.ars-grin.gov/gringlobal/taxonomybrowse.aspx.

[12]http://www.gbif.org/.

[13]http://www.fao.org/wiews-archive/wiews.jsp.

[14]http://www.fao.org/faostat/en/#home.

[15]http://eurisco.ipk-gatersleben.de/apex/f?p=103:1.

[16]http://ealldb.ipk-gatersleben.de/861/.

Gatersleben. EURISCO and EALLDB use standardized passport datasets to describe accessions. They are also step by step completed by characterization and evaluation data.

Additionally to these general databases, special databases provide information, often supported by comprehensive sets of images. IPK Gatersleben provides an image database on its most important garlic and shallot accessions, the Garlic and Shallot Core Collection (GSCC)[17] as well as a database on wild *Allium* species Taxonomic *Allium* Reference Collection Database.[18] GSCC comprises passport data, infraspecific grouping, pictures of field plots, bulb and inflorescences in and, in some cases, ontogeny of the given accession (Colmsee et al. 2012).

In order to increase the international comparability of crops in the diverse collections, European Cooperative Programme for Plant Genetic Resources (ECPGR) initiated the so-called descriptor lists covering passport, characterization and evaluation data. In the course of a European project on genetic resources, a subset of 100 accessions of garlic was characterized by minimum descriptors in IPK (Keller and Senula 2001). These descriptors were elaborated and refined together with other working groups and were finally condensed in the International Plant Genetic Resources Institute (IPGRI) minimum descriptor list of *Allium* (IPGRI et al. 2001). This characterization covered 16 characters of garlic which were used to identify the material such as the shapes and structures of bulbs, their colour, ability to flower, leaf attitude, etc. All these characters were used to formulate infraspecific groups (Messiaen et al 1993) and can be set into relation with existing classifications which are based on molecular markers (Maaß and Klaas 1995; García Lampasona et al. 2003; Ipek et al. 2003, 2005; Volk et al. 2004a; Pinto da Cunha et al. 2014).

IPK hosts one of the largest crop species herbaria. Herbarium voucher specimens have been sampled during collecting missions and from the living Taxonomic *Allium* Reference Collection (8500 specimens including older specimens at Gatersleben). The Crop *Allium* Collection is documented by 2500 herbarium vouchers from the Gatersleben field propagation (Fritsch and Pistrick, pers. comm.). The others come from the wild species collection. All this material is a part of the Herbarium Gatersleben. Recently, a project was funded to digitalize these herbaria specimen in order to make the herbarium more easily accessible. In the frame of the ongoing IPK Herbarium digitalization (Pistrick and Knüpffer 2013), 7260 scans with the standardized collecting and nomenclatural data have been provided via the international database system Virtual Herbaria JACQ[19] and combined with data from the Genebank Information System (GBIS)[20] of the IPK in the Herbarium Gatersleben.[21] Image databases complement the herbarium (IPK GSCC and IPK Taxonomic *Allium* Reference Collection).

3.4.1 Global Crop Portals

A large improvement in the conservation and utilization of plant genetic resources since the 2000s has been the development of global genetic resources portals on the internet, which allow for the first time online searches, in which the precision of the analysis depends, of course, on the passport data included in the databases. These global portals did not exist when Astley wrote his chapter on the conservation of genetic resources in the Onions and *Allium* Crops Vol I book (Rabinowitch and Brewster 1990), as at that time only paper versions existed of plant genetic resources (e.g. Frison and Serwinski 1995). As different catalogs were not easy to obtain and no common format was used, it was difficult to analyze these catalogs in combination. Fortunately, this has changed dramatically in the past years.

According to the global portal on Plant Genetic Resources for Food and Agriculture GENESYS,[22]

[17] http://www.ipk-gatersleben.de/databases/gscc.

[18] http://www.ipk-gatersleben.de/databases/allium.

[19] http://herbarium.univie.ac.at/database/collections.htm.

[20] http://gbis.ipk-gatersleben.de/.

[21] http://herbarportal.ipk-gatersleben.de/.

[22] www.genesys-pgr.org.

there are 20,627 *Allium* accessions conserved and according to the global portal of the Botanical Gardens Conservation International PLANTSEARCH[23] 6270 *Allium* accessions are being conserved. These numbers are, however, minimum estimates as *Allium* accessions from genebanks of countries like PR China (*n* = 1030; H Wang, pers. comm.), India (*n* = 2050; Khosa et al. 2016) and Japan (*n* = 1352; FAO 2014) are not yet included in these databases. Without any doubt, a significant improvement would be that countries upload and refresh their *Allium* genetic resources in both databases which could help the protection of these valuable genetic resources.

Around 1000 *Allium* species are currently recognized (FO Khassanov, pers. comm.; Table 3.5). The most recent classification of *Allium* species worldwide has appeared in 2006 and was written by Friesen, Fritsch and Blattner for the whole genus and also for several subgenera (Friesen et al. 2006; Fritsch 2015, 2016; Fritsch and Gurushidze 2009; Fritsch et al. 2010; Fritsch and Abbasi 2013; Gurushidze et al. 2008). They recognized 14 subgenera and most *Allium* species occur in three subgenera namely *Allium, Amerallium* and *Melanocrommyum* (see Table 3.5).

In genebanks, most accessions are being conserved namely 20,627, whereas in botanical gardens 6270 accessions are maintained. Five hundred and eighty-one (*n* = 581) species are conserved worldwide, which is 56% of the number of species (*n* = 1038) recognized. Slightly more *Allium* species are being conserved by genebanks (*n* = 445) compared to botanical gardens (*n* = 409). This is for a large part due to the large number of species conserved in genebanks of subgenus *Melanocrommyum* (*n* = 58) and *Allium* (*n* = 54) compared to botanical gardens (*n* = 9, respectively, *n* = 41). The percentage of species present in genebanks and botanical gardens versus the total number of species recognized per subgenus is on average 56% and varies for the different subgenera in between 33% (*Caloscordum*) to 100% (*Butomissa, Microscordum, Nectaroscordum, Porphyroprason* and *Vvedenskya*).

[23]https://www.bgci.org.

On the basis of the statistical overview of GENESYS, the number of holding institutes for *Allium* species is 86. The countries with the highest numbers of accessions conserved are USA, Germany, Spain, Russia and United Kingdom: all these countries conserve over 2000 accessions in their genebanks (Fig. 3.5).

The composition of the genebank collections worldwide in terms of the biological status of an accession (SAMPSTAT descriptor,[24] Multi-Crop Passport Descriptors v2, 2012) is presented in Fig. 3.6. Landraces, improved/advanced cultivars and wild material are the categories for which most accessions are present. No GMO accessions are reported in the genebank collections worldwide.

On the basis of the data from GENESYS, most *Allium* accessions in genebanks worldwide are conserved only as seed (*n* = 10,938) or in combination with other conservation methods like field, in vitro, cryo and other (*n* = 1361). Also, a considerable number of accessions are cultivated only in the field (*n* = 4417) or in combination with other measures (*n* = 1954). Other single methods of storage like in vitro, cryo and DNA storage are less used. The GENESYS data can be confirmed for Europe, where a recent overview about maintenance methods and their duplication was provided by EURISCO (Weise pers. comm. 3. 3. 2017; Table 3.6).

As GENESYS reports that 'no historic records are present (=false)', it can be assumed that all 20,627 accessions are present in genebanks worldwide. However, the availability of these *Allium* accessions is far from clear as for 17,053 accessions it is not specified whether they are available. Only 1999 accessions are reported available for distribution and 1571 are reported not to be available at present for distribution.

Safety duplication of seed accessions has taken place to a large extent as 18,511 accessions (90%) have been duplicated in 16 sites as reported to GENESYS. Safety duplication of *Allium* accessions at the Global Seed Vault in

[24]http://www.bioversityinternational.org/fileadmin/user_upload/online_library/publications/pdfs/FAOBioversity_multi_crop_passport_descriptors_V_2_Final_rev_1526.pdf).

Table 3.5 An overview of the *Allium* genetic resources maintained worldwide on the basis of the global portals GENESYS (genebanks: gb) and PLANTSEARCH (botanical gardens: bg), #: number, acc: accession

Subgenus	Total # of species	# unique species in bg	# unique species in gb	# species in common	Total # species in bg and gb	% # species in bg and gb and # species	# acc bg	# acc gb
Allium	434	41	54	79	174	40	1442	7244
Amerallium	180	38	19	52	109	61	1341	524
Anguinum	14	3	2	1	6	43	115	45
Butomissa	4	0	0	4	4	100	196	223
Caloscordum	3	0	0	1	1	33	3	5
Cepa	32	5	7	17	29	91	873	10031
Cyathophora	5	1	0	2	3	60	78	18
Melanocrommyum	170	9	58	57	124	73	1080	526
Microscordum	1	1	0	0	1	100	7	0
Nectaroscordum	3	0	0	3	3	100	34	8
Polyprason	60	16	13	25	54	90	310	222
Porphyroprason	1	0	0	1	1	100	65	16
Reticulato-bulbosa	85	13	10	15	38	45	203	117
Rhizirideum	45	9	9	15	33	73	408	353
Vvedenskya	1	0	0	1	1	100	0	3
?							115	1292
Total	1038	136	172	273	581	56	6270	20,627

Subgenera according to Friesen et al. (2006)

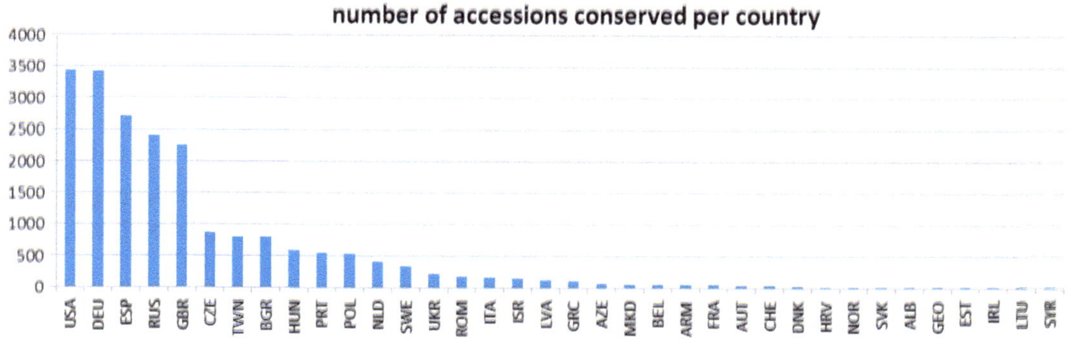

Fig. 3.5 The number of *Allium* accessions conserved per country based on GENESYS

Svalbard has also taken place to a large extent as 19,998 accessions (97%) have been duplicated at the Vault. Safety duplication of vegetatively propagated *Allium* species has also taken place. In this context, a part of the *Allium* field collection of the Czech genebank at Olomouc was duplicated in IPK in 1993 and 1994 (Maggioni et al. 2002). Cryopreserved samples have also been duplicated. In the frame of the European *Allium* project EURALLIVEG,[25] a trilateral

[25]http://euralliveg.ipk-gatersleben.de/.

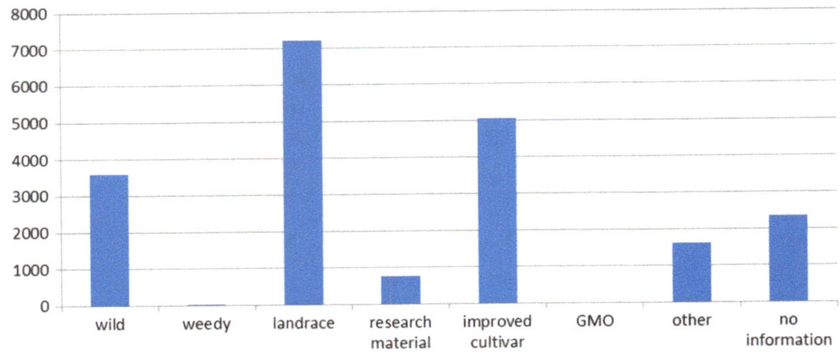

Fig. 3.6 The number of *Allium* accessions per biological status of the material held in genebanks worldwide based on GENESYS

Table 3.6 The method of conservation of *Allium* accessions as provided by EURISCO

	Seed	Field	In vitro	Cryo[a]	Sum
Seed	7661	1224	16	89	8990
Field		4468	442	258	5168
In vitro			475	136	611
Cryo				340	340
Sum	7661	5692	933	823	15,109

[a]The cryopreservation figure does not include the samples stored as pollen

safety duplication was established between the Czech Republic, Germany and Poland covering 200 accessions (Keller et al. 2012c).

The amount of ordinary duplication among accessions cannot be determined via GENESYS. The European PGR database EURISCO[26] allows for such an analysis as the fields DONORCODE and DONOR DESCRIPTION can be used as a proxy for duplication. In Table 3.7 an overview is presented on the percentage of duplication that has taken place in European genebanks. This percentage of duplication varies between 1 and 100% with an average of 47%. However, one needs to be carefully interpreting these data as they might give an overestimation of the percentage of duplication. This can, among others, be due to the fact that collections from which accessions were duplicated do not exist anymore and as such the accessions that were originally duplicated are not duplicated anymore. In case of DEU, this applies to 79 accessions and in case of NLD, this applies to 247 accessions, which would lead to a percentage of duplication of 26 and 42% in case of DEU and NLD, respectively.

Eighty-nine percent (=18,409/20,627) of the *Allium* genetic resources held worldwide are cultivated species (Table 3.8).

The number of accessions of cultivated *Allium* species (see for overview Fritsch and Friesen 2002) occupy, as expected, the major share (91%) in genebanks compared to botanical gardens (9%). The number of cultivated species held in both genebanks and botanical gardens is 23 for genebanks and 22 for botanical gardens.

3.5 Characterization, Evaluation, Utilization and Distribution

To stimulate the use of PGR, it is of significant importance to characterize (description of phenotypic traits) and evaluate (description of disease and pest resistances, biochemical traits, molecular markers, etc.) the material present in a collection.

CGN and to a lesser extent the German IPK genebank have been chosen as an example in this context. At CGN, ample consideration is given to the characterization and evaluation of its collections. The total number of *Allium* accessions in

[26]http://eurisco.ipk-gatersleben.de.

Table 3.7 Duplication of *Allium* accessions within Europe based on EURISCO data (dump: March 28 2017), using donor code and donor description as a proxy for duplication; data kindly provided by S Weise (IPK, Germany)

Holding country	No. of *Allium* acc held	No. of *Allium* access with DONORCODE/DONORDESCR	% of duplication
ARM	57	1	2
AUT	44	15	34
AZE	40	1	3
BGR	816	374	46
CHE	50	48	96
CZE	817	666	82
DEU	2772	789	28
ESP	1882	262	14
FRAU	54	33	61
GBR	2153	1784	83
HUN	590	230	39
ITA	170	1	1
LTU	1	1	100
NGB	312	120	38
NLD	428	428	100
POL	548	320	58
ROU	176	15	9
SVK	5	5	100
UKR	229	20	9

the collection is modest namely 428. In total, 42 phenotypic traits were characterized and six disease and one pest resistance trait(s) were evaluated in the past (Table 3.9).

The characterization of the material was carried out by CGN only, whereas the majority of the disease and pest evaluations were carried in combination with breeding companies. In the latter case, CGN coordinated the evaluation taking care of the distribution of accessions to breeding companies and coordinating the data management, whereas breeding companies did the evaluations. The data obtained were subjected to an embargo period, which did not last longer than 5 years after an accession was evaluated. For many genebanks, however, the cooperation with breeding companies is not feasible due to various reasons and this limits their genebank operation (Frese et al 2016a, b). IPK published the characterization data of their garlic and shallot core collections in their GSCC database (Colmsee et al. 2012).

A policy focussed on the utilization of germplasm together with a strict policy concerning the online availability of the material, resulted in an annual distribution rate in the last decade of 280 samples of *Allium* accessions from the CGN genebank and 380 samples from the IPK genebank. Therefore, over the last 10 years, both genebanks accounted for the distribution of around 6600 *Allium* samples in total (Fig. 3.7).

3.6 Exploitation of Wild Relatives

The species belonging to genepool 1 for the various *Allium* crops are often not really known as few species crossability studies have been made to identify these genepools (e.g. Phillips and Hubstenberger 1987; Van Raamsdonk et al

Table 3.8 The number of accessions of cultivated *Allium* species (sensu Fritsch and Friesen 2002) conserved per subgenus and in botanical gardens and genebanks. The total number of accessions held is 18539, of which 1679 are held in botanical gardens and 16860 in genebanks

Subgenus/species	# accessions	Botanical garden	Genebank	Subgenus/species	# accessions	Botanical garden	Genebank
Allium				***Butomissa***			
ampeloprasum	2013	161	1852	*ramosum*	103	45	58
sativum	4634	130	4504	*tuberosum*	299	142	157
macrostemon	23	14	9	***Cepa***			
rotundum	127	30	97	*fistulosum*	975	105	870
Amerallium				*altaicum*	136	42	94
canadense	49	38	11	*cepa*	8660	215	8445
hookeri	15	11	4	*chinense*	33	16	17
kunthii	12	7	5	*oschaninii*	48	18	30
neapolitanum	117	76	41	*pskemense*	66	38	28
ursinum	177	103	74	*schoenoprasum*	607	256	351
wallichii	30	26	4	*x proliferum*	81	0	81
Anguinum				***Polyprason***			
victorialis	150	104	46	*obliquum*	63	42	21
				Rhizirideum			
				nutans	121	60	61

Table 3.9 The number of traits, accessions and observations characterized and evaluated at the CGN *Allium* collection

Type of data	Number of traits observed	Mean (min-max) number of accessions	Mean (min-max) number of observations
Phenotypic	42	137 (25–619)	244 (25–327)
Disease	6	38 (8–110)	44 (14–110)
Pest	1	49	73

2000, 2003). The only crop/genepool 1 complex that has been studied in some detail is *A. cepa* and its wild relatives. In this context, *A. galanthum, A. roylei, A. vavilovii, A. fistulosum, A. oschaninii, A. pskemense* and *A. altaicum* are wild relatives that have been successfully crossed with onion. However, as only part of the 32 species belonging to subgenus *Cepa* have been studied, it would be of interest to carry out a more in-depth study comprising more species from this subgenus as wild relatives have proven to be unique gene reservoirs for the improvement of crops. This not only is true for onion, but also for the other *Allium* crops (Kik 2002). However, the number of crop wild relatives accessions in genebank collections worldwide is small (e.g. for onion less than 200 accessions) and it would be of great interest to collect more of these species not only for breeding new cultivars but also for fundamental research.

For example, crosses between *A. cepa* and *A. galanthum* showed that *galanthum* mtDNA in a nuclear background of *A. cepa* resulted in cytoplasmic male sterility (CMS). This maybe of importance for F_1 hybrid breeding in onions as an additional source of CMS next to the S and T cytoplasm which are already known (Shigyo and Kik 2008). Not only species crosses should be considered within subgenera, but also between subgenera as is the case for *A. cepa* (*Cepa*) and *A. roylei* (*Polyprason*) (Khrustaleva and Kik 1998, 2000). This species cross proved to be of

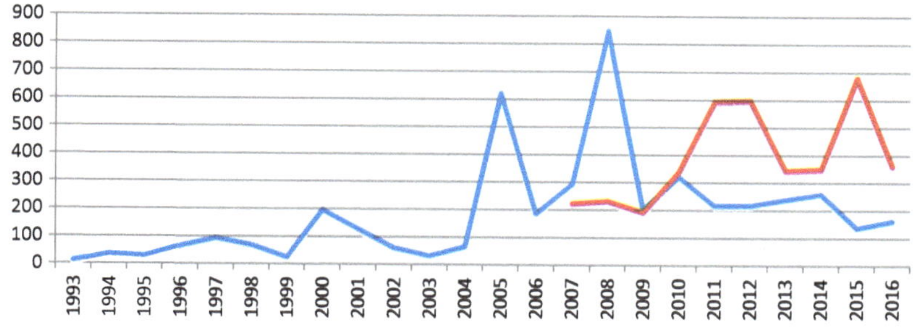

Fig. 3.7 The annual distribution of *Allium* samples from the IPK (red) and CGN (blue) collections since distribution was recorded at each genebank

high interest for breeders due to the complete resistance to *Peronospora destructor*, the causal agent of downy mildew, which is present in *A. roylei* (Kofoet et al. 1990; Scholten et al. 2007). Currently, onion cultivars resistant to downy mildew are used primarily in organic cultivation. As *A. roylei* is also highly resistant to *Botrytis squamosa*, the causal agent of onion leaf blight, the breeding of new cultivars resistant to both downy mildew and onion leaf blight seems to be within reach for the near future (de Vries et al. 1992; Scholten et al. 2016).

Some of the widely used ornamental alliums have also hybrid origin (Friesen et al. 1997). Crosses with more or less distant species may be facilitated by in vitro culture using embryo rescue techniques. Their principle consists of the following facts: in angiospermous plants, the fertilization is a two-way process. One of the two nuclei of the male gamete fuses with the egg cell, the other one with the diploid endosperm. Since the second process fails in many distant crosses, the embryo does not have nutritive tissue available. Cultivation on artificial media may support the embryo development, therefore it is called 'embryo rescue'. The first investigations were performed by Doležel et al. (1980) on hybrids with onion. Embryo rescue was used for crosses of *A. chinense* with distant species (Nomura and Oosawa 1990; Nomura et al. 1994). Crosses of onion with 19 species of the subgenera *Rhiziridеum* and *Allium* were performed resulting in primary hybrids in 19 combinations (Keller et al. 1996). Unfortunately, after colchicination chromosome doubling no fertility restoration could be reached (data unpublished). Genes from related species may also be introduced by protoplast fusion or gene transfer. *Allium* seems to be difficult in this respect, due to complications in the in vitro culture (Fellner and Havranek 1992; Fellner 1993; Fellner et al. 1996), but some first indications of successful culture and regeneration were published (van der Valk et al. 1992; Song and Peffley 1994; Karim and Adachi 1997; Barandiaran et al. 1999). In an effort to introduce CMS into leek, Buiteveld and Creemers-Molenaar (1994), Buiteveld et al. (1998a, b) produced interspecific hybrid plants based on somatic hybridization between onion and leek, however, these plants proved to be sterile. Ohsumi et al. (1992) produced interspecific plants between onion and garlic, however, also these plants proved to be sexually sterile.

Although not strictly fitting into the exploitation of wild *Allium* relatives, it is worth mentioning in this context the extensive research carried out in genetic transformation of *Allium* crops. Using *Agrobacterium tumefaciens* as a vector, Zheng et al. (2004, 2005) and Eady et al. (2008) managed to genetically transform *A. cepa* and *A. sativum*. Introducing, for example resistance to beet armyworm (Bt) into a background of onion and garlic and silencing the lachrymatory factor (LF) in onion.

3.7 Phytosanitary Aspects

In the EU phytosanitary guidelines[27], the seeds and bulbs intended for planting of the cultivated species *A. cepa* Common Onion group, *A. cepa* Aggregatum group,[28] *A. schoenoprasum* and *A. porrum* are subjected to control for *Ditylenchus dipsaci*. For all the other *Allium* species no control on *Ditylenchus dipsaci* is needed. The control of the material is needed for imports to and often exports from the EU and for transfer of material within the EU.

3.7.1 Virus

In collections of vegetatively propagated crops, which concern garlic, shallot and some minor *Allium* germplasm, one of the most important phytosanitary aspects is the colonization of the plants by viruses. The infections are transmitted by vectors like aphids and nematodes. In the life cycle of plants, the distribution of most viruses is blocked by the seed phase. In vegetatively propagated germplasm, this phase is absent, and, therefore, viruses may accumulate in the crops to a dangerous level, not only reducing vigor and yield but even endanger the existence of the plants itself (Havranek 1974; Conci et al. 2003). The first measure to manage the impact of viruses is to record the virus incidence by indexing them with biochemical (ELISA) or molecular (RT-PCR) techniques (Karlová et al. 2009; Lunello et al. 2005; Leisova-Svobodova and Karlova-Smekalova 2011). In the field of IPK Gatersleben, onion yellow dwarf (OYDV), leek yellow stripe (LYSV), garlic common latent (GLCV) and shallot latent (SLV) viruses as well as various allexiviruses (GarV-A, GarV-B, GarV-C, GarV-D, shallot virus X) were found in garlic (Keller et al. 1995, 2012a, c). Virus distribution was also screened in the Czech garlic collection (Smékalová et al. 2010). Eradication of viruses may be performed by meristem culture, thermo-, chemo- and cryotherapy. No one of these complementary techniques will per se result in 100 percent virus-free plants. Thus, after the treatments, repeated indexing is needed to prove the 'virus-free' state keeping in mind that particles may persist below the detection limit and provide the initial point for later new outbreaks. Furthermore, different virus species may be more or less easily eradicated. On the other hand, the effect of various viruses may be very different reaching from unimportant latent to very influential ones. In genebanks, the impact of these various viruses on the value of the germplasm needs to be thoroughly assessed to decide whether the germplasm can be distributed or not. Finally, it is necessary to separate the cleaned germplasm from the still infected material. Isolation of them in tightly closed field cages against aphids, as in vitro cultures or in cryopreservation is needed.

In *Allium*, the main way of virus elimination has so far been meristem culture. Since the growth of the plant tissue is mostly faster than the virus multiplication, the meristem proper and the first leaf primordia may be free of viruses or may contain a reduced level of them. If the isolation succeeds to separate the freed from the infected tissues, meristem culture may result in healthy tissue which can be cultivated for regenerating virus-free plants. First meristem culture of *Allium* was published by a Czech group (Havranek 1972). Later on, laboratories in UK, Argentina, France, Germany and the Netherlands followed (Walkey et al. 1987; Conci

[27]Concerning EU phytosanitary guidelines the following directives are of importance:

1. Council Directive 2000/29/EC of 8 May 2000 on protective measures against the introduction into the Community of organisms harmful to plants or plant products and against their spread within the Community (http://eur-lex.europa.eu/oj/direct-access.html; 2000, L169),
2. Commission Directive 2008/61/EC of 17 June 2008 establishing the conditions under which certain harmful organisms, plants, plant products and other objects listed in Annexes I–V to Council Directive 2000/29/EC may be introduced into or moved within the Community or certain protected zones thereof, for trial or scientific purposes and for work on varietal selections (http://eur-lex.europa.eu/oj/direct-access.html; 2008, L158/41).

[28]For historical reasons, this taxon is still mentioned as *Allium ascalonicum* in the EU directive, however, this name is invalid because it refers to another species.

and Nome 1991; Keller et al. 1995; Verbeek et al. 1995; Lot et al. 1998). In the Czech Republic, Argentina and Brazil, intense activities are running also recently (Fajardo et al. 2001; Karlová et al. 2009; Smékalová et al. 2010; Celli et al. 2016). In France and Argentina, virus-free material is subject to marketing. Activities of successful elimination of garlic viruses by means of cryotherapy are reported from South Korea, Japan and from Brazil (Ayabe and Sumi 2001; Kim et al. 2012; Vieira et al. 2015). The principle of cryotherapy is based on the fact, that meristems better survive cryopreservation than differentiated tissue around the meristem (Wang et al. 2009), thus acting in the same direction as manual manipulations for tissue excision (it is the so-called 'cryo-knife' Zámečník, EUR-ALLIVEG). Activities to reinforce the effect of meristem culture by heat treatments (thermotherapy) or chemicals (chemotherapy) were also reported (Conci and Nome 1991; Senula et al. 2000). In genebanks, virus-free material requires isolation in all types of culture to protect it from re-infection.

Since the treatments to free an accession of viruses are costly and time consuming, their benefit needs to be assessed. In genebank collections, it may be the case that only the most valuable material can be treated and not the main bulk (Senula and Keller 2002). On the other hand, in regular field production the higher income earned by higher yields of virus-free material in comparison to infected material, one may justify consistent treatment of large quantities of material in commercial companies (Conci et al. 2005) in conjunction with virus-free declarations in the frame of a certification procedure as it is the case in France (PROSEMAIL).[29]

3.8 *Allium* Networks

As for most crops, also for *Allium*, numerous joint activities exist around the world in order to bundle efforts for mutual benefit. In the 70s of the last century, the Food and Agriculture Organization of the United Nations (FAO) was aware of the needs for consolidation of the activities related to plant genetic resources and founded the International Board of Plant Genetic Resources (IBPGR) in 1974.[30] Currently, after transient re-naming (International Plant Genetic Resources Institute (IPGRI), 1991–2006 this organization is called Bioversity International, and located in Rome, Italy. The organization functions since 1994 under the umbrella of the Consultative Group on International Agricultural Research (CGIAR; founded in 1971)[31] Fifteen research centres exist, which have manifold cooperation with each other. Several other institutions network internationally. Amongst them, the World Vegetable Center, founded 1971 in Taiwan under the name 'Asian Vegetable Research and Development Center' AVRDC, a member of the Association of International Research and Development Centers AICRA,[32] coordinates research projects on *Allium* (Ebert 2011).[33] It holds also a collection of tropical and subtropical *Allium* germplasm. In Europe, the establishment of ECPGR[34] took place in 1980 and is funded by national governments. ECPGR consists at present of 21 working groups. One of the first was the *Allium* Working Group,[35] established in 1982 as one of the original six Working Groups developed during the first Phase of ECPGR. Its first Chairman was Dave Astley, Institute of Horticultural Research, Wellesbourne, UK. He was its Chairman until 2011, followed by E.R. Joachim Keller, IPK, Gatersleben, Germany (2011–2014) and, until today, by Helena Stavělíková (CRI, Dept. Vegetables and Special Crops, Olomouc, Czech Republic). The *Allium* Working Group has at present (February 2017) 66 members out of 36 countries.[36]

[29] http://plant-certifie-ail.org/en/index.php.

[30] www.bioversityinternational.org/about-us/who-we-are/history/.

[31] www.cgiar.org/.

[32] http://www.airca.org/.

[33] https://avrdc.org/.

[34] http://www.ecpgr.cgiar.org/about-ecpgr/overview/.

[35] http://archive-ecpgr.cgiar.org/working_groups/allium.html.

[36] http://www.ecpgr.cgiar.org/working-groups/allium/.

Various non-governmental and private entities should not be forgotten. Amongst them, the Global Crop Diversity Trust,[37] collecting and bundling donations of private persons and organizations in favour of crop diversity, is of major impact for plant genetic resources financing international collaborative projects.

3.8.1 EU-Funded *Allium* Projects

Allium-related projects were and are the main field of concrete activities. The direction of Agriculture of the EU funded, in its first call for the regulation 1467/94, a project entitled 'Protecting future European Community crops: a programme to conserve, characterize, evaluate and collect *Allium* crops and wild species' (1996–2001). In the course of the project implementation, two sub-groups were formed as the vegetatively propagated *Allium* genetic resources required priorities different from the seed-propagated ones. In the sub-project on vegetative *Allium* germplasm, a preliminary garlic core collection was established and material was micropropagated using meristem culture in the start phase. A number of 95 accessions were freed from 5 viruses, the status of which was proven by ELISA tests. Elaborate characterization took place, which contributed to the development of the *Allium* IPGRI descriptor list (IPGRI et al. 2001). On the basis of 16 minimum descriptors also a European garlic core collection of 50 accessions was developed. The conclusions of this project were further discussed in a special follow-up meeting of the ECPGR *Allium* Working group in 2001. The main technical subjects were characterization, virus elimination and cryopreservation (Maggioni et al. 2002). In the second call of regulation 870/2004, the Directorate on Agriculture granted the project EURALLIVEG ('Vegetatively propagated alliums, Europe's core collection with higher maintenance safety at lower cost and better health conditions') from 2007 until 2011 (Keller et al. 2012b). Apart from the activities to use molecular markers for describing the diversity and excluding undesired duplicates, the main subjects were cryopreservation and virus elimination. The main outcome of this project was the establishment of an initial European Core Collection of garlic covering 200 accessions which fulfill the criteria of Most Appropriate Accessions set by the European Genebank Integrated System AEGIS.[38] This core collection is preserved by a tripartite cryobank situated in the Czech, German and Polish genebanks. This safety duplication has been implemented in March 2010. Safety duplication is now an important constituent of the FAO genebank standards (FAO 2014; Hanson 2017). Thus, all the 200 accessions are safety-duplicated between the three partners so that all of them are situated in two places. Another joint action in this context was a small project financed by ECPGR/AEGIS, in the course of which three European partners (genebanks from Germany, Poland and Portugal) successfully adapted the droplet-vitrification technique to cryopreserve bases of young garlic inflorescences, which had been published by Kim et al. (2006), to the European conditions (Keller et al. 2011c).

Apart from these projects, directly focussed on genetic resources of *Allium*, several research projects had a substantial impact on *Allium* genebanking. In the years 1996–2001, a project was funded by the General Directorate for Research of EU named 'Tailoring the onion crop for the twenty-first century. The development of high quality fresh and processed onions' (FAIR CT95-465). Though it was mainly focussed on the organo-sulphur and carbohydrate metabolism, it also contributed to the establishment of a working collection of onion which proved to be very useful for the evaluating the onion germplasm for organo-sulphur and carbohydrate compounds (e.g. Kahane et al. 2001). From 2000 to 2004, another EU project on *Allium* was granted entitled 'Garlic & Health, the development of high-quality garlic and its influence on biomarkers of atherosclerosis and cancer in humans for disease prevention' (QLK1-CT-1999-498). This project

[37] https://www.croptrust.org/.

[38] http://www.ecpgr.cgiar.org/aegis/about-aegis/.

focussed on the potential health beneficial effects of garlic, and produced amongst others detailed data on the organo-sulphur metabolism in garlic germplasm (Kik et al. 2001; Kamenetsky et al. 2007; Zheng et al. 2007; Huchette et al. 2007; Jones et al. 2007; Espirito Santo et al. 2007; Siess et al. 2007). From 2004 to 2006, a research grant from the EU was obtained in the framework of the SEAFOODplus project which was carried out in the sixth framework programme of the EU. In this research grant, research was carried out to analyse the interaction between garlic germplasm and arbuscular mycorrhizal fungi in the uptake of selenium and conversion into organo-selenium compounds (Larsen et al 2006).

For special research topics, such as cryopreservation of *Allium*, a COST grant was obtained from the European Science Foundation (ESF). The COST action 871 CRYOPLANET ('Cryopreservation of Crop Species in Europe'—2007–2010) brought together many researchers from Europe to further develop cryopreservation by scientific meetings and training courses including *Allium* cryopreservation (Keller et al. 2011b).

3.9 Conclusions

An update is given on the progress that has been made in the management of *Allium* genetic resources over the past 25 years. A significant development during the past twenty-five years since Astley wrote his chapter on *Allium* genetic resources in the book entitled Onions and Allied Crops (Rabinowitch and Brewster 1990) has been the introduction and further development of web-based information systems (e.g. GENESYS, PLANTSEARCH), which facilitated the exchange of data to a large extent between *Allium* collection holders worldwide. These information systems made it possible to obtain an overview of the *Allium* genetic resources managed worldwide. However, these overviews are at the moment not completely accurate as many collection holders did not yet contribute to these systems. Nevertheless, it became already clear, that still, large gaps are present in *Allium* collections worldwide, especially with respect to the wild crossable relatives of the various *Allium* crops. This warrants future collecting expeditions to the centres of biodiversity of the various species.

A second significant area of progress in *Allium* genetic resources is the development of new methods for the maintenance of *Allium* germplasm. An important development in this area was the introduction of cryopreservation. This method has made it possible to maintain *Allium* accessions in a cheap and effective way. The method is especially important for the conservation of vegetatively maintained germplasm. The more 'traditional' methods, such as ex situ field and seed management were refined in the past 25 years, but not essentially revolutionized.

Other developments in *Allium* genetic resources are the improvement of the health status of the germplasm kept in the collections and furthermore the ongoing characterization and evaluation of accessions, stimulating the utilization of the *Allium* genetic resources.

Fundamental changes could also be observed with respect to acquisition and exchange of plant genetic resources everywhere in the world due to many and complex new regulations on the legal and organizational levels due to the adoption of the CBD and IT-PGRFA by many countries. It makes the handling of the plant accessions safer and more consistent but also more circumstantial.

Finally, we need to underline that in a more and more changing world with all the threats of genetic erosion and extinction due to disappearance of traditional cultivation methods, devastation of our environment and climatic change, the conservation of genetic resources remains of prime importance for agriculture. Especially for breeders and researchers, this is true as a highly diverse gene pool is for them an invaluable treasure. The importance to keep this treasure will no doubt become even more important in the future.

Acknowledgements The authors gratefully acknowledge the contributions they received from many colleagues working in the field of *Allium* genetic resources, mentioned in the text. Many of them provided information which added to our knowledge of activities occurring worldwide, especially for the data about their collections. More, in particular, CK wants to thank Dr. F. O. Khassanov for his help with respect to the evaluation of the taxonomic data as presented in Tables 3.5 and 3.8. Furthermore, the authors are also highly appreciative to all the donors of funds who made it possible to conduct projects and to carry out the necessary tasks to manage the precious *Allium* genetic resources which are conserved globally.

References

Astley D (1990) Conservation of genetic resources. In: Rabinowitch HD, Brewster JL (eds) Onions and allied crops, vol 1. Botany physiology and genetics. CRC Press, Boca Raton, Florida, pp 177–198

Ayabe M, Sumi S (2001) A novel and efficient tissue culture method—"stem-disc dome culture"—for producing virus-free garlic (*Allium sativum* L.). Plant Cell Rep 20:503–507

Baek H, Kim H, Cho E, Chae Y, Engelmann F (2003) Importance of explant size and origin and of preconditioning treatments for cryopreservation of garlic shoot apices by vitrification. CryoLetters 24:381–388

Barandiaran X, Martin N, Rodriguez-Conde MF, Di Pietro A, Martin J (1999) An efficient method for callus culture and shoot regeneration of garlic (*Allium sativum* L.). HortScience 34:348–349

Bhojwani SS (1980) *In vitro* propagation of garlic by shoot proliferation. Sci Hortic 13:47–52

Bockish T, Saranga Y, Altman A (1997) Garlic micropropagation by somatic embryogenesis. Acta Hortic 447:241–242

Brewbaker JL, Kwack BH (1963) The essential role of calcium in pollen germination and pollen tube growth. Am J Bot 50:859–865

Buiteveld J, Creemers-Molenaar J (1994) Plant regeneration from protoplasts isolated from suspension cultures of leek (*Allium ampeloprasum* L.). Plant Sci 100:203–210

Buiteveld J, Kassies W, Geels R, van Lookeren Campagne MM, Jacobsen E, Creemers-Molenaar J (1998a) Biased chloroplast and mitochondrial transmission in somatic hybrids of *Allium ampeloprasum* L. and *Allium cepa* L. Plant Sci 131:219–228

Buiteveld J, Suo Y, Lookeren Campagne M, Creemers-Molenaar J (1998b) Production and characterization of somatic hybrid plants between leek (*Allium ampeloprasum* L.) and onion (*Allium cepa* L.). Theor Appl Genet 96:765–775

Celli MG, Perotto MC, Buraschi D, Conci VC (2016) Biological and molecular characterization of Garlic Virus D and its effects on yields in garlic. Acta Hortic 1143:193–200

Colmsee C, Keller ERJ, Zanke C, Senula A, Funke T, Oppermann M, Weise S, Scholz U (2012) The garlic and shallot core collection image database of IPK presenting two vegetatively maintained crops in the Federal ex situ genebank for agricultural and horticultural crops at Gatersleben, Germany. Genet Resour Crop Evol 59:1407–1415

Conci VC, Nome SF (1991) Virus free garlic (*Allium sativum* L.) plants obtained by thermotherapy and meristem tip culture. J Phytopathol 132:186–192

Conci VC, Canavelli A, Lunello P, Rienzo JD, Nome SF, Zumelzu G, Italia R (2003) Yield losses associated with virus-infected garlic plants during five successive years. Plant Dis 87:1411–1415

Conci VC, Perotto MC, Cafrune E, Lunello P (2005) Program for intensive production of virus-free garlic plants. Acta Hortic 688:195–200

Cromarty AS, Ellis RH, Roberts EH (1982) rev. 1985, 1990. The design of seed storage facilities for genetic conservation. Handbooks for genebanks. No 1. IBPGR, Rome

de Vries JN, Wietsma WA, de Vries T (1992) Introgression of leaf blight resistance from *Allium roylei* Stearn into onion (*A. cepa* L.). Euphytica 62:127–133

Doležel J, Novák FJ, Lužný J (1980) Embryo development and *in vitro* culture of *Allium cepa* and its interspecific hybrids. Z Pflanzenzücht 85:177–184

Dulloo ME, Ebert AW, Dussert S, Gotor E, Astorga C, Vasquez N, Rakotomalala JJ, Rabemiafara A, Eira M, Bellachew B, Omondi C, Engelmann F, Anthony F, Watts J, Qamar Z, Snook L (2009) Cost efficiency of cryopreservation as a long-term conservation method for coffee genetic resources. Crop Sci 49:2123–2138

Dunstan DI, Short KC (1979) Shoot production from the flower head of *Allium cepa* L. Sci Hortic 10:345–356

Dussert S, Engelman F, Noirot M (2003) Development of probabilistic tools to assist in the establishment and management of cryopreserved plant germplasm collections. CryoLetters 24:149–160

Eady CC, Kamoi T, Kato M, Porter NG, Davis S, Shaw M, Kamoi A, Imai S (2008) Silencing onion lachrymatory factor synthase causes a significant change in the sulfur secondary metabolite profile. Plant Physiol 147:2096–2106

Ebert AW (2011) Vegetable germplasm conservation and utilization at AVRDC—The World Vegetable Center. Acta Hortic 898:89–96

Ellis D, Skogerboe D, Andre C, Hellier B, Volk G (2006) Implementation of garlic cryopreservation techniques in the national plant germplasm system. CryoLetters 27:99–106

Ellis RH, Hong TD, Roberts EH (1985a), repr. 1989. Principles and methodology. Handbooks for genebanks No. 2, Handbook of seed technology for genebanks, vol 1, IBPGR, Rome

Ellis RH, Hong TD, Roberts EH (1985b) repr. 1989. Compendium of specific germination information and test recommendations. Handbooks for genebanks. No. 3. Handbook of seed technology for genebanks, vol 2. IBPGR, Rome

Engelmann F (2012) Germplasm collection, storage, and conservation. In: Altman A, Hasegawa PM (eds) Plant biotechnology and agriculture. Prospects for the 21st century. Elsevier Inc., pp 255–267

Espirito Santo S, Keiss H-P, Meyer K, Buytenhek R, Roos T, Dirsch V, Buniatian G, Ende C, Günther J, Heise K, Kellert D, Lerche K, Pavlica S, Struck F,

Usbeck E, Voigt J, Zellmer S, Princen JMG, Vollmar AM, Gebhardt R (2007) Garlic and cardiovascular diseases. Med Aromat Plant Sci Biotechnol 1:25–30

Etoh T (1986) Fertility of the garlic clones collected in Soviet Central Asia. J Jpn Soc Hortic Sci 55:312–319

Etoh T, Keller ERJ, Senula A (2001) Fertile garlic clones in the Gatersleben collection. Mem Fac Agric Kagoshima Univ 37:29–35

Fajardo TVM, Nishijima M, Buso JA, Torres AC, Ávila AC, Resende RO (2001) Garlic viral complex: Identification of *Potyviruses* and *Carlavirus* in Central Brazil. Fitopatol Brasil 26:619–626

FAO (2014) Genebank Standards for Plant Genetic Resources for Food and Agriculture. Food and Agriculture Organization of the United Nations, Commission on Genetic Resources for Food and Agriculture, Rome

Fellner M (1993) Problems of garlic protoplast culture—literature review, speculations and hypothesis. IAPTC Newsl 71:4–13

Fellner M, Havranek P (1992) Isolation of *Allium* pollen protoplasts. Plant Cell Tiss Org 29:275–279

Fellner M, Kneifel W, Gregorits D, Leonhardt W (1996) Identification and antibiotic sensitivity of microbial contaminants from callus cultures of garlic *Allium sativum* L. and *Allium longicuspis* Regel. Plant Sci 113:193–201

Féréol L, Chovelon V, Causse S, Michaux-Ferriere N, Kahane R (2002) Evidence of a somatic embryogenesis process for plant regeneration in garlic (*Allium sativum* L.). Plant Cell Rep 21:197–203

Féréol L, Chovelon V, Causse S, Kalumvueziko ML, Kahane R (2005) Embryogenic cell suspension cultures of garlic (*Allium sativum* L.) as method for mass propagation and potential material for genetic improvement. Acta Hortic 688:65–74

Frankel OH, Bennett E (1970) Genetic resources in plants—their exploration and conservation. IBP handbook 11. Blackwell Scientific Publications, Oxford, UK, p 554

Frese L, Palmé A, Neuhaus G, Bülow L, Maxted N, Poulsen G, Kik C (2016a) On the conservation and sustainable use of plant genetic resources in Europe: a stakeholder analysis. In: Maxted N, Dulloo ME, Ford-Lloyd BV (eds) Enhancing crop genepool use, capturing wild relative and landrace diversity for crop improvement. CABI Publishing, Wallingford, UK, pp 388–400

Frese L, Palmé A, Bülow L, Kik C (2016b) Towards an improved European plant germplasm system. In: Maxted N, Dulloo ME, Ford-Lloyd BV (eds) Enhancing crop genepool use, capturing wild relative and landrace diversity for crop improvement. CABI Publishing, Wallingford, UK, pp 401–411

Friesen N, Fritsch R, Bachmann K (1997) Hybrid origin of some ornamental *Alliums* of subgenus *Melanocrommyum* verified with GISH and RAPD. Theor Appl Gen 95:1229–1238

Friesen N, Fritsch RM, Blattner FR (2006) Phylogeny and new intrageneric classification of *Allium* (Alliaceae) based on nuclear ribosomal DNA ITS sequences. Aliso 22:372–395

Frison EA, Serwinski J (1995) Directory of European institutions holding plant genetic resources, 4th edn, vols 1 and 2. International Plant Genetic Resources Institute, Rome, Italy, p 499

Fritsch RM (2015) Checklist of ornamental Allium species and cultivars currently offered in the trade. www.ipk-gatersleben.de/fileadmin/content-ipk/content-ipk-institut/Downloads/2015/150312_OrnamAllium Checkl_FritschMR.pdf, pp 1–64

Fritsch RM (2016). A preliminary review of Allium subg. Melanocrommyum in Central Asia. IPK Gatersleben. www.ipk-gatersleben.de/DOI/3a44112c-340e-442e-ac65-720ee85c38df/833a14d4-3a8e-4cf9-8a0b-50a40 144e490/2

Fritsch RM, Abbasi M (2013) A taxonomic review of *Allium* subg. Melanocrommyum in Iran. IPK, Gatersleben. ISBN 978-3-9813096-3-8

Fritsch RM, Friesen N (2002) Evolution, domestication and taxonomy. In: Rabinowitch HD, Currah L (eds) Allium crop science: recent advances. CAB International, pp 5–30

Fritsch RM, Gurushidze M (2009) Phylogenetic relationships of ornamental species in *Allium* L. subg. *Melanocrommyum* (Webb et Berthel.) Rouy (Alliaceae). Israel J Plant Sci 57:287–295

Fritsch RM, Blattner FR, Gurushidze M (2010) New classification of *Allium* L. subg. *Melanocrommyum* (Webb & Berthel.) Rouy (Alliaceae) based on molecular and morphological characters. Phyton 49:145–220

Ganeshan S (1986) Viability and fertilizing capacity of onion pollen *Allium cepa* stored in liquid nitrogen. Trop Agric (St Augustine) 63:46–48

Ganeshan S, Rajasekharan PE, Shashikumar S, Decruze W (2008) Cryopreservation of pollen. In: Reed BM (ed) Plant cryopreservation: a practical guide. Springer, New York, pp 443–464

García Lampasona S, Martínez L, Burba JL (2003) Genetic diversity among selected Argentinean garlic clones (*Allium sativum* L.) using AFLP (Amplified Fragment Length Polymorphism). Euphytica 132:115–119

Gregory M, Fritsch RM, Friesen NW, Khassanov FO, McNeal DW (1998) Nomenclator Alliorum. *Allium* names and synonyms—a world guide. Royal Botanical Gardens Kew, UK

Gurushidze M, Fritsch RM, Blattner FR (2008) Phylogenetic analysis of *Allium* subg. *Melanocrommyum* infers cryptic species and demands a new sectional classification. Mol Phylogenet Evol 49:997–1007. https://doi.org/10.1016/j.ympev.2008.09.003 Epub 2008 Sep 14

Hajjar R, Hodgkin T (2007) The use of wild relatives in crop improvement: a survey of developments over the last 20 years. Euphytica 156:1–13

Hammer K, Hanelt P, Tittel C (1977) A collection of indigenous cultivated plants from the territory of GDR. Kulturpflanze 25:89–99 (in German)

Hannan R, Garoutte D (1996) Cryopreservation of garlic germplasm as meristems in LN for long term storage. In: Havey MJ (eds) Proceedings of the national onion research conference, USA. Madison, WI, pp 101–103

Hanson J (2017) Safety duplication. Available via Crop Genebank knowledge base. http://cropgenebank.sgrp.cgiar.org/index.php?option=com_content&view=article&id=58&Itemid=207&lang=english

Havranek P (1972) Virus-free garlic clones obtained from meristem cultures. Ochr Rostl 8:291–298

Havranek P (1974) The effect of virus diseases on the yield of common garlic. Ochr Rostl 10:251–256

Havranek P, Novak FJ (1973) The bud formation in the callus cultures of *Allium sativum* L. Z Pflanzenphysiol 68:308–318

Hong TD, Linington S, Ellis RH (1998) Compendium of information on seed storage behaviour. Handbook for genebanks No. 4, IPGRI, Rome

Huchette O, Auger J, Arnault I, Barandiaran X, Chovelon V, Kahane R (2007) Garlic cultivation for high health-value. Med Aromat Plant Sci Biotechnol 1:16–20

Ipek M, Ipek A, Simon PW (2003) Comparison of AFLPs, RAPD markers, and isozymes for diversity assessment of garlic and detection of putative duplicates in germplasm collections. J Amer Soc Hortic Sci 128:246–252

Ipek M, Ipek A, Almquist SG, Simon PW (2005) Demonstration of linkage and development of the first low-density genetic map of garlic, based on AFLP markers. Theor Appl Genet 110:228–236

IPGRI, ECP/GR, AVRDC (2001) Descriptors for *Allium* (*Allium* spp.). International Plant Genetic Resources Institute, Rome, Italy; European Cooperative Programme for Crop Genetic Resources Networks (ECP/GR), Asian Vegetable Research and Development Center, Taiwan

Jenderek MM, Hannan RM (2000) Seed producing ability of garlic (*Allium sativum* L.) clones from two public US collections. In: Randle WM (ed) Proceedings of the third international symposium on edible Alliaceae. University of Georgia, Athens, GA, USA, pp 73–75

Johri BM, Vasil IK (1961) Physiology of pollen. Bot Rev 27:325–381

Jones MG, Collin HA, Tregova A, Trueman L, Brown L, Cosstick R, Hughes J, Milne J, Wilkinson MC, Tomsett AB, Thomas B (2007) The biochemical and physiological genesis of alliin in garlic. Med Aromat Plant Sci Biotechnol 1:21–24

Kahane R, Rancillac M, Teyssendier de la Serve B (1992) Long-term multiplication of onion (*Allium cepa* L.) by cyclic shoot regeneration *in vitro*. Plant Cell Tiss Org 28:281–288

Kahane R, Vialle-Guerin E, Boukema I, Tzanoudakis D, Bellamy C, Chamaux C, Kik C (2001) Changes in non-structural carbohydrate composition in high solid and sweet onions in a multi-location trail. Environ Exp Bot 45:73–83

Kamenetsky R, Rabinowitch HD (2001) Floral development in bolting garlic. Sex Plant Reprod 13:235–241

Kamenetsky R, London Shafir I, Khassanov F, Kik C, van Heusden AW, Vrielink-van Ginkel M, Burger-Meijer K, Auger J, Arnault I, Rabinowitch HD (2005) Diversity in fertility potential and organo-sulphur compounds among garlics from Central Asia. Biodivers Conserv 14:281–295

Kamenetsky R, Khassanov F, Rabinowitch HD, Auger J, Kik C (2007) Garlic biodiversity and genetic resources. Med Aromat Plant Sci Biotechnol 1:1–5

Kästner U, Klahr A, Keller ERJ, Kahane R (2001) Formation of onion bulblets in vitro and viability during medium-term storage. Plant Cell Rep 20:137–142

Karim MA, Adachi T (1997) Cell suspension, isolation and culture of protoplasts of *Allium cepa*. Plant Cell Tiss Org 51:43–47

Karlová K, Dušek K, Stavělíková H (2009) Virus diseases in collection of genetic resources of garlic in the Czech Republic. Agriculture (Pol'nohospodárstvo) 55:58–60

Keller ERJ (2002) Cryopreservation of *Allium sativum* L. (garlic). In: Towill LE, Bajaj YPS (eds) Biotechnology in agriculture and forestry, Cryopreservation of plant germplasm II, vol 50. Springer, Heidelberg, pp 37–47

Keller ERJ (2005) Improvement of cryopreservation results in garlic using low temperature preculture and high-quality in vitro plantlets. CryoLetters 26:357–366

Keller ERJ (2006) Die Erhöhung von Effektivität und Sicherheit bei der Erhaltung permanent vegetativer Genbank-Akzessionen durch In-vitro-Kultur und Kryokonservierung. Vortr Pflanzenzücht 70:16–26

Keller ERJ, Senula A (2001) Progress in structuring and maintaining the garlic (*Allium sativum*) diversity for the European GenRes Project. Acta Hort 555:189–193

Keller ERJ, Senula A (2013) Micropropagation and cryopreservation of garlic (*Allium sativum* L.). In: Lambardi M, Ozudogru EA, Jain SM (eds) Protocols for micropropagation of selected economically-important horticultural plants. Methods in molecular biology vol 11013, Springer Science Business Media, New York, pp 353–368

Keller ERJ, Senula A (2016) Recent aspects of *Allium* cryopreservation in the Federal German Genebank. Acta Hortic 1143:35–44

Keller ERJ, Lesemann DE, Maaß HI, Meister A, Lux H, Schubert I (1995) Maintenance of an in vitro collection of *Allium* in the Gatersleben genebank—problems and use. In: Terzi M, Cella R, Falavigna A (eds) Current issues in plant molecular and cellular biology. Proceedings of VIIIth international congress plant tissue cell culture, Florence, June 12–17, 1994, pp 347–352

Keller ERJ, Schubert I, Fuchs J, Meister A (1996) Interspecific crosses of onion with distant *Allium* species and characterization of the presumed hybrids by means of flow cytometry, karyotype analysis and genomic in situ hybridization. Theor Appl Genet 92:417–424

Keller ERJ, Lesemann D-E, Lux H, Maaß HI, Schubert I (1997) Application of in vitro culture to onion and garlic for the management and use of genetic resources at Gatersleben. Acta Hortic 433:141–150

Keller ERJ, Kaczmarczyk A, Senula A (2008) Cryopreservation for plant genebanks—a matter between high expectations and cautious reservation. CryoLetters 29:53–62

Keller ERJ, Senula A, Zanke C (2011a) Alliaceae in cryopreservation, achievements and constraints. Acta Hortic 908:495–508

Keller ERJ, Senula A, Zanke C, Grübe M, Kaczmarczyk A, Nukari A, Teyssedre D, Kremer Morales C, Edesi J, Pelc M, Olas-Sochacka M (2011b). Ways of collaboration—COST short-term scientific missions on three crops and their outcomes—potato, garlic and mint. In: Grapin A, Keller ERJ, Lynch PT, Panis B, Revilla Bahillo A, Engelmann F (eds) Cryopreservation of crop species in Europe. COST Office, Brussels, pp 110–115

Keller ERJ, Zanke CD, Kotlinska T, Olas-Sochacka M, Reis A, Barata AM (2011c) Broadening the spectrum of usable organ sources for cryopreservation of garlic —an AEGIS project report. Bioversity Newsl Europe 53:9

Keller ERJ, Blattner FR, Fritsch R, Pistrick K, Senula A, Zanke CD (2012a) The genus *Allium* in the Gatersleben plant collections—progress in germplasm preservation, characterization and phylogenetic analysis. Acta Hortic 969:273–287

Keller ERJ, Panis B, Engelmann F (2012b) In vitro storage and cryopreservation as substantial complements in concerted actions to better maintain and use crop germplasm. Acta Hortic 961:35–50

Keller ERJ, Zanke C, Blattner FR, Stavělíková H, Zámečník J, Kotlińska T, Miccolis V, Kik C, Esnault F, Solberg S (2012c) EURALLIVEG: establishment of a European core collection by cryopreservation and virus elimination in garlic. Acta Hortic 969:319–327

Keller ERJ, Senula A, Höfer M, Heine-Dobbernack E, Schumacher HM (2013a) Cryopreservation of plant cells. In: Flickinger MC (eds) Encyclopedia of industrial biotechnology: bioprocess, bioseparation, and cell technology. Wiley. https://doi.org/10.1002/9780470054581.eib244.pub2, pp 1–15

Keller ERJ, Zanke CD, Senula A, Breuing A, Hardeweg B, Winkelmann T (2013b) Comparing costs for different conservation strategies of garlic (*Allium sativum* L.) germplasm in genebanks. Genet Resour Crop Evol 60:913–926

Keller ERJ, Senula A, Grübe M, Diekmann K, Dehmer KJ (2014) Fifteen years of cryopreservation in the IPK genebank—experience, conclusions and outlook. Acta Hortic 1039:249–263

Keller ERJ, Grübe M, Hajirezaei M-R, Melzer M, Mock H-P, Rolletschek H, Senula A, Subbarayan K (2016) Experience in large-scale cryopreservation and links to applied research for safe storage of plant germplasm. Acta Hortic 1113:239–249

Khosa JS, McCallum J, Dhatt AS, MacKnight RC (2016) Enhancing onion breeding using molecular tools. Plant Breeding 135:9–20

Khrustaleva LI, Kik C (1998) Cytogenetical studies in the bridge cross *Allium cepa* × (*A. fistulosum* × *A. roylei*). Theor Appl Genet 96:8–14

Khrustaleva LI, Kik C (2000) Introgression of *Allium fistulosum* into *A. cepa* mediated by *A. roylei*. Theor Appl Genet 100:17–26

Kik C (2002) Exploitation of wild relatives for the breeding of cultivated *Allium* species. In: Rabinowitch HD, Currah L (eds) *Allium* crop science: recent advances. CABI Publishing Oxon UK, pp 81–100

Kik C, Kahane R, Gebhardt R (2001) Garlic and health. Nutr Metab Cardiovasc Dis 11:57–65

Kim HH (2004) Analysis of factors affecting the cryopreservation of garlic shoot tips. Dissertations Department of Agronomy, Graduate School, Seoul National University

Kim H-H, Cho E-G, Baek H-J, Kim C-Y, Keller ERJ, Engelmann F (2004a) Cryopreservation of garlic shoot tips by vitrification: effects of dehydration, rewarming, unloading and regrowth conditions. CryoLetters 25:59–70

Kim H-H, Kim J-B, Baek H-J, Cho E-G, Chae Y-A, Engelmann F (2004b) Evolution of DMSO concentration in garlic shoot tips during a vitrification procedure. CryoLetters 25:91–100

Kim J-B, Kim H-H, Baek H-J, Cho E-G, Kim Y-H, Engelmann F (2005a) Changes in sucrose and glycerol content in garlic shoot tips during freezing using PVS3 solution. CryoLetters 26:103–112

Kim H-H, Yoon J-W, Kim J-B, Engelmann F, Cho E-G (2005b) Thermal analysis of garlic shoot tips during a vitrification procedure. CryoLetters 26:33–44

Kim H-H, Lee J-K, Yoon J-W, Ji J-J, Nam S-S, Hwang H-S, Cho E-G, Engelmann F (2006) Cryopreservation of garlic bulbil primordia by the droplet vitrification procedure. CryoLetters 27:143–153

Kim H-H, Lee J-K, Hwang H-S, Engelmann F (2007a) Cryopreservation of garlic germplasm collections using the droplet-vitrification technique. CryoLetters 28:471–482

Kim H-H, Lee Y-K, Kim T-S, Cho E-G, Lee J-K, Ji J-J, Nam S-S, Engelmann F (2007b) Implementation of cryopreservation for garlic genetic resources by the droplet vitrification procedure. Acta Hortic 760:209–215

Kim H-H, Lee Y-G, Park S-U, Lee S-C, Baek H-J, Cho E-G, Engelmann F (2009a) Development of alternative loading solutions in droplet-vitrification procedures. CryoLetters 30:291–299

Kim H-H, Lee Y-G, Shin D-J, Ko H-C, Gwag J-G, Cho E-G, Engelmann F (2009b) Development of alternative plant vitrification solutions in droplet-vitrification procedures. CryoLetters 30:320–334

Kim H-H, Popova E, Shin D-J, Yi J-Y, Kim C-H, Lee J-S, Yoon M-K, Engelmann F (2012) Cryobanking of Korean *Allium* germplasm collections: results from a 10 year experience. CryoLetters 33:45–57

Kofoet A, Kik C, Wietsma WA, de Vries JN (1990) Inheritance of resistance to downy mildew (*Peronospora destructor* [Berk.]Casp.) from *Allium roylei* Stearn in the backcross *Allium cepa* L. × (*A. roylei A. cepa*). Plant Breeding 105:144–149

Larsen EH, Lobinski R, Burger-Meijer K, Hansen M, Ruzik R, Mazurowska L, Rasmussen PH, Sloth JJ, Scholten O, Kik C (2006) Uptake and speciation of selenium in garlic cultivated in soil amended with symbiotic fungi (mycorrhiza) and selenate. Anal Bioanal Chem 385:1098–1108

Leisova-Svobodova L, Karlova-Smekalova K (2011) Detection of garlic viruses using SYBR Green real-time reverse transcription-polymerase chain reaction. J Phytopathol 159:429–434

Lot H, Chovelon V, Souche S, Delecolle B (1998) Effects of onion yellow dwarf and leek yellow stripe viruses on symptomatology and yield loss of three French cultivars. Plant Dis 82:1381–1385

Lunello P, Ducasse D, Conci V (2005) Improved PCR detection of potyviruses in *Allium* species. Eur J Plant Pathol 112:371–378

Lynch P, Souch G, Trigwell S, Keller J, Harding K (2011) Plant cryopreservation: from laboratory to genebank. Asia Pac J Mol Biol Biotechnol 18:239–242

Lynch PT, Souch GR, Harding K (2012) Effects of post-harvest storage of *Allium sativum* bulbs on the cryopreservation of stem-discs by encapsulation/dehydration. J Hortic Sci Biotechnol 87:588–592

Lynch P, Souch GR, Zámečník J, Harding K (2016) Optimization of water content for the cryopreservation of *Allium sativum* in vitro cultures by encapsulation-dehydration. CryoLetters 37:308–317

Maaß HI, Klaas M (1995) Infraspecific differentiation of garlic (*Allium sativum* L) by isozyme and RAPD markers. Theor Appl Genet 91:89–97

Maggioni L, Keller J, Astley D (2002) European collections of vegetatively propagated *Allium*. Report of a workshop, 21–22 May 2001, Gatersleben, Germany. International Plant Genetic Resources Institute, Rome, Italy

Makowska Z, Keller J, Engelmann F (1999) Cryopreservation of apices isolated from garlic (*Allium sativum* L.) bulbils and cloves. CryoLetters 20:175–182

Mazur P (1965) The role of cell membranes in the freezing of yeast and other single cells. Ann NY Acad Sci 125:658–676. https://doi.org/10.1111/j.1749-6632.1965.tb45420.x

Mazur P, Schmidt JJ (1968) Interactions of cooling velocity, temperature, and warming velocity on the survival of frozen and thawed yeast. Cryobiology 5:1–17

Messiaen CM, Cohat J, Leroux JP, Pichon M, Beyries A (1993) Les allium alimentaires reproduits par voie vegetative. INRA, Paris

Moriconi DN, Conci VC, Nome SF (1990) Rapid multiplication of garlic (*Allium sativum* L.) in vitro. Phyton (Argentina) 51:145–151

Nirmal Babu K, Yamuna G, Praveen K, Minoo D, Ravindran PN, Peter KV (2012) Cryopreservation of spices genetic resources. In: Katkov I (ed) Current frontiers in cryobiology, pp 457–484, InTech, Available from http://www.intechopen.com/books/current-frontiers-in-cryobiology/cryopreservationof-spices-genetic-resources

Niwata E (1995) Cryopreservation of apical meristems of garlic (*Allium sativum* L.) and high subsequent plant regeneration. CryoLetters 16:102–107

Nomura Y, Oosawa K (1990) Production of interspecific hybrids between *Allium chinense* and *A. thunbergii* by *in ovulo* embryo culture. Jpn J Breed 40:531–535

Nomura Y, Maeda M, Tsuchiya T, Makara K (1994) Efficient production of interspecific hybrids between *Allium chinense* and edible *Allium* spp. through ovary culture and pollen storage. Breed Sci 44:151–155

Ohsumi C, Kojima A, Hinata K, Etoh T, Hayashi T (1992) Interspecific hybrid between *Allium cepa* and *Allium sativum*. Theor Appl Genet 85:969–975

Petrzelova I, Doležalová I, Dušková E, Hýbl M, Kopecký P, Stavělíková H, Dušek K (2016) Collection of landraces of vegetables and wild ecotypes of medicinal, aromatic and culinary plants (MAPs) In: Papoušková L (ed) Racionální rozšiřování kolekcí v rámci Národního programu rostlin. CRI, Prague, pp 65–76 (in Czech)

Phillips GC, Hubstenberger JF (1987) Plant regeneration in vitro of selected *Allium* spp. and interspecific hybrids. HortScience 22:124–125

Pinto da Cunha C, Vilela Resende F, Zucchi MI, Pinheiro JB (2014) SSR-based genetic diversity and structure of garlic accessions from Brazil. Genetica 142:419–431

Pistrick K, Knüpffer H (2013) DFG project on IPK Herbarium digitalization. IPK J 22:12

Popova E, Shukla M, Kim H-H, Saxena PK (2015) Plant cryopreservation for biotechnology and breeding. In: Al-Khayri JM, Jain SM, Johnson DV (eds) Advances in plant breeding strategies: breeding, biotechnology and molecular tools. Springer International Publishing Switzerland, https://doi.org/10.1007/978-3-319-22521-0_3, pp 63–93

Rabinowitch HD, Brewster JL (eds) (1990) Onions and allied crops, botany physiology and genetics, vol 1. CRC Press, Boca Raton, Florida

Rao NK, Hanson J, Dulloo ME, Ghosh K, Nowell D, Larinde M (2006) Manual of seed handling in genebanks. Handbooks for genebanks No 8. Bioversity International, Rome, Italy. ISBN 978-92-9043-740-6

Reed B (ed) (2008) Plant cryopreservation. A practical guide. Springer Science + Business Media LLC, ISBN 978-0-387-72275-7

Sackville Hamilton NR, Chorlton KH (1997) Regeneration of accession in seed collections: a decision guide. Handbooks for genebanks No. 5, IPGRI, Rome

Schäfer-Menuhr A, Müller E, Mix-Wagner G (1996) Cryopreservation: an alternative for the long-term storage of old potato varieties. Potato Res 39:507–513

Scholten OE, van Heusden AW, Khrustaleva LI, Burger-Meijer K, Mank RA, Antonise RGC, Harrewijn JL, Van Haecke W, Oost EH, Peters RJ, Kik C

(2007) The long and winding road leading to the successful introgression of downy mildew resistance into onion. Euphytica 156:345–353

Scholten OE, Van Kaauwen MPW, Shahin A, Hendrickx PM, Keizer LCP, Burger K, Van Heusden AW, Van der Linden CG, Vosman B (2016) SNP-markers in *Allium* species to facilitate introgression breeding in onion. BMC Plant Biol 16:187–196

Senula A, Keller ERJ (2002) Germplasm exchange of vegetative alliums—minimum phytosanitary requirements. In: Maggioni L, Keller J, Astley D (eds) European collections of vegetatively propagated *Allium*. Report Workshop Gatersleben, Germany, May 21–22, 2001. IPGRI, Rome, pp 66–71

Senula A, Keller ERJ (2014) Pollen cryopreservation to support maintenance of a wild species collection of the genus *Allium*. Acta Hortic 1039:289–296

Senula A, Keller ERJ, Lesemann DE (2000) Elimination of viruses through meristem culture and thermotherapy for the establishment of an *in vitro* collection of garlic (*Allium sativum*). Acta Hortic 530:121–128

Senula A, Büchner D, Keller ERJ (2010) Dehydrating pretreatment, water relations and survival of shallot shoot tips from field and in vitro culture (Abstr.). Cryobiology 61:400

Siess MH, Le Bon AM, Teyssier C, Belloir C, Singh V, Bergès R (2007) Garlic and cancer. Med Aromat Plant Sci Biotechnol 1:31–36

Shemesh-Mayer E, Ben-Michael T, Rotem N, Rabinowitch HD, Doron-Faigenboim A, Kosmala A, Perlikowski D, Sherman A, Kamenetsky R (2015) Garlic (*Allium sativum* L.) fertility: transcriptome and proteome analyses provide insight into flower and pollen development. Front Plant Sci 6:271. https://doi.org/10.3389/fpls.2015.00271

Shigyo M, Kik C (2008) Onion. In: Prohens J, Nuez F (eds) The handbook of plant breeding—vegetables II: Fabaceae, Liliaceae, Solanaceae and Umbelliferae. Springer, New York, pp 121–159

Singh AK, Varaprasad KS, Venkateswara K (2012) Conservation costs of plant genetic resources for food and agriculture: seed genebanks. Agric Res 1:223–239. https://doi.org/10.1007/s40003-012-0029-3

Smékalová K, Stavlíkova H, Dušek K (2010) Distribution of viruses in the garlic germplasm collection of the Czech Republic. J Plant Pathol 92:273–274

Song P, Peffley EB (1994) Plant regeneration from suspension cultures of *Allium fistulosum* and an *A. fistulosum* × *A. cepa* interspecific hybrid. Plant Sci 98:63–68

Souch GR (2006) Long-term conservation of *Allium* germplasm resources; the cryopreservation of *A. sativum* using an encapsulation/dehydration approach. Dissertations University of Derby, UK

Sudarmonowati E (2001) Cryopreservation of garlic (*Allium sativum*) cv. Lumbu Hijau using vitrification technique. Ann Bogor 8:39–47

van der Valk P, Scholten OE, Verstappen F, Jansen RC, Dons JJM (1992) High frequency somatic embryogenesis and plant regeneration from zygotic embryo-derived callus cultures of three *Allium* species. Plant Cell Tiss Org 30:181–191

van Hintum T, van Treuren R (2012) Reliability of germination testing of ex situ conserved seeds: a genebank case study on outsourced analyses. Plant Genet Resour 10:134–136

van Raamsdonk LWD, Vrielink-van Ginkel M, Kik C (2000) Phylogeny reconstruction and hybrid analysis in *Allium* subgenus *Rhizirideum*. Theor Appl Genet 100:1000–1009

van Raamsdonk LWD, Ensink W, van Heusden AW, Vrielink-van Ginkel M, Kik C (2003) Biodiversity assessment based on cpDNA and crossability analysis in selected species of *Allium* subgenus *Rhizirideum*. Theor Appl Genet 107:1048–1058

van Treuren R, de Groot EC, van Hintum TJL (2013) Preservation of seed viability during 25 years of storage under standard genebank conditions. Genet Resour Crop Evol 60:1407–1421

Verbeek M, Van Dijk P, Van Well PMA (1995) Efficiency of eradication of four viruses from garlic (*Allium sativum*) by meristem-tip culture. Eur J Plant Pathol 101:231–239

Vieira RL, da Silva AL, Zaffari GR, Steinmacher DA, Pacheco de Freitas Fraga H, Guerra MP (2015) Efficient elimination of virus complex from garlic (*Allium sativum* L.) by cryotherapy of shoot tips Acta Physiol Plant 37:1733. https://doi.org/10.1007/s11738-014-1733-3

Volk GM, Henk AD, Richards CM (2004a) Genetic diversity among U.S. garlic clones as detected using AFLP methods. J Am Soc Hortic Sci 129:559–569

Volk GM, Maness N, Rotindo K (2004b) Cryopreservation of garlic (*Allium sativum* L.) using plant vitrification solution 2. CryoLetters 25:219–226

Volk GM, Walters C (2006) Plant vitrification solution 2 lowers water content and alters freezing behavior in shoot tips during cryopreservation. Cryobiology 52:48–61

Walkey DGA, Webb MJ, Bolland CJ, Miller A (1987) Production of virus-free garlic (*Allium sativum* L.) and shallot (*A. ascalonicum* L.) by meristem-tip culture. J Hortic Sci 62:211–220

Walters C (2004) Principles for preserving germplasm in gene banks. In: Guerrant Jr, EO, Havens K, Maunder M (eds) Ex situ plant conservation. Supporting species survival in the world. Island Press, Washington, pp 113–138

Wang QC, Panis B, Engelmann F, Lambardi M, Valkonen JPT (2009) Cryotherapy of shoot tips: a technique for pathogen eradication to produce healthy planting materials and prepare healthy plant genetic resources for cryopreservation. Ann Appl Biol 154:351–363

Weise S, Oppermann M, Maggioni L, van Hintum T, Knüpffer H (2017) EURISCO: the European search catalogue for plant genetic resources. Nucleic Acids Res 45(D1):D1003–D1008. https://doi.org/10.1093/nar/gkw755

Zámečník J (2002) Cryopreservation at the research institute of crop production, Czech Republic. In: Maggioni L, Keller J, Astley D (eds) European collections of vegetatively propagated *Allium*. Report Workshop Gatersleben, Germany, May 21–22, 2001, IPGRI, Rome, pp 82–87

Zámečník J, Faltus M (2014) Physiological and biophysical responses of plants under low and ultralow temperatures. In: Pessarakli M (ed) Handbook of plant and crop physiology, 3rd edn. CRC Press, Boca Raton, London, pp 557–584

Zámečník J, Šesták J (2011) Constrained states occurring in plants cryo-processing and the role of biological glasses. In: Šesták J, Mareš JJ, Hubík P (eds) Glassy, amorphous and nano-crystalline materials, Hot topics in thermal analysis and calorimetry, vol 8. Springer Science + Business Media BV, https://doi.org/10.1007/978-90-481-2882-2_18, pp 291–310

Zámečník J, Grospietsch M, Bilavčík A (2001) Dehydration and glass transition of encapsulated garlic shoot tips. Cryobiology 43:328–329

Zámečník J, Faltus M, Bilavčík A (2007) Cryoprotocols used for cryopreservation of vegetatively propagated plants in the Czech cryobank. Adv Hortic Sci 21:247–250

Zámečník J, Faltus M, Bilavčík A (2012) Comparison of cryopreservation methods of vegetatively propagated crops based on thermal analysis. In: Katkov I (ed) Current frontiers in cryopreservation. InTech Open Science, Rijeka, Croatia, pp 333–358

Zanke C, Zámečník J, Kotlińska T, Olas M, Keller ERJ (2011) Cryopreservation of garlic for the establishment of a European core collection. Acta Hortic 908:431–438

Zheng S-J, Henken B, Ahn Y-K, Krens FA, Kik C (2004) The development of a reliable transformation protocol for garlic (*Allium sativum* L.) and the production of transgenic garlic resistant to beet armyworm. Mol Breed 14:293–307

Zheng S-J, Henken B, de Maagd RA, Purwito A, Krens FA, Kik C (2005) Two different *Bacillus thuringiensis* toxin genes confer resistance to beet armyworm (*Spodoptera exigua* Hübner) in transgenic *Bt*-shallots (*Allium cepa* L.). Transgenic Res 14:261–272

Zheng S-J, Kamenetsky R, Féréol L, Barandiaran X, Rabinowitch HD, Chovelon V, Kik C (2007) Garlic breeding system innovations. Med Aromat Plant Sci Biotechnol 1:6–15

Classical Genetics on Gene Mapping

4

Ali Fuat Gökçe

Abstract

While onion (*Allium cepa* L.) plants have significantly different morphologies in height, leaves, scape, and inflorescence, their bulbs differ in pungency, color, size, firmness, shape, tightness of bulb scales and neck. As in other vegetables, resistance to diseases, better quality, high yield, and uniformity are the major breeding achievements in onion. However, being a biennial plant with alternative seed-to-bulb and bulb-to-seed generations, the genetic progression of any desired trait in onion is relatively slow. This chapter reviews general breeding objectives, genetics and morphological markers found in seed, seedling, leaf, scape, bulb scale, and flower, genetics of some biotic and abiotic stresses, and cautions for marker-assisted selection to develop maintainer lines.

4.1 Breeding Goals for Edible Onion

The *Allium* genus widely dispersed over temperate and warm-temperate zones of the northern hemisphere in Turkey, Iran, Iraq, Afghanistan, Kazakhstan, and western Pakistan (Hanelt 1990). The Alliums are very diverse with more than 500 species including important vegetable and ornamental types and typically grow in fairly infertile places, in steppes, dry mountain slopes, stony or stony open sites, or summer-dry flora (Hanelt 1990; Jones and Mann 1963; Traub 1968). The edible *Alliums* include several cultivated economically and dietarily important species such as the bulb onion (*A. cepa*), the closely related shallot (*A. cepa* var. *ascalonicum*) and potato onion (*A. cepa* var. *aggregatum*), chive (*A. schoenoprasum*), Chinese chive (*A. tuberosum*), Japanese bunching onion (*A. fistulosum*), garlic (*A. sativum*), leek (*A. ampeloprasum* var. *porrum* syn. *A. porrum*), and rakkyo (*A. chinense*) (Havey 1995a).

Common onion (*A. cepa*) is the most important among the genus along with garlic (*A. sativum*). Vavilov (1951) indicated its origin belongs to the Central Asiatic Center, including northwest India, all of Afghanistan, the former Soviet republics of Tajikistan and Uzbekistan, and Western Tian-Shan. Secondary centers for onion are the Near-Eastern Center, the interior of Asia Minor, Trans-Caucasus, the high ground of Turkmenistan and Iran, whereas the Mediterranean center was hypothesized to be the origin of large-sized onions (Vavilov 1951). Common onion is believed to be domesticated in fields of the Afghanistan, Takistan, Tajikistan, or northern Iran (Brewster 1994). They have been cultivated for many centuries, and today, they are grown,

A. F. Gökçe (✉)
Department of Agricultural Genetic Engineering, Ayhan Sahenk Faculty of Agricultural Sciences and Technologies, Niğde Ömer Halisdemir University, 51240 Niğde, Turkey
e-mail: gokce01@yahoo.com

M. Shigyo et al. (eds.), *The Allium Genomes*, Compendium of Plant Genomes,
https://doi.org/10.1007/978-3-319-95825-5_4

traded, and consumed almost worldwide, and traditionally maintained as open-pollinated populations. The bulb onion, Chinese chive, chive, Japanese bunching onion, and leek, on the other hand, are mainly seed propagated (Havey 1995a).

Bulb formation in onion is affected by both day length and temperature. Cultivated onions that require 14 h or more to bulb are classified as long-day, and those that bulb as day length exceeds 11–12 h are considered to be short-day onions (Pike 1986). Other than these two major types, there is also a third group with bulb formation in between these two major groups, called as an intermediate day length type. Apart from bulbing requirements, onion plants also come with significantly different height, leaves, the morphology of scapes and inflorescence. Furthermore, onion bulbs differ in pungency, color, size, firmness, shape, single center, and tightness of bulb scales and neck. As in many other vegetables, breeding for resistance to diseases, better quality, high yield, and uniformity are the major goals for agricultural production in onion. The consumer choices, on the other hand, are also very important for onion breeding considering wide differences in traditional cooking preferences of this worldwide grown vegetable in every culture. With the growing concern on the adverse effects of pesticide usage in agriculture for human health and the environment and the need to reduce the cost of using pesticide sprays, breeding disease-resistant onion cultivars has become one of the utmost onion breeding goals. It is also important to mention that the pesticides commonly used for the control of disease spreading are no longer registered and, currently, there are relatively few chemicals available for onion growers. Unfortunately, being a biennial plant with alternative seed-to-bulb and bulb-to-seed generations, the genetic progression is relatively slow in onion, and the breeding studies require a double time period compared to annual crops, which limits the achievement of any desired trait in onion.

4.2 Overview of Basic Breeding Traits

Bulb color, size, shape, pungency, amount of total soluble solids, firmness, dormancy, resistance to biotic stresses, and tolerance to abiotic stresses are the most important bulb quality traits for onion breeding. McCollum (1966, 1968, 1971) reported that bulb diameter and weight were highly affected by environmental conditions indicating low heritability, high heritability of soluble solids (62–82%) with a negative genetic correlation between soluble solids, and bulb size, while the soluble solids, bulb shape, and size were characterized by quantitative phenotypic variation suggesting quantitative inheritance. The QTL correlated with traits of pungency, solids, and antiplatelet activity was later identified by Galmarini et al. (2001).

The bulb color is another important factor for onion breeding which highly depends on the consumer preferences in different locations of the world. The first extensive study of the bulb color inheritance identified an incompletely dominant factor (*I* over *i*) as the inhibiting factor for the expression of color (Rieman 1931). Later, Clarke et al. (1944) proposed a three independent factor scheme for bulb color development. Here, the *C* gene that was completely dominant over *c* was hypothesized to be definitely required for any pigment formation in the onion bulb, and the *R* gene was responsible for the development of red color. A dominant allele at the *G* locus was shown to be responsible for the golden bulb color without any pink, and the red pigment is conditioned by dominant *L* (El-Shafie and Davis 1967).

Fusarium basal rot (*Fusarium oxysporum* f. sp. *cepae*), neck rot (*Botrytis allii*), white rot (*Sclerotium cepivorum*), pink root (*Pyrenochaeta terrestris*), downy mildew (*Peronospora destructor*), purple blotch (*Alternaria porri*), and leaf blight (*Botrytis squamosa*) are still among the most common onion diseases causing significant losses as reported by Havey (1993b).

Resistances to some, but not all of these diseases have been reported, including purple blotch (Riollano 1943), downy mildew (Warid and Tims 1952), *Botrytis* neck rot (van der Meer et al. 1970; Miyaura et al. 1985), and white rot (Utkhede and Rahe 1978).

4.3 Classical Genetics of Onion

The biennial nature of the plant limits the study of onion genetics compared to that of annual crops requiring less time to understand the genetics of most important traits. An outstanding review of the biochemical, classical, and molecular markers of onion emphasizing sources of each was published by Cramer and Havey (1999). Nevertheless, the availability of onion germplasms possessing the genes nominated in this report is currently unknown.

4.3.1 Morphological Markers Found in Onion Seed, Seedling, Leaf, and Scape

Table 4.1 summarizes the morphological markers identified in onion seed, seedling, leaf, and scape.

The common onion seed coat color is black controlled by the *b* locus with a genotype of *B* (Davis 1966). However, homozygous recessive genotypes with *bb* locus are characterized by a brown seed coat (Davis 1966).

The morphological markers in onion seedling, such as albino, pale green, yellow lethal, and virescent, are mostly characterized by chlorophyll-deficient mutants and, therefore, under the natural selection pressure. For example, albinism (*a*) is a recessive gene not interfering with the growth of the plant, whereas homozygous recessive (*aa*) plants cannot even survive seedling stage (Jones et al. 1944; Rasmusson 1920). The pale-green-colored onion seedlings with the lethal locus pale green (*pg*) cause the death of the newly emerging seedling possessing homozygous recessive (*pgpg*) locus (Jones et al. 1944). Jones et al. (1944) suggested two yellow lethal genes with the observation that some seedlings with yellow lethal allele also specified with a morphological change in leaves, glossy foliage. Seedlings with light-yellow leaves that die shortly after germination were identified as homozygous recessive (*y1y1y2y2*) in terms of the yellow lethal loci, *y1* and *y2* (Jones et al. 1944; Rasmusson 1920). Two recessive (*vv*) alleles, virescent (*v*), were shown to result in seedlings having pale-green foliage at low temperatures (Jones et al. 1944). Virescent plants, as opposed to other chlorophyll-deficient mutants, can survive and the color of the foliage becomes similar to that of nonvirescent foliage by development (Jones et al. 1944). They are, however, characterized by reduced bulb yield with smaller in size and lower seed set, and therefore, are still considered to be under the natural selection pressure.

The yellow lethal locus (*y1*) linked glossy foliage, having no or weak wax on the surface of the leaf, is caused by a recessive allele at the *gy* locus. Glossy foliage locus was correlated with the thrips (*Thrips tabaci* Lindeman) resistance (Jones et al. 1944; Molenaar 1984). Scape glossiness, however, is governed by separate loci, which is conditioned by homozygous recessive genotypes at both loci (*gls1* and *gls2*) (Molenaar 1984). The *gls1* locus was reported to be epistatic to *gls2* and double glossy scapes were characterized by two recessive alleles at both loci (Molenaar 1984). A dominant allele at the *gls2* locus leads to an intermediary phenotype between double glossy and single glossy scapes, whereas a dominant allele at the *gls1* locus produced single glossy scapes, irrespective of the alleles at the *gls2* locus (Molenaar 1984). These three loci for glossiness were suggested to be linked by Molenaar (1984).

Rabinowitch et al. (1984, 1991) described the dwarf scape locus inheritance, *dw1*, as a single gene, whereas the F_3 segregation of families from crosses between normal and dwarf cultivars suggested the presence of several other loci governing the variation in the scape height (Horobin 1986).

Table 4.1 Morphological marker loci of onion seed, seedling, leaf and scape

Loci	Genotypes	Phenotypes	Clarifications	References
B	*bb*	Brown	Brown seed coat. Wild-type is black	Davis (1966)
	Bb	Black		
	BB	Black		
A	*aa*	Albino	Albino. Chlorophyll-deficient seedlings. Albino plants die soon after germination. Under natural selection pressure, only carried by heterozygous genotypes to the subsequent generation	Jones et al. (1944), Rasmusson (1920)
	Aa	Green		
	AA	Green		
Pg	*pgpg*	Pale	Pale green foliage. Seedlings soon die. Under natural selection pressure, only carried by heterozygous genotypes to the subsequent generation	Jones et al. (1944)
	Pgpg	Green		
	PgPg	Green		
Y1	*y1y1*	Yellow	Yellow lethal. Linked with glossy foliage	Jones et al. (1944)
	Y1y1	Green		
	Y1Y1	Green		
Y2	*y2y2*	Yellow	Yellow lethal. Not linked with glossy foliage	Jones et al. (1944)
	Y2y2	Green		
	Y2Y2	Green		
	y1y1/y2y2	Yellow	Double gene action with a complete dominance	Jones et al. (1944)
	y1_1/_2_2	Green		
	_1_1/_2y2	Green		
	Y1Y1/Y2Y2	Green		
V	*vv*	Virescent	Virescent foliage. Virescent onion plants are not as vigorous in growth as plants with nonvirescent foliage	Jones et al. (1944)
	Vv	Vigorous		
	VV	Vigorous		
Gl	*glgl*	Glossy	Glossy foliage. Correlated with resistance to thrips	Jones et al. (1944)
	Glgl	Waxy		
	GlGl	Waxy		
Gls1	*gls1gls1*	Glossy	Glossy scape 1. Reported as epistatic to *gls2*. Correlated with thrips resistance.	Molenaar (1984)
	Gls1gls1	Waxy		
	Gls1Gls1	Waxy		
Gls2	*gls2gls2*	Glossy	Glossy scape 2. Correlated with thrips resistance	Molenaar (1984)
	Gls2gls2	Waxy		
	Gls2Gls2	Waxy		
	gls1gls1/gls2gls2	Double Glossy	Double gene action with an epistatic action of the gls1 to gls2	Molenaar (1984)
	__1__1/Gls2gls2	Mid Glossy		
	Gls1gls1/__2__2	Single Glossy		
	Gls1__1/Gls2_2	Waxy		
Dw1	*dw1dw1*	Dwarf	Dwarf seed stalk. Some other genes may also be responsible in seed stalk height	Horobin (1986), Rabinowitch et al. (1984), Rabinowitch et al. (1991)
	Dw1dw1	Tall		
	Dw1Dw1			

4.3.2 Morphological Markers Found in Onion Scale

Onion bulbs with various scale colors are preferred by consumers in different locations of the world and their genetics are summarized in Table 4.2. The first extensive study of the bulb color inheritance dates back to Rieman (1931), where an incompletely dominant factor (*I* over *i*) was described as the inhibiting factor for the expression of color. Bulbs are white when *I* homozygous dominant (*II*), regardless of the other genes. Clarke et al. (1944) proposed three independent factors governing the pigmentation of onion bulb;

(a) the *I* gene that was Rieman's inhibiting factor,
(b) the *C* gene completely dominant over *c* and needed for the production of any pigment, and
(c) the bulb color is governed by a independent third gene, where a dominant allele at the *G* locus resulting a golden color without any pink. The red pigment is governed by the fourth *L* locus with a dominant *L* allele, and the fifth locus *R* gene results in red color (El-Shafie and Davis 1967).

Accordingly, when homozygous recessive at *I* and a dominant allele at *C*, bulbs with a dominant allele at *R* are red (*iiC_R_*), and with recessive *r* alleles are yellow (*iiC_rr*). The genotype of *IiCCrr* is buff white, of recessive (*iiccR_*) and of dominant (*II__*) are white, and of *iiC-G-L-rr* is golden (El-Shafie and Davis 1967). Heterozygosity at *L* or *R* loci results in light-red or pink scales. A complementary light-red bulb color described by Jones and Peterson (1952) was hypothesized to be conditioned by the interaction of the *R* and *L* loci (El-Shafie and Davis 1967). Although light-red bulb color was later assigned to be governed by *Crb-1* loci (King et al. 1998a, b), the relationship among *Crb-1*, *R*, and *L* are still unknown. Homozygous recessive (*iiC_ggllrr*) bulbs are chartreuse, where a dominant allele at the third locus, *G*, produces golden bulbs (*iiC_G_llrr*) (El-Shafie and Davis 1967). The homozygous dominancy at the fourth locus *L* and the fifth locus *R* leads to dark-red bulbs and, their heterozygosity at either locus results in light-red or pink scales (El-Shafie and Davis 1967). Since the fifth locus *R* controls the quantity of red pigmentations produced by the epidermal cells of the scale, light-red or pink bulbs are governed by a dominant *R* allele (*iiC_R_*) and brown or yellow bulbs by the recessive *r* allele (*iiC_rr*) (El-Shafie and Davis 1967).

4.3.3 Morphological Markers Found in Onion Flower

Some loci governing floral morphology and predominantly male sterility in onion flower have been well studied due to its importance for hybrid seed production (Table 4.3).

The trait observed in some plants, where the perianth does not fully develop around the reproductive organs and appear to be protruding from the bud, is referred as exposed anthers (*ea*) and governed by the recessive genotype at the locus (Jones et al. 1944). Anthers can be both green and yellow, and yellow anthers are conditioned by the presence of two recessive alleles of the *ya* locus (Jones et al. 1944). Although there is no segregation data available, Davis (1960) suggested white perianth to be recessive and controlled by a single locus.

Cytoplasmic-genic male sterility (CgMS) is formerly reported as CMS (cytoplasmic male sterility). However, "cytoplasmic male sterility" and "cytoplasmic-genic male sterility" are two different types of male sterilities, both are abbreviated as "CMS" causing confusion. Herein, CMS will be used for cytoplasmic male sterility and CgMS for cytoplasmic-genic male sterility. The male-sterile (S) cytoplasm was first described by Jones and Emsweller (1936). The male sterility is governed by the interaction of the cytoplasm, normal (N) male-fertile versus male-sterile (S) cytoplasm, and a single nuclear gene (*Ms*) (Jones and Clarke 1943). The dominant allele at the nuclear male fertility locus *Ms* restores the male fertility in plants with S cytoplasm (Jones and Clarke 1943). The recessive

Table 4.2 Morphological marker loci of onion scale color

Loci	Genotypes	Phenotypes	Clarifications	References
I	*ii* *Ii*	Depends on the other loci	Inhibitor of scale color. Must be homozygous recessive to form colored scales	Rieman (1931)
	II	Dominant white		
C	*cc*	Recessive white	Colored bulb scales. Must be dominant to form colored onion bulbs. All bulbs white when the dominant allele is absent	Clarke et al. (1944)
	Cc	Depends on the other loci		
	CC			
Crb1		See the R and L loci	Complementary red bulb. May be either *R* or *L*	Jones and Peterson (1952)
G	*gg*	Depends on the other loci	Golden scale color. Recessive allele gives chartreuse. Must have a dominant allele for red bulb color	El-Shafie and Davis (1967)
	Gg			
	GG			
L	*ll*	Depends on the other loci	Red scale color 2. Results in red onion bulbs when dominant, particularly when the *R* locus is recessive	
	Ll			
	LL			
R	*rr*	Depends on the other loci	Red scale color 1. Recessive allele results in yellow or brown-colored onion bulbs	Clarke et al. (1944), El-Shafie and Davis (1967)
	Rr			
	RR			
	II_ _ _ _	White		
	_cc _ _ _	White		
	IiCc _ _ _	Buff white		
	IiCC _ _ _	Pale pink		
	iicc_ _ _	White		
	iiC_ggllrr	Chartreuse		
	iiCCG_llrr	Yellow		
	*iiC_G_ rrL_*	Golden		
	iiCcggL_rr	Pink		
	iiCCggL_Rr	Pink		
	iiCCggllRr	Pink		
	iiCCggLLRR	Red		

allele (*ms*) conditions male sterility in plants with S cytoplasm (S-*msms*), which is called as cytoplasmic-genic male sterility (CgMS) (Jones and Clarke 1943). All the plants with N cytoplasm, on the other hand, are male-fertile irrespective of the genotype at the *Ms* locus, and plants having S-*MsMs* and S-*Msms* genotypes are male-fertile owing to the presence of a dominant allele (Jones and Clarke 1943). Since the cytoplasm of onion shows maternal inheritance, CgMS can be maintained by crossing plants that are genetically similar but differing in the cytoplasm, i.e., S-*msms* plants with an N-*msms* maintainer line (Table 4.3).

The second source of CgMS in onion, the T-cytoplasm, was discovered in the French cultivar Jaune Paille des Vertus by Berninger (1965). Here, as opposed to S cytoplasm, male sterility is controlled by the interaction between the T-cytoplasm and three independently

Table 4.3 Morphological marker loci of onion flower

Loci	Genotypes	Phenotypes	Clarifications	References
Ea	*eaea*	Exposed	Exposed anther locus. In the exposed phenotype, Perianth does not fully develop around anthers	Jones et al. (1944)
	Eaea	Normal		
	EaEa			
Ya	*yaya*	Yellow anther	Yellow anther locus	Jones et al. (1944)
	Yaya	Green anther		
	YaYa			
Ms	*S-msms*	Male Sterile	S cytoplasm with recessive alleles leads cytoplasmic-genic male sterility (S-CgMS). Either N-cytoplasm or a nuclear dominant allele at *Ms* locus leads male fertility. *N-msms* is used to seed propagate S-*msms*	Jones and Clarke (1943)
	S-Msms	Male fertile		
	S-MsMs			
	N-msms			
	N-Msms			
	N-MsMs			
Ta	*T-aa*	Depends on the loci B&C	Nuclear male fertility restorer locus in T-cytoplasm independent from B and C loci	Schweisguth (1973)
	T-Aa	Male fertile		
	T-AA			
Tb	*T-bb*	Depends on the loci A	Nuclear male fertility restorer locus in T-cytoplasm complementary with the C locus and independent from A locus	
	T-Bb	Depends on the loci C		
	T-BB			
Tc	*T-cc*	Depends on the loci A	Nuclear male fertility restorer locus in T-cytoplasm complementary with the B locus and independent from A locus	
	T-Cc	Depends on the loci B		
	T-CC			
Ta *Tb* *Tc*	*T-aabbcc*	Male sterile T1	T-cytoplasm with homozygous recessive alleles at A locus and at either B or C locus leads cytoplasmic-genic male sterility (T-CgMS)	Schweisguth (1973)
	*T-aabbC_*	Male sterile T2		
	T-aaB_cc	Male sterile T3		
	T-A_ __ __	Male Fertile	A dominant allele at either A locus alone or at both B and C loci lead male fertility	
	*T-aa B_ C_*			
	N-_ _ _			
	N-aabbcc	Maintains T1, T2, and T3	All N cytoplasmic individuals are male fertile	
	N-aabb__	Maintains T1 and T3		
	N-aa__cc	Maintains T1 and T2		
__	*G-__*	Male sterile	*Galantum* cytoplasm leads the cytoplasmic male sterility (CMS). Any male fertile can be used to seed propagate	Havey (1999)

inherited recessive nuclear genes. These three loci were later renamed by Cramer and Havey (1999) as T*a*, T*b*, and T*c*, as the names *a*, *b*, and *c* was previously assigned in onion. Among these three loci, the T*a* gene operates independently of T*b* and T*c*, where all T-*aa* plants with the T-cytoplasm are sterile given that plant has T-*bbcc*. Loci T*b* and T*c*, on the other hand, act complementarily, where all T-aa*bbcc* plants are sterile (Schweisguth 1973). Dominance at the *A* locus or at both the *B* and *C* loci is required for restoration of male fertility (Table 4.3).

Besides the S- and T-cytoplasmic-genic male sterility (CgMS) sources, the third type of male sterility, cytoplasmic male sterility (CMS), was developed by Havey (1999) while crossing *A. galantum* Kar. et Kir. with *A. cepa L* and back-crossing the hybrid to *A. galanthum*. In this cross, the male-sterile backcross progenies possess the *Galantum* cytoplasm and the nuclear genome of the *A. cepa* L. Thus, the *Galantum* cytoplasm completely lacking the anthers leads to the male sterility. Unlike S- and T-CgMS, no nuclear locus is needed to have male-sterile individuals, and as expected, there is no report on the male fertility restorer locus for this CMS system.

4.3.4 Morphological Markers of Biotic and Abiotic Stresses in Onion

Fusarium basal rot (*Fusarium oxysporum* f. sp. *cepae*), neck rot (*Botrytis allii*), white rot (*Sclerotium cepivorum*), pink root (*Pyrenochaeta terrestris*), downy mildew (*Peronospora destructor*), purple blotch (*Alternaria porri*), and leaf blight (*Botrytis squamosa*) are among the diseases causing significant losses in onion production, whereas the resistance of only a number of them (Table 4.4) have been reported to date (Havey 1993b; Cramer and Havey 1999).

Resistance to Fusarium basal rot was reported to be conditioned by either one locus (Tsutsui 1991), or two loci, designated as *Foc1* and *Foc2* (Bacher 1989; Bacher et al. 1989). Bacher et al. (1989) suggested that genes partially dominant for both loci (*Foc1foc1 Foc2foc2*) showed resistant genotype and that the gene effects (*Foc1-Foc1Foc2Foc2*) at each locus were additive.

Resistance to pink root was reported to be inherited as a single locus with a recessive allele, designated as *pr1* (Jones and Perry 1956; Nichols et al. 1965; Cramer and Havey 1999). Based on the observation of some crosses, pink root resistance was also speculated to be multigenic with additional loci required for the presence of resistance (Nichols et al. 1965).

Resistance to downy mildew was proposed to be controlled by two recessive alleles (*s1* and *s2*) at two dominant loci (Warid 1952; Warid and Tims 1952). The highest resistance to downy mildew was observed when both loci were homozygous (*s1s1s2s2*) recessive (Warid 1952; Warid and Tims 1952). The transfer of a single dominant locus (*Pd1*) for downy mildew resistance was reported from *A. roylei* Stearn to *A. cepa L.* and was proposed by Kofoet et al. (1990).

In addition, resistance to purple blotch was reported to be governed by a single gene resistance when it is recessive (Ekanayake and Ewart, 1997). Walker (1923) reported pigmented onion bulbs are resistant to onion smudge, whereas nonpigmented onions are susceptible, and white onion bulbs are susceptible to diplodia stain while red and yellow bulbs are resistant (Sumner 1995). Based on these observations, resistances to onion smudge and diplodia stain were believed to be connected with another trait, onion bulb pigmentation (Walker 1923; Sumner 1995).

Except the resistance to ozone, there is very limited information on the resistance to abiotic stresses as expected from such a complex trait. Engle and Gabelman (1966) described resistance to ozone observed in several onion inbred lines to be dominant and proposed that the resistance was associated with the closure of leaky stomata which prevents inner tissues damage by ozone. Since Engle and Gabelman (1966) failed to designate this locus, Cramer and Havey (1999) proposed *Oz* locus for resistance to ozone (Cramer and Havey 1999).

Table 4.4 Morphological marker loci of biotic and abiotic stresses in onion

Loci	Genotypes	Phenotypes	Clarifications	References
Foc1	*foc1foc1*	Susceptible	Fusarium basal plate rot resistance locus 1. Partially dominant with *Foc2* shows additive resistance with *Foc2*	Bacher (1989), Bacher et al. (1989)
	Foc1foc1	Partial resistant		
	Foc1Foc1	Resistant		
Foc2	*foc2foc2*	Susceptible	Fusarium basal plate rot resistance locus 2. Partially dominant with *Foc1* shows additive resistance with *Foc1*	
	Foc2foc2	Partial resistant		
	Foc2Foc2	Resistant		
Pr1	*pr1pr1*	Resistant	Pink root resistance locus. Reported that other modifier genes may be involved in resistance	Jones and Perry (1956), Nichols et al. (1965)
	Pr1pr1	Susceptible		
	Pr1Pr1			
Pd1	*pd1pd1*	Susceptible	Downy mildew resistance locus 1. Dominant gene transferred from *A. roylei*	Kofoet et al. (1990)
	Pd1pd1	Resistant		
	Pd1Pd1			
S1	*s1s1*	Resistant	Downy mildew resistance locus 1. Reported in Calred onion. Shows better resistance when both loci homozygous recessive (*s1s1 s2s2*)	Warid (1952), Warid and Tims (1952)
	S1s1	Susceptible		
	S1S1			
S2	*s2s2*	Resistant	Downy mildew resistance locus 2. Reported in Calred onion. Shows better resistance when both loci homozygous recessive (*s1s1 s2s2*)	
	S2s2	Susceptible		
	S2S2			
Oz	*ozoz*	Susceptible	Ozone tolerance locus. Individuals with recessive alleles at the *Oz* locus susceptible to ozone stress.	Engle and Gabelman (1966), Cramer and Havey (1999)
	Ozoz	Tolerant		
	OzOz			

4.4 Molecular Marker Assisted Selection

Plant breeding to develop the desired trait requires quite a long time, especially in biennial plants such as onion; therefore, development of genetic maps and markers are essential to speed up the process. Traits that are difficult to phenotype under field conditions due to environmental effect can easily be selected in the laboratory, depending on the presence of a closely linked marker, since alleles tend to be inherited together.

Genetic markers can be classified as morphological, isozyme, or DNA markers. Morphological markers are often undesirable in breeding programs, due to their deleterious pleiotropic effect. DNA-based molecular markers, on the other hand, are often common and in sufficient numbers enabling generation of detailed genetic maps within single species. The earlier molecular marker based genetic maps were reported both in onion (King et al. 1998a, b) and between *A. cepa* and *A. roylei* (van Heusden et al. 2000a, b).

4.4.1 Marker Development Linked to *Ms* Locus

Although several morphological markers have been described for onion (Tables 4.1, 4.2 and 4.3), most of the work on marker development is quite specific to the identification of molecular markers neighboring the nuclear *Ms* locus

(Table 4.4) enabling the selection of maintainer lines for hybrid seed production. Such markers are quite important since they allow onion breeders to retain only the maintainer plants (N-*msms*), or to easily develop maintainer lines (N-*Msms*) by reducing the number of plants to be test crossed to a male-sterile line.

The identification of polymorphisms in the chloroplast and mitochondrial DNAs that differentiate N and S cytoplasms of onion allowed breeders to identify the cytoplasm without crosses (de Courcel et al. 1989; Holford et al. 1991; Havey 1993a; Satoh et al. 1993). Havey (1995b) and Sato (1998) developed molecular markers to establish the cytoplasm of an individual onion plant by the polymerase chain reaction (PCR) which takes only days, as opposed to years. Cytoplasmic determinations are especially imperative given that open-pollinated populations may completely possess S cytoplasm (Havey 1993a). That is especially important in countries where hybrid-onion cultivar seed has been reserved and used to improve open-pollinated onion populations.

Segregation at the *Ms* locus was reported by King et al. (1998a). However, since male fertility restoration can be affected by the environmental factors (van der Meer and van Bennekom 1969 and 1972) some families appearing male-sterile at early stages can still shed pollen after over half of the flowers in the umbel had dehisced. Gökçe et al. (2002) suggested that the consistent appearance of too few male-fertile plants may be due to scoring the restoration of male fertility too early, incomplete dominance, and/or reduced penetrance of the dominant allele at *Ms*. Later Melgar and Havey (2010) demonstrated that the dominant *Ms* allele shows incomplete penetrance, requiring scoring male fertility restoration in S cytoplasm over different environments and over the entire pollination period of each umbel for years to more confidently assign genotypes at *Ms*.

4.4.2 Cytoplasmic-Genic Male Sterility (CgMS)

The hybrid-onion seed production becomes economically more feasible using CgMS. Male-sterile plants possess S cytoplasm and are homozygous recessive at the nuclear *Ms* locus (S-*msms*). Male-sterile inbred lines can only be seed propagated by crossing with a maintainer line possessing normal (N) male-fertile cytoplasm and the homozygous recessive genotype at the nuclear restorer *Ms* locus (N-*msms*). Therefore, production of hybrid-onion seed is reliant on the availability of maintainer lines (N-*msms*) to seed-propagate the male-sterile (S-*msms*) lines (Jones and Davis 1944). Development of maintainer lines (*N-msms*) from onion populations with a high prevalence of the dominant allele at *Ms* or the frequency of S cytoplasm, on the other hand, has not been successful (Davis 1957; Havey 1993a; Havey and Randle 1996; Little et al. 1944; Satoh et al. 1993).

Sterile cytoplasm used to produce hybrid onion in the USA traces back to an individual plant discovered in the cultivar Italian Red at the University of California Davis in 1925 (Jones and Emsweller 1936). Bulbs from Italian Red were selected from a commercial lot, planted, and self-pollinated. One of those bulbs, Italian Red 13-53, flowered normally, but failed to produce seed and was only propagated by small bulbs in the inflorescence (Jones and Emsweller 1936). The plants of genotype S-*msms* were used as the female line in hybrid production and CgMS was maintained by crossing S-*msms* plants with a maintainer line (N-*msms*). The male-sterile character of Italian Red 13-53 was later incorporated into important commercial onion inbreds by backcrossing (Jones and Davis 1944).

Numerous molecular markers have been established to date to identify different cytoplasm types to reduce time and labor required to identify cyto types by test crossing taking 4–8 years due to the biennial cycle of onion (Havey 1995b; Sato 1998; Engelke et al. 2003). Similarly, many efforts have been made to identify molecular markers closely linked to the *Ms* locus (Gökçe et al. 2002; Bang et al. 2011, 2013; Havey 2013), but their linkage phase validation in an uncharacterized population is needed before applying them in molecular marker-assisted selection of maintainer lines.

4.4.3 Use of Cytoplasmic and Nuclear DNA Markers Linked to the *Ms* Locus

The nuclear and cytoplasmic onion genotypes can be classified as N-*msms*, N-*Msms*, or S-*Msms* by test crossing to a male-sterile plant (S-*msms*), self-pollinating the male-fertile plant, and separately scoring the male fertility of S_1 and test-cross progenies. The same procedure will produce all male-fertile S_1 and test cross phenotypes for N-*MsMs* and S-*MsMs*. For these genotypes, a maintainer plant (N-*msms*) must be crossed with N-*MsMs* and S-*MsMs* and progenies from these crosses should be self-pollinated and scored for male fertility to establish their cytoplasms. The most commonly used source of cytoplasmic-genic male sterility (CgMS) in onion is conditioned by the interaction of male-sterile (S) cytoplasm and the homozygous recessive (S-*msms*) genotype at a nuclear male fertility restoration (*Ms*) locus (Jones and Clarke 1943; Jones and Emsweller 1936).

Due to the biennial nature of onion, 4 to 8 years are necessary to determine if maintainer lines (N-*msms*) can be selected from uncharacterized populations or segregating families (Havey 1995b). Fortunately, molecular markers flanking the *Ms* locus is available, the sequence can be used to reveal polymorphism among individuals or populations using the polymerase chain reaction (PCR). These nuclear and organellar markers can be used to select maintainer genotypes from onion populations or segregating families and should reduce the number of test crosses necessary to pinpoint maintainer individual onion plants. Maintainer lines are used to seed propagate male-sterile (*S-msms*) lines and possess normal (N) male-fertile cytoplasm and the homozygous recessive (N-*msms*) genotype at *Ms* locus. Fortunately, PCR markers were developed to distinguish N and S cytoplasms (Havey 1995b; Sato 1998; Engelke et al. 2003) and polymorphism flanking the *Ms* locus was reported (Gökçe et al. 2002; Bang et al. 2011, 2013; Havey 2013). Identification of nuclear markers closely linked to the *Ms* locus facilitates the molecular-assisted selection of maintainer lines. With the presence of these PCR markers, one can define the cytoplasm or nucleus of an individual onion plant in a day, as opposed to years. However, breeders should consider two possible cases while using molecular marker-facilitated selection. The first is the possibility that *Ms* linked marker is not the *Ms* gene to the genotype. The molecular markers and *Ms* must not be at or near linkage equilibrium to facilitate molecular marker-assisted selection in a given onion population and also that the marker and Ms alleles might be repulsion phase of what reported in the marker developed population (Tekeli, 2015). The other case is that the bulb onion is a diploid species with natural outcrossing rates ranging from 70 to 100% in the field. Male-sterile plants must receive the male gametes from male-fertile plants in order to set seed. Therefore, the frequency of the *ms* allele will change over generations in onion populations possessing S cytoplasm or mixtures of N and S cytoplasms due to the natural selection since male-sterile (S-*msms*) plants fail to produce male gametes (Gökçe and Havey 2006).

Gökçe et al. (2002) evaluated the segragation of the mapping families previously described by King et al. (1998a) for linkage to *Ms* locus and reported a closely linked *Ms* marker. Gökçe and Havey (2002) used *Ms* linked molecular marker identified in their study in three different onion populations to test the efficient use of this marker. After evaluating the frequency of genotypes at the molecular marker and *Ms* among plants from three different onion populations, chi-square analyses demonstrated that these two loci are at linkage equilibrium indicating that adequate generations of random mating had occurred to reach linkage equilibrium assuming single *Ms* locus (Gökçe and Havey 2002). Since DNA-based markers linked to *Ms* locus are not the alleles of the male sterility gene, they could not be directly used to predict genotypes at *Ms* in linkage phase unknown onion populations without determination of allelic phases of the *Ms* locus and the molecular marker loci. Tekeli (2015) used the *Ms* linked markers (Bang et al. 2013) in *Ms* genotype known onion inbred lines and reported that although markers and *Ms* alleles were at complete disequilibrium, they were at repulsion phase of what reported by Bang et al. (2013).

Therefore, it is important to note that molecular markers are useful when the breeder generates linkage disequilibrium by crossing among restorer and maintainer onion genotypes. To use the DNA markers flanking the *Ms* locus, one should determine the alleles revealed by the flanking DNA markers for the parents used in the fertile-by-fertile cross. Coupling- or repulsion-phase linkage with the recessive *ms* allele can be identified. The segregation of the dominant and recessive alleles at *Ms* can then be predicted using these linked phase-known markers. Although molecular characterization of nuclear genotypes and the cytoplasm does not reduce the onion breeding cycle (2 years per seed to seed generation are still required), molecular identification represents a more proficient use of resources by decreasing the total number of crosses and test crosses prerequisite to select maintainer onion plants.

References

Bacher JW (1989) Inheritance of resistance to *Fusarium oxysporum* f. sp. *cepae* in cultivated onions. Dissertation, Michigan State University

Bacher JW et al (1989) Inheritance of resistance to *Fusarium oxysporum* f. sp. *cepae* in cultivated onions. In: Jensen L (ed) Proceedings of national onion research conference, Idaho, pp 85–91

Bang H, Cho DY, Yoo K-S, Yoon M-K, Patil BS, Kim S (2011) Development of simple PCR-based markers linked to the *Ms* locus: a restorer-of-fertility gene in onion (*Allium cepa* L.). Euphytica 179:437–449

Bang H, Kim S, Park SO, Yoo KS, Patil BS (2013) Development of a codominant CAPS marker linked to the *Ms* locus controlling fertility restoration in onion (*Allium cepa* L.). Sci Hortic 153:42–49

Berninger E (1965) Contribution a l'etude de la sterilite-male de l'oignon (*Allium cepa* L.) (Contribution to the study of male sterility in the onion). Ann Amelior Plantes 15:183–199

Brewster JL (1994) Onions and other vegetable Allies. CAB International, Wallingford, UK

Clarke AE, Jones HA, Little TM (1944) Inheritance of bulb color in onion. Genetics 29:569–575

Cramer CS, Havey MJ (1999) Morphological, biochemical, and molecular markers in onion. HortScience 34:589–593

Davis E (1957) The distribution of the male-sterility gene in onion. J Amer Soc Hort Sci 70:316–318

Davis EW (1960) Marker genes to facilitate rouging onion-seed fields. Seed World 87:4–6

Davis EW (1966) An improved method of producing hybrid-onion seed. J Hered 57:55–57

de Courcel AGL, Vedel F, Boussac JM (1989) DNA polymorphism in *Allium cepa* cytoplasms and its implications concerning the origin of onions. Theor Appl Genet 77:793–798

Ekanayake N, Ewart LC (1997) Inheritance of resistance to *Alternaria porri* (Ellis) Cife in cultivated onion (*Allium cepa* L.). Allium Improv Nswl 7:20–22

El-Shafie M, Davis G (1967) Inheritance of bulb color in the onion (*Allium cepa* L.). Hilgardia 9:607–622

Engelke T, Terefe D, Tatlioglu T (2003) A PCR-based marker system monitoring CMS-(S), CMS-(T) and (N)-cytoplasm in the onion (*Allium cepa* L.). Theor Appl Genet 107:162–167

Engle RL, Gabelman WH (1966) Inheritance and mechanism for resistance to ozone damage in onion, *Allium cepa* L. J Amer Soc Hort Sci 89:423–430

Galmarini CR, Goldman IL, Havey MJ (2001) Genetic analyses of correlated solids, flavor, and health-enhancing traits in onion (*Allium cepa* L.). Mol Genet Genomics 265:543–551

Gökçe AF, Havey MJ (2002) Linkage equilibrium among tightly linked RFLPs and the *Ms* locus in open-pollinated onion populations. J Amer Soc Hort Sci 127:944–946

Gökçe AF, Havey MJ (2006) Selection at the *Ms* locus in open pollinated onion (*Allium cepa* L.) populations possessing S-cytoplasm or mixtures of N- and S-cytoplasms. Genet Resour Crop Evol 53:1495–1499

Gökçe AF, McCallum J, Sato Y, Havey MJ (2002) Molecular tagging of the *Ms* locus in onion. J Amer Soc Hort Sci 127:576–582

Hanelt P (1990) Taxonomy, evolution, and history. In: Rabinowitch H, Brewster JL (eds) Onion and allied crops. CRC Press, Florida, pp 1–26

Havey MJ (1993a) A putative donor of S-cytoplasm and its distribution among open-pollinated populations of onion. Theor Appl Genet 86:128–134

Havey MJ (1993b) Onion (*Allium cepa* L.). In: Kalloo G, Bergh BO (eds) Genetic improvement of vegetable crops. Pergamon Press, New York, pp 35–49

Havey MJ (1995a) Onion and other cultivated alliums spp. (*Liliaceae*). In: Smartt J, Simmonds NW (eds) Evolution of crop plants, 2nd end. Longman Scientific and Technical, Wiley, New York, pp 344–350

Havey MJ (1995b) Identification of cytoplasms using the polymerase chain reaction to aid in the extraction of maintainer lines from open-pollinated populations of onion. Theor Appl Genet 90:263–268

Havey MJ (1999) Seed yield, floral morphology, and lack of male-fertility restoration of male-sterile onion (*Allium cepa*) populations possessing the cytoplasm of *Allium galanthum*. J Amer Soc Hort Sci 124:626–629

Havey MJ (2013) Single nucleotide polymorphisms in linkage disequilibrium with the male-fertility restoration (*Ms*) locus in open-pollinated and inbred populations of onion. J Amer Soc Hort Sci 138:306–309

Havey MJ, Randle WT (1996) Combining abilities for yield and bulb quality among long- and intermediate-day open-pollinated onion populations. J Amer Soc Hort Sci 121:604–608

Holford P, Croft J, Newbury N (1991) Differences between, and possible origins of, the cytoplasms found in fertile and male-sterile onions (*Allium cepa* L.). Theor Appl Genet 82:737–744

Horobin JF (1986) Inheritance of a dwarf seed stalk character in onion. Ann Appl Biol 108:199–204

Jones HA, Clarke AE (1943) Inheritance of male sterility in the onion and the production of hybrid seed. Proc Amer Soc Hort Sci 43:189–194

Jones HA, Davis GN (1944) Inbreeding and heterosis and their relation to the development of new varieties of onions. USDA Tech Bull 874:1–28

Jones HA, Emsweller SL (1936) A male sterile onion. Proc Am Soc Hortic Sci 34:582–585

Jones HA, Mann LK (1963) Classification and identification. Onions and their allies. Interscience, New York, pp 24–46

Jones HA, Perry BA (1956) Inheritance of resistance to pink root in the onion. J Hered 47:33–34

Jones HA, Peterson CE (1952) Complementary factors for light-red bulb color in onions. Proc Amer Soc Hort Soc 59:457

Jones HA, Clarke AE, Stevenson FJ (1944) Studies in the genetics of the onion (*Allium cepa* L.). Proc Amer Soc Hort Sci 44:479–484

King JJ, Bradeen JM, Bark O, Havey MJ (1998a) A low-density genetic map of onion reveals the role of duplication in the evolution of an extremely large diploid genome. Theor Appl Genet 96:52–62

King JJ, Bradeen JM, Havey MJ (1998b) Variability for RFLPs and relationships among elite commercial inbred and virtual hybrid populations of onion. J Amer Soc Hort Sci 123:1034–1037

Kofoet A, Kik C, Wietsma WA, De Vries JN (1990) Inheritance of resistance to downy mildew [*Peronospora destructor* (Berk.) Casp.] from *Allium roylei* Stearn in the backcross *Allium cepa* L. x (*A. roylei* x *A. cepa*). Plant Breed 105:144–149

Little T, Jones HA, Clarke A (1944) The distribution of the male-sterility gene in varieties of onion. Hebertia 11:310–312

McCollum GD (1966) Heritability and genetic correlation of some onion bulb traits. J Hered 57:105–110

McCollum GD (1968) Heritability and genetic correlation of soluble solids, bulb size and shape in white sweet Spanish onion. Can J Genet Cytol 10:508–514

McCollum GD (1971) Heritability of onion bulb shape and size. Estimates from half-sib families. J Hered 62:101–104

Melgar S, Havey MJ (2010) The Dominant *Ms* allele in onion shows reduced penetrance. J Amer Soc Hort Sci 135:49–52

Miyaura K, Shinada Y, Gabelman WH (1985) Selection for resistance of onions to *Botrytis allii* by scale inoculation method. HortScience 20:769–770

Molenaar ND (1984) Genetics, thrips (*Thrips tabaci* L.) resistance and epicuticular wax characteristics of non glossy and glossy onions (*Allium cepa* L.). Dissertation, University of Wisconsin

Nichols CG, Gabelman WH, Larson RH, Walker JC (1965) The expression and inheritance of resistance to pink root in onion seedlings. Phytopathology 55:752–756

Pike LM (1986) Onion Breeding. In: Bassett MJ (ed) Breeding vegetable crops. AVI Publishing Company, Connecticut, pp 357–394

Rabinowitch HD, Friedlander B, Peters RJ (1984) Inheritance and characterization of dwarf scape in onions: a progress report. Proceedings in EUCARPIA 3rd Allium Symposium. Wageningen, The Netherlands, pp 83–89

Rabinowitch HD, Friedlander B, Peters R (1991) Dwarf flower stalk in onion: characterization, genetic control, and physiological response to Ethephon and gibberellic acid. J Amer Soc Hort Sci 116:574–579

Rasmusson J (1920) Mendelnde chlorophyllfaktoren bei *Allium cepa*. Hereditas 1:128–134

Rieman GH (1931) Genetic factors for pigmentation in the onion and their relation to disease resistance. J Ag Res 42:251–278

Riollano A (1943) Louisiana Red Creole, a new promising onion variety for Puerto Rico. Proc Amer Soc Hort Sci 43:272–274

Sato Y (1998) PCR amplification of CMS-specific mitochondrial nucleotide sequences to identify cytoplasmic genotypes of onion (*Allium cepa* L.). Theor Appl Genet 96:367–370

Satoh Y, Nagai M, Mikami T, Kinoshita T (1993) The use of mitochondrial DNA polymorphism in the classification of individual plants by cytoplasmic genotypes. Theor Appl Genet 86:345–348

Schweisguth B (1973) Etude díun nouveau type de sterilite male chez l'oignon, *Allium cepa* L. (A new type of male-sterility in onion, *Allium cepa* L.). Ann Amelior Plantes 23:221–233

Sumner DR (1995) Diplodia stain. In: Schwartz HF, Mohan SK (eds) Compendium of onion and garlic diseases. APS Press, St. Paul, Minnesota, pp 28–29

Tekeli FÖ (2015) Detection of linkage phase of the molecular markers and nuclear male sterility gene alleles in onion (*Allium cepa* L.) genotypes to be used for hybrid cultivar development. Dissertation, Niğde Ömer Halisdemir University

Traub HP (1968) Orientation of vascular bundles in Allium leaves. Plant Life 24:143–146

Tsutsui K (1991) Inheritance of resistance to *Fusarium oxysporum* in onion. Dissertation, University Wisconsin

Utkhede RS, Rahe JE (1978) Screening world onion germplasm collection for resistance to white rot. Can J Plant Sci 58:819–822

van der Meer QP, van Bennekom JL (1969) Effect of temperature on the occurrence of male sterility in onion (*Allium cepa* L.). Euphytica 18:389–394

van der Meer QP, van Bennekom JL (1972) Influence of the environment on the percentage of self-fertilization in onions and some consequences for breeding. Euphytica 21:450–453

van der Meer QP, van Bennekom JL, van der Giessen AC (1970) Testing onions (*Allium cepa* L.) and other Allium species for resistance to *Botrytis allii* Munn. Euphytica 19:152–162

van Heusden AW, van Ooijen JW, Vrielink-van Ginkel R, Verbeek WHJ, Wietsma WA, Kik C (2000a) A genetic map of an interspecific cross in Allium based on amplified fragment length polymorphism (AFLP) markers. Theor Appl Genet 100:118–126

van Heusden AW, Shigyo M, Tashiro Y, Vrielink-van Ginkel R, Kik C (2000b) AFLP linkage group assignment to the chromosomes of *Allium cepa* L. via monosomic addition lines. Theor Appl Genet 100:480–486

Vavilov NI (1951) The origin, variation, immunity and breeding of cultivated plants. Chron Bot 13:1–366

Walker JC (1923) Disease resistance to onion smudge. J Agr Res 24:1019–1040

Warid WA (1952) Inheritance studies in the onion, *Allium cepa* L. Dissertation, Louisiana State University

Warid WA, Tims EC (1952) Studies on the inheritance to downy mildew in onion incited by Peronospora destructor. Phytopathology 42:22

Cytological Details of Genome

5

Ludmila Khrustaleva

Abstract

This chapter presents a molecular cytogenetic progress in the study of the *Allium* chromosomes and its application in onion breeding. A crucial milestone in modern cytogenetics has been the development of fluorescence in situ hybridization (FISH) technology. Thanks to FISH and numerous other technical improvements it became possible to visualize on chromosome repetitive and single-copy DNA sequences that allowed to construct cytogenetic maps and integrate them with recombination maps. A complete set of alien monosomic addition lines (AMALs) of *A. fistulosum* with an extra chromosome from *A. cepa* played an essential role for alignment of *Allium* genetic maps to physical chromosomes. The recent discovery of unusual *Allium* telomere and the first achievement in the study of the onion centromere provide insight into the biology of this important functional structure of chromosomes. The molecular cytogenetic analysis of the *Allium* genome has revealed a great species-specific diversity of the chromosome structure, which once again convinces us in the complexity of the plant genome evolution. For the future, much more molecular cytogenetic markers are needed to assist whole-genome assembly in the ongoing SEQUON—Onion Genome Sequencing project. High-density cytogenetic maps will fill the gaps between genome sequencing data and the sub-chromosomal level.

L. Khrustaleva (✉)
Molecular Biotechnology Center, Russian State Agrarian University-Moscow K. A. Timiryazev Agricultural Academy, 49 Timiryazevskaya Str, 127550 Moscow, Russia
e-mail: khrustaleva@rgau-msha.ru

5.1 Introduction

Each cell's nucleus contains a certain number of DNA molecules in each species. DNA molecule is tightly coiled many times around proteins in chromosome structure. An individual chromosome becomes visible under a microscope only during cell division. The organization of genes, tandem repeats, retrotransposons, satellite DNA, and other chromosomal sequences related to the frequency of recombination can be established on the cytological study. In situ chromosome analysis of the gene and other sequence locations can provide a great deal of information for many aspects of genetics and genome assembly that can fully assist plant breeding.

Most *Alliums* are diploids ($2n = 2x$), with basic chromosome numbers $x = 8$ (Eurasia and Mediterranean basin), $x = 7$ (North America), or $x = 9$ (Eurasia) (Havey 2002). Polyploids such as triploids (*A. rupestre*, *A. scordoprasum*), tetraploids (*A. ampeloprasum, A. chinense, A. nutans*),

M. Shigyo et al. (eds.), *The Allium Genomes*, Compendium of Plant Genomes,
https://doi.org/10.1007/978-3-319-95825-5_5

pentaploids (*A. splendens*), hexaploids (*A. lineare*), and octoploids (*A. nutans*) also occur (Jones and Rees 1968; Jones 1990; Konishi et al. 2011). Genome sizes display 4.5-fold differences among *Allium* species from 7 pg/1 C in *A. altyncolicum* (Ricroch et al. 2005) to 31.49 pg/1 C in *A. ursinum* (Ohri 1998). Bulb onion (*Allium cepa*, $2n = 2x = 16$) has one of the largest genomes among crop plants, with its 17 pg (16 Gb) haploid nuclear genome being more than 100-fold larger than that of *Arabidopsis*. The huge genome of bulb onion is placed in 8 large chromosomes and that made this plant a good model in classical cytogenetics. However, from positions of modern molecular cytogenetics, onion is a difficult subject for scientific study. The mean condensation of an onion mitotic metaphase chromosome has been estimated at 249.6 Mb/μm, assuming uniform condensation along the entire chromosome (Khrustaleva and Kik 2001). This is much greater than the condensation of the human chromosome at 26.6 Mb/μm (Alberts et al. 1989) or tomato chromosomes at 40.6 Mb/μm (Anderson et al. 1985). On an average, one onion chromosome possesses the same amount of DNA as the entire diploid genome of tomato. Chromatin condensation in onion chromosomes remains a riddle in biology.

5.2 *Allium* Chromosome Nomenclature and Homoeology Relationships

Participants in a workshop on the onion chromosome nomenclature held at the University of Warwick, Coventry, U.K. on 8 September 1988 during the Eucarpia 4th *Allium* Symposium agreed with Kalkman's nomenclature (de Vries 1990). Since then, the nomenclature system proposed by Kalkman (1984) has provided the standard identification procedure of *A. cepa* chromosomes. Cytogeneticists who have studied the karyotype of *A. cepa* in several varieties of LD, SD and multiplier genotypes including shallots largely agreed to those of Kalkman (Peffley and Currah 1988; de Putter and van de Vooren 1988; de Vries and Jongerius 1988).

For six *Allium* species, de Vries and Jongerius (1988) produced karyotype idiograms based on morphometry and C-banding with the chromosome nomenclature according to Kalkman's standard system (Table 5.1). For these species, chromosomes with the same rank order number are not necessarily homoeologues. The chromosomes known to be homoeologous are the nucleolar organizing *A. cepa* and *A. fistulosum* chromosomes 6 because these satellite chromosomes form bivalents clearly sharing one nucleolus in pachytene and early diplotene of an interspecific hybrid between *A. cepa* and *A. fistulosum* (Jones and Rees 1968). *Allium cepa* and *A. roylei* possess two 5S rDNA loci, both localized on the short arm of the smallest metacentric chromosome 7 (Shibata and Hizume 2002; Fredotovic et al. 2014). *Allium fistulosum* possesses one 5S locus on the short arm of the smallest metacentric chromosome 7 (Son et al. 2012; Kirov et al. 2017). Thus, the smallest metacentric chromosomes 7 of *A. cepa, A. roylei* and *A. fistulosum* are homoeologues. The homoeology relationships between the remaining six chromosomes should be proved in further cytogenetic studies of the genus *Allium*.

5.3 Genes, Genetic Markers, and Linkage Group Assignment to the Chromosomes Via Alien Monosomic Addition Lines

Alien monosomic addition lines (AMALs) contain only one single chromosome of an alien donor species in addition to the entire chromosome complement of the recipient species. The genetic analyses of AMALs are an effective approach to allocate genes and genetic markers to physical chromosomes in wheat.

A unique resource—a complete set of monosomic alien addition lines, representing the eight different chromosomes of shallot (*A. cepa* L. Aggregatum group) in an *A. fistulosum* background ($2n = 17$, 16FF +C1-8) have been developed (Shigyo et al. 1996, 1998). Chromosome-specific genetic markers (ten

Table 5.1 Relative chromosome length and centromeric index of the six Allium species (produced from the data published by de Vries and Jongerius 1988)

Species	Chromosome number															
	1		2		3		4		5		6		7		8	
	Lr, M ± m	CI, M ± m	Lr, M ± m	CI, M ± m	Lr, M ± m	CI, M ± m	Lr, M ± m	CI, M ± m	Lr, M ± m	CI, M ± m	Lr, M ± m	CI, M ± m	Lr, M ± m	CI, M ± m	Lr, M ± m	CI, M ± m
A. cepa	14.8 ± 0.3	46.9 ± 3.2	14.4 ± 0.3	33.9 ± 1.2	13.8 ± 0.1	41.7 ± 4.5	12.6 ± 0.7	39.3 ± 2.1	12.7 ± 1.0	48.7 ± 0.7	12.2 ± 0.4	21.9 ± 4.1	10.4 ± 0.6	47.3 ± 4.2	9.2 ± 0.1	36.4 ± 2.8
A. fistulosum	15.2 ± 0.5	46.6 ± 2.6	14.4 ± 0.4	36.1 ± 0.3	12.0 ± 0.7	42.8 ± 6.2	11.5 ± 1.5	47.9 ± 1.9	11.9 ± 0.3	47.9 ± 1.9	11.9 ± 0.3	18.2 ± 2.6	10.3 ± 0.5	47.1 ± 0.7	10.3 ± 0.3	38.2 ± 3.0
A. roylei	15.3 ± 0.2	47.5 ± 0.5	13.9 ± 0.7	38.7 ± 4.7	13.6 ± 1.0	45.6 ± 2.1	13.3 ± 0.6	42.3 ± 4.1	12.1 ± 0.1	47.4 ± 3.1	11.3 ± 0.3	21.9 ± 7.3	10.7 ± 0.7	48.0 ± 0.4	9.8 ± 0.7	40.1 ± 4.2
A. vavilovii	15.0 ± 0.1	46.2 ± 2.4	14.0 ± 0.5	37.9 ± 0.6	13.1 ± 0.3	42.6 ± 0.3	13.0 ± 0.2	39.2 ± 2.4	12.9 ± 0.1	48.1 ± 0.2	11.9 ± 1.1	25.2 ± 0.8	10.6 ± 0.4	49.4 ± 4.7	9.9 ± 0.1	41.8 ± 3.5
A. galanthum	14.6 ± 0.5	43.8 ± 2.9	14.2 ± 0.6	37.2 ± 1.7	13.6 ± 0.7	44.2 ± 2.4	12.7 ± 1.0	45.4 ± 0.2	13.7 ± 0.4	48.4 ± 1.2	11.5 ± 0.7	20.2 ± 2.4	10.8 ± 0.1	44.3 ± 1.1	9.6 ± 0.1	40.0 ± 2.0
A. pskemense	14.0 ± 0.6	46.5 ± 4.8	13.3 ± 0.1	37.5 ± 0.1	13.3 ± 0.1	44.3 ± 2.2	17.8 ± 0.6	43.0 ± 1.1	12.4 ± 0.3	47.1 ± 1.0	12.1 ± 0.2	24.6 ± 1.7	10.7 ± 0.1	46.2 ± 0.4	10.3 ± 0.1	40.7 ± 1.5

Lr—the relative chromosome length (% of length of chromosome/total length of the complement)
CI—centromeric index (% of length of short arm/length of chromosome)

isozymes, 5S rDNA and 16 RAPD markers) (Shigyo et al. 1994, 1995a, b, 1996, 1997), genes related to flavonoid and anthocyanin production (Shigyo et al. 1997), genes involved in flavonoid biosynthesis (partial sequence of candidate genes CHS-A, CHS-B, CHI, F3H, DFR, ANS), and 3GT gene for glycosylation of anthocyanidin (Masuzaki et al. 2006) have been assigned to the individual shallot chromosomes (Table 5.2).

Using the AMALs, linkage groups based on *A. cepa*-markers developed via the genetic analysis of an interspecific cross between *A. cepa* and *A. roylei* with the AFLP marker were assigned to the individual physical chromosomes of *A. cepa* (van Heusden et al. 2000). The onion and shallot accessions used in this study were genetically not very different. More than 90% of their AFLP fragments were equally sized in both the accessions. Fifty-one AFLP markers of *A. cepa* that were absent in *A. roylei* were distributed over all eight linkage groups, thereby allowing the assignment of these groups to chromosomes. These 51 markers with their known chromosomal origin, determined with the set of AMALs, were used to find any correspondence between the linkage groups and the physical chromosomes. The availability of a complete set of AMALs allowed the detection of 186 *A. cepa* chromosome-specific AFLP markers. Further 74 codominant onion EST-derived markers were evaluated in the *A. cepa* x *A. roylei* interspecific population, enabling the merging of the AFLP-based maps (McCalum et al. 2012).

The AMALs were used to assign the onion linkage groups constructed on the intraspecific segregating population to physical chromosomes (Martin et al. 2005). The low-density intraspecific onion genetic map derived from the BYG15-23 × AC43 segregating family based primarily on restriction fragment length polymorphisms (RFLPs) composed of 116 markers was developed (King et al. 1998). Later 100 new expressed sequence tag (EST)-derived markers were added to the onion genetic map of King et al. (1998) to produce the most detailed intraspecific map to date, encompassing 1907 cM, and the linkage groups were assigned to onion chromosomes (Martin et al. 2005).

Thus the AMALs developed by Shigyo and his colleagues is the key resource that has enabled alignment of *Allium* genetic maps to physical chromosomes and facilitated a comparative study among species (McCalum et al. 2012). The AMALs were also used to anchor SSR-based maps in *A. fistulosum* to physical chromosomes (Tsukazaki et al. 2008). Using the AMALs and allotriploid-bunching onion single alien deletions, 513 EST-derived markers were assigned to the eight chromosomes of *A. fistulosum* and *A. cepa* (Tsukazaki et al. 2010). Data concerning markers assigned in multiple studies to the *Allium* physical map using AMALs have been compiled and these data reveal extensive synteny between *A. cepa* and *A. fistulosum* (McCalum et al. 2012).

Table 5.2 Chromosome specific genetic markers of *Allium cepa* L. determined via AMALS

Chromosome	Marker[a]	Source
1C	*Lap-1*	Shigyo et al. (1995a)
2C	*Got-1, 6-Pgdn-2, CHS-A*	Shigyo et al. (1995a), Masuzaki et al. (2006)
3C	*Tri-1, CHI, F3H*	Shigyo et al. (1996), Masuzaki et al. (2006)
4C	*Mdh-1, CHS-B, ANS, 3GT*	Shigyo et al. (1996), Masuzaki et al. (2006)
5C	*Idn-1, Pgi-1*	Shigyo et al. (1995b)
6C	*Adh-1, Got-2*	Shigyo et al. (1994, 1995a)
7C	5S rDNA-3, *DFR*	Shigyo et al. (1996), Masuzaki et al. (2006)
8C	*Gdh-1*	Shigyo et al. (1995a)

[a]RAPD markers (see Shigyo et al. 1997) are not included in the table

AMALs of *A. cepa* carrying extra chromosomes from *A. roylei* (RR, $2n = 2x = 16$) were constructed and all eight possible types of *A. cepa*—*A. roylei* monosomic addition lines (CC+R–CC+8R) were identified using the analyses of isozymes, EST markers, and karyotypes (Hoa et al. 2012). The availability of this AMALs for scientists and breeders will extend our knowledge about the *A. roylei* genome and genetics and will improve the introgression of desirable genes from *A. roylei* into *A. cepa*.

5.4 In Situ Direct Mapping Genes/Markers on *Allium* Chromosomes

How to visualize genes/markers and other DNA sequences on physical chromosomes? Fluorescence in situ hybridization (FISH) provides researchers with a way to visualize and map the DNA sequence in a specific position of an individual chromosome. FISH combines the chromosome preparation of classical cytogenetics with recombinant DNA technology. The hybridization site of the labeled DNA probe on chromosomal DNA is visualized using a fluorescent reporter molecule directly or indirectly bound to the probe.

5.4.1 rRNA Encoding Genes Organized in Tandem Arrays

The ribosomal RNA (rRNA) genes are arranged in hundreds or thousands of tandem repeats and are highly conserved between species. This chromosome region is the most widely investigated by FISH since the positions of these sites are well known in the hundreds of plant species. In higher eukaryotes, rRNA genes are organized as two distinct multigene families comprised of tandemly arrayed repeats. One family is represented by the 45 s rDNA which consists of a transcriptional unit that codes for the 18S, 5.8S, and 28S rRNAs, and an intergenic spacer (IGS). Multiple copies of 45 s rDNA correspond to the nucleolar organizer regions (NORs). The other family codes for the 5S rRNA and consists of a highly conserved coding sequence of 120 base pairs (bp) which is separated from each coding unit by an IGS. The 5S rDNA is not normally associated with NORs (reviewed in Long and David 1980). The 45S rDNA sequences were shown to be on the NOR of the satellite chromosomes and the 5S rDNA to be located, generally, on one of the other chromosomes (Gerlach and Dyer 1980; Leitch and Heslop-Harrison 1992). Due to FISH technology, it was found that the 45S rRNA genes are located not only on the site of secondary constriction (NORs) but on others sites where they were not previously detected (Guerra et al. 1996). It is known that the secondary constrictions represent only the expression of rRNA genes which were active during the last interphase and that other functional sites may not form secondary constrictions, especially if located too close to the terminal end of the chromosomes (Vanzela et al. 2003; Roa and Guerra 2012).

In *A. cepa*, the 45S rDNA were located on short arm of NOR-bearing chromosome 6 and at the distal region of the short arm of the smallest chromosome 8 (Ricroch et al. 1992; Do et al. 2001; Mancia et al. 2015) (Fig. 5.1a). In *A. fistulosum,* FISH probing with the 18S rDNA (Hizume 1994) and 45S rDNA (Kudryavtseva and Khrustaleva 2018 unpublished data) (Fig. 5.1b) revealed the hybridization signal only on the short arm of NOR-bearing chromosome 6. In *A. wakegi* ($2n = 2x = 16$), a natural allodiploid hybrid between *A. cepa* and *A. fistulosum,* probing with 18S rDNA (Hizume 1994) and 45S rDNA (Kirov and Khrustaleva 2018 unpublished data) (Fig. 5.1c) revealed the fluorescent signals on three chromosomes, at secondary constrictions of chromosomes 6 in *A. cepa* and *A. fistulosum* and on the short arm of chromosome 8 of *A. cepa.* The parental origin of chromosomes was proved by genomic in situ hybridization (GISH) (Hizume 1994). The 5S rDNA genes have been mapped to two loci (proximal and interstitial) on the short arm of chromosome 7 of *A. cepa* (Hizume 1994; Lee and Seo 1997) (Fig. 5.1d). Shibata and Hizume (2002) analyzed the structure and chromosomal location of different 5S rDNA subunit of *A. cepa* using microdissection and FISH. The authors dissected separately the proximal and distal segments of 5S rDNA and used them as templates for PCR. They showed that the long 5S rDNA unit was only present distally and the short unit was predominantly located proximally on the short arm of chromosome 7. In *A. fistulosum,* the 5S rDNA genes have been mapped to one locus on the short arm of chromosome 7 (Hizume 1994; Do and Seo 2000; Kirov et al. 2017) (Fig. 5.1e).

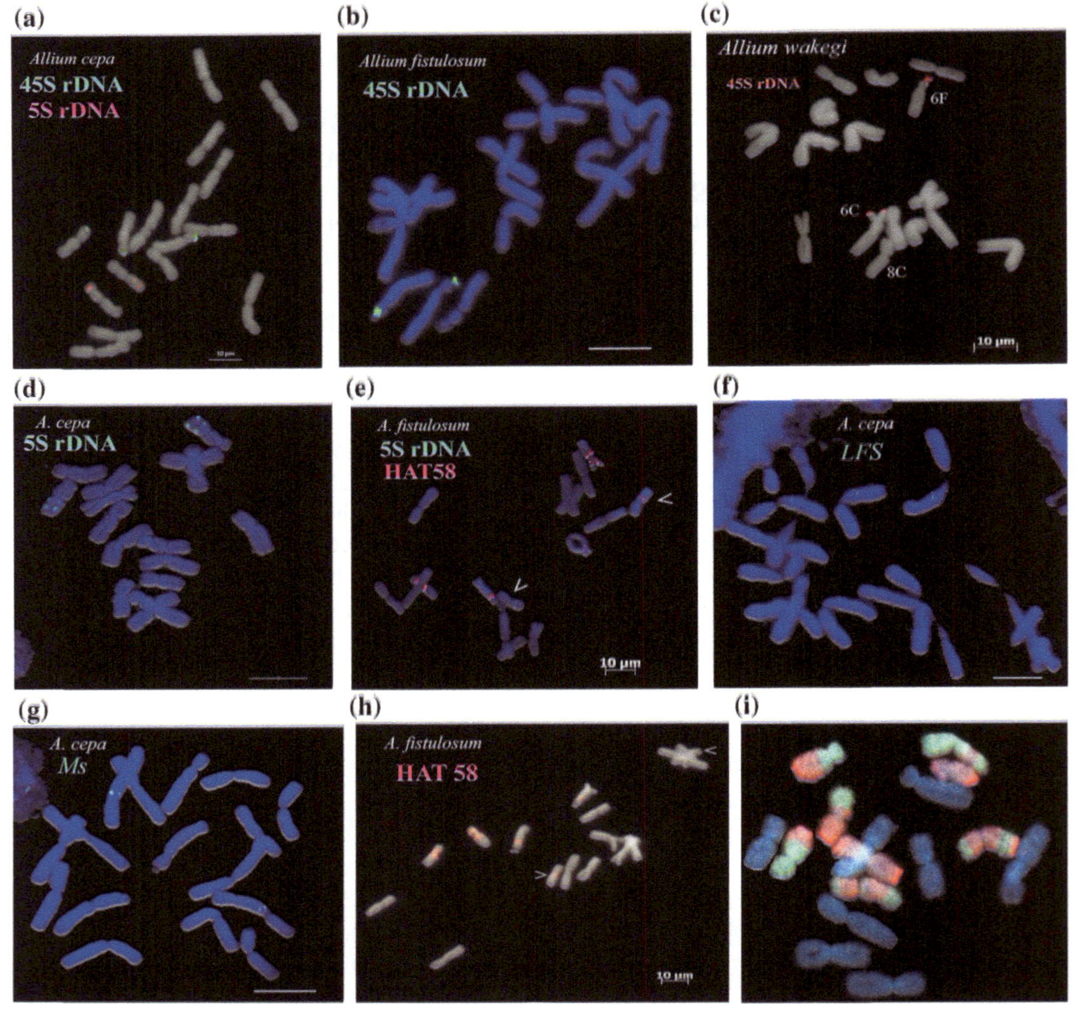

Fig. 5.1 Application of FISH in the *Allium* genome research. **a** Two-color FISH probing with 45S rDNA (green) and 5S rDNA (red) in *A. cepa*. Photo by I. Kirov. **b** FISH probing with 45S rDNA (green) in *A. fistulosum*. Photo by N. Kudryavtseva. **c** FISH probing with 45S rDNA (red) in *A.* x *wakegi*. Photo by I. Kirov. **d** FISH probing with 5S rDNA (green) in *A. cepa*. Confocal microscopy photo by D. Romanov. **e** Two-color FISH probing with 5S rDNA (green) and HAT58 (red), the arrows indicate polymorphic sites of HAT58 on the long arm of chromosome7. Photo by I.Kirov **f** BAC-FISH probing with the clone possessing of the *LFS* gene insert. C_0t-100 fraction was used as a block. Photo by L. Khrustaleva. **g** Tyr-FISH probing with cocktail of SNPs markers tightly linked to the *Ms* locus. Photo by L. Khrustaleva. **h** FISH probing with HAT58, the arrows indicate polymorphic sites of HAT58 on the long arm of chromosome7. Photo by I. Kirov **(i)** Multicolor GISH in the first generation of bridge-cross (*A. cepa* x (*A. roylei* x *A. fistulosum*)), *A. roylei* (green), *A. fistulosum* (red) and *A. cepa* (block DAPI, blue). Photo by L. Khrustaleva

The FISH study of 15 *Allium* species showed that the 5S rDNA were primarily located on chromosomes 5 and/or 7 in diploid species and various chromosomes in alloploid species (Do and Seo 2000).

5.4.2 Unique Genes Mapping on Highly Condensed *Allium* Chromosomes

5.4.2.1 BAC-FISH Mapping

While rRNA genes organized in tandemly arrayed repeats can be easily localized by FISH, unique genes have not been readily mapped on physical chromosomes due to the technical challenge of visualizing a small target DNA. In plant species with small gene-rich genomes, such as *Arabidopsis thaliana* (Koornneef et al. 2003) or rice (Jiang et al. 1995), detection of specific loci can be accomplished by FISH with large genomic Bacterial Artificial Chromosome (BAC) clones as probes. Unfortunately, the onion genome contains families of abundant repetitive elements (Stack and Comings 1979; Pearce et al. 1996). So, any clone that includes a copy of a repetitive DNA will hybridize across the genome, making it unsuitable as a FISH probe. Overcoming the technical difficulties for the first time, BAC clones possessing the lachrymatory factor synthase (LFS) gene were mapped on the *A. cepa* chromosome (Masamura et al. 2012) (Fig. 5.1f). The C_0t-100 fraction was used to block the repetitive sequence on a target DNA. Two totally sequenced BAC clones, 2E8/10 and 4F10/155, were hybridized to the mitotic metaphase chromosome. Sequence comparison of two BAC clones bearing LFS genes, LFS amplicons from diverse germplasm, and expressed sequences from a doubled haploid line revealed variation consistent with duplicated LFS genes. The BAC-FISH study showed that these BAC clones are co-localized in the proximal region of the long arm of the chromosome 5. However, the clones can be distant from each other up to 25 Mbp and still be located in the same position because of high compactization of mitotic metaphase chromosome and the resolution limit of a conventional light microscope. The results suggested that LFS in *A. cepa* is transcribed from at least two loci and that they are localized on chromosome 5. Genetic mapping of polymorphisms detected by heteroduplex analysis of LFS amplicons in the *A. cepa* x *A. roylei* interspecific cross revealed co-segregation with markers linked to chromosome 5 (Masamura et al. 2012). The position of LFS on the genetic map was linked with its position on the physical chromosome (Fig. 5.2b).

5.4.2.2 ESTs Mapping

ESTs are attractive candidates for chromosomal gene mapping because they possess protein-coding sequences and often do not contain dispersed repetitive DNA sequences that may complicate FISH signals. These probes do not require blocking of dispersed repeat sequences by C_0t fractions. Unfortunately, detection of such small unique sequences on plant chromosomes has been difficult because the length of the target chromosomal DNA that can be routinely visualized by FISH is 10 kb (Jiang et al. 1995; Jiang and Gill 2006), which is longer than the average gene length of 2.5–4.0 kb. To overcome the FISH sensitivity limitation, Raap et al. (1995) introduced the use of fluorescent tyramide conjugates as substrates for Horse Radish Peroxidase (HRP) into FISH technology. The technique combines the advantage of an enzymatic procedure that provides signal amplification due to the deposition of many substrate molecules, and that of fluorescence-based detection, which is higher than absorbency used in enzymatic detection. With this method, the detection sensitivity can be increased up to 100 times compared to the conventional FISH procedure. Khrustaleva and Kik (2001) have adapted for plant cytogenetics this ultrasensitive FISH method termed tyramide-FISH (Tyr-FISH). The authors were able to visualize the position of T-DNA inserts as small as 710 bp in transgenic shallots. Another problem faced by researchers is the presence of a cell wall in plants that hampers the availability of the target DNA sequence for the hybridization probe. The chromosome preparation procedure has a very strong impact on chromatin accessibility

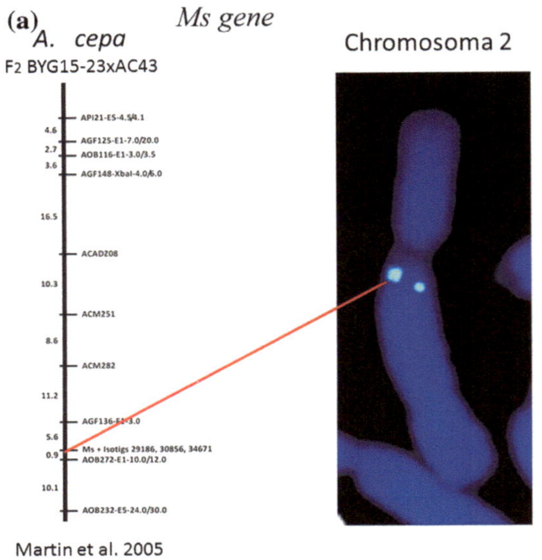

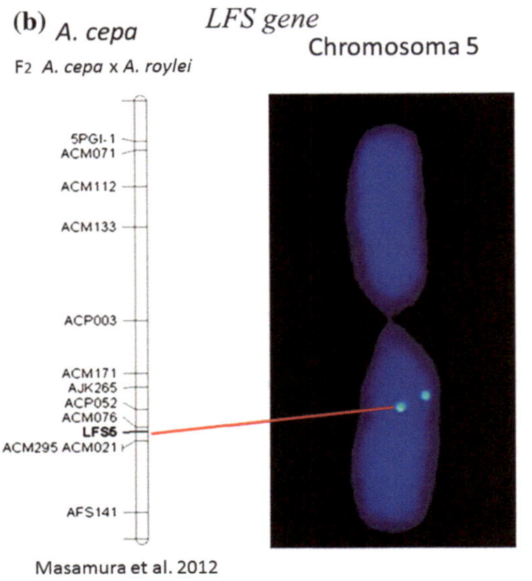

Fig. 5.2 Integration of the gene position on physical chromosome into recombination map. **a** *Ms* locus visualized on the long arm of chromosome 2 of *A. cepa* using tyr-FISH. The chromosome extracted from metaphase on Fig. 5.1g. **b** *LFS* locus visualized on the long arm of chromosome 5 of *A. cepa* using BAC-FISH. The chromosome extracted and straightened from metaphase on Fig. 5.1f

and short probe detection. Recently, a novel method named "SteamDrop" for the preparation of high-quality well-spread mitotic and pachytene chromosomes of plants was developed (Kirov et al. 2014, video of "SteamDrop" method available at www.plantgen.com). The sequence information about expressed sequence tags (ESTs) and genes that have been sequenced or partially sequenced are available publicly from GenBank at the NCBI (http://www.ncbi.nlm.nih.gov/genbank/) for the construction of primers or oligonucleotides to produce a DNA probe for in situ mapping. Using Tyr-FISH, the EST-clones API15 (GenBank Accession Number: BE205550.1), API59 (BE205590.1), API23 (BE205556.1), API92 (BE205605.1), and API66 (BE205593.1) were visualized on chromosome 5 of *A. cepa*, which carries several QTLs and desirable genes (Romanov et al. 2015). These EST clones were previously mapped in the same linkage group that was assigned to chromosome 5 (King et al. 1998; Martin et al. 2005). The position of these markers on the genetic map was integrated with their physical position on a chromosome. Through the integration of genetic and cytogenetic maps, the distribution of recombination events along onion chromosome 5 was estimated. The highest base pair/centimorgan estimates were (0.8 Mb/cM) between the markers API59-H3-15.0/9.5, API92-E1-11.0/12.0, and API66-E5-6.7/9.5 located in the interstitial region and over 20 times less (21.8 Mb/cM) for API23-H3-12.0/6.5 and API15-E1-3.0 markers located in the proximal region (Romanov et al. 2015).

5.4.2.3 Markers Tightly Linked to the Male Fertility Restoration Locus (*Ms*) of Onion

The importance of knowledge about the gene position on the physical chromosome was clearly demonstrated in the work on the Tyr-FISH mapping of markers linked to Male Fertility Restoration Locus (*Ms*) in *A. cepa* (Khrustaleva et al. 2016). Hybrid onion seed is commonly produced using cytoplasmic male sterility (CMS). Seed propagation of male-sterile plants (S *msms*) is possible by crossing with maintainer (N *msms*) plants (Jones and Davis 1944) and selection of superior maintainer lines is a primary focus of hybrid onion breeding programs. To more quickly determine genotypes at *Ms*, several research groups have identified molecular markers from the genome or transcriptome showing linkage to *Ms* (Gökçe et al. 2002; Gökçe and Havey 2002; Huo et al. 2012; Yang et al. 2013; Bang et al. 2013; Havey 2013; Park et al. 2013; Kim 2014). Many markers linked to *Ms* have been identified even though in some cases relatively few clones or primers were screened. It is known that genetically close markers may actually be far apart in terms of base pairs (or vice versa) due to differences in the frequency of recombination along the length of a chromosome. Thus, in chromosome regions experiencing relatively low recombination showing tight genetic linkage may be quite physically distant from each other. Tyr-FISH probing with the markers linked to *Ms* (SNPs and RFLP) revealed the proximal locations of these markers close to the centromere on the long arm of chromosome 2 (Khrustaleva et al. 2016) (Fig. 5.1g), a region of lower recombination (Albini and Jones 1988). The position of *Ms* on the genetic map was linked with its position on the physical chromosome (Fig. 5.2a). Four markers were co-localized at a relative position from the centromere -0.1 ± 0.02 on high condense mitotic metaphase chromosome. On super-stretched pachytene chromosomes, four markers were visualized as a linear string of fluorescent signals measuring 7.4 ± 0.6 μm. If the lengths of super-stretched pachytene chromosome are assumed to be 20 times longer than regular pachytene chromosomes and correspond to 1.5 Mb/μm (Koo and Jiang 2009), the markers would be located across a 10-Mb region. This does not diminish the usefulness of these molecular markers to predict genotypes at *Ms*; however, it does indicate that eventual map-based cloning of *Ms* may be arduous. As the cost of DNA sequencing continues to decline, the nuclear genome of onion will eventually be sequenced and assembled. Nevertheless, identification of candidates for *Ms* may be difficult because flanking markers may not locate onto a single contig.

5.5 Chromosomal Organization of Repetitive DNA Sequences

5.5.1 Tandem Repeats

A tandem repeat in DNA is two or more adjacent, approximate copies of a pattern of nucleotides. Tandem repeats are associated with important chromosomal landmarks such as centromeres, telomeres, subtelomeric, and other heterochromatic regions and have been widely studied during the last decades (Henikoff et al. 2001; Jiang et al. 2003; Koo et al. 2011). The first publication on chromosomal distribution of tandem repeats in Alliums was written by Barnes et al. (1985). A 375-bp fragment was isolated from the *Bam*HI digest of *A. cepa* genomic DNA that was cloned and sequenced. The tritiated plasmid DNA was hybridized to metaphase chromosomes and detected on autoradiographs. The 375-bp tandem repeats were located at the telomeric ends of all chromosomes except on the short arm of NOR-bearing chromosomes. The 375-bp tandem repeats constitute about 4% of the *A. cepa* genome. Later, Irifune et al. (1995) isolated a 380 bp DNA sequence in EcoRV digests of the total genomic DNA of *A. fistulosum* and showed with FISH that this tandem repeat had the same chromosomal localization as the 375 bp repeat in *A. cepa*. Moreover, the 380 bp DNA sequence of *A. fistulosum* had 82% homology with the 375 bp repeat of *A. cepa*. A copy number of the 380 bp tandem repeat was estimated about 2.8×10^6 per haploid genome of *A. fistulosum* (Irufine et al. 1995) Using FISH and PCR analysis of the *A. fistulosum* genome, Fesenko et al. (2002) showed that the 380 bp repeat arrays possess inversions and are interspersed with microsatellite and Ty1-copia retrotransposon sequences. A 314-bp tandemly repeated DNA sequence, named p*Ac*074, with high homology to the 375 bp repeat sequence was isolated by PCR with a set of random primers and FISH located it at the telomeric end of the *A. cepa* chromosomes (Do et al. 2001). FISH with the 375 bp repeat sequence probe derived from *A. cepa* on the chromosomes of 27 species (in 37 accessions) belonging to 14 sections of four subgenera in *Allium* showed that the analyzed closely related species possessed a very similar satellite sequence at the telomeric end of their chromosomes (Pich et al. 1996a). Taking all things together, we may conclude that this repeat had evolved already in progenitor forms of Alliums and remained unusually well conserved during speciation.

FISH analysis of BAC clones containing about 100 kb inserts of genomic DNA of *A. cepa* allowed to select, among the 91 randomly selected clones, nine clones showing FISH signals at centromeric and proximal regions and three clones showing telomeric signals (Suzuki et al. 2001). Considering that they did not use C_0t-1 as a competitor DNA in the BAC-FISH experiments, the BAC clones with distinct localized signals might possess large arrays of tandem repeats.

Tandem repeats are a valuable source of cytogenetic markers for distinguishing individual chromosomes (Albert et al. 2010). The repeatome analysis of *A. fistulosum* genome allowed to identify novel tandem repeats in *A. fistulosum* genome that can be used as cytogenetic markers (Kirov et al. 2017). Using next-generation sequencing data, authors identified two novel tandem repeats HAT58 and CAT36, which together occupy 0.25% of the *A. fistulosum* genome with 160,000 copies of HAT58 and 93,000 copies of CAT36 per haploid genome. FISH analysis showed that CAT36 is located in the pericentromeric regions of chromosomes 5 and 6 of *A. fistulosum*. HAT58 occupied intercalary heterochromatin of chromosome 6, 7, and 8 associated with C-banding patterns (Fig. 5.1h). Moreover, FISH with HAT58 revealed that this tandem repeat is polymorphic because plants with three type location patterns of this repeat were observed: both homologous chromosome 7 with signals, only one homolog with signals and missing signals on both homologs. Thus, HAT 58 might quickly spread to new genomic regions resulting in polymorphic sites. This finding suggests that the rapid evolution of the HAT58 repeat is still ongoing. HAT58 and CAT36 are species-specific tandem repeats that were shown for two closely related species *A. cepa* and *A. fistulosum*, and *A. wakegi* ($2n = 2x = 16$), a natural allodiploid hybrid between *A. fistulosum* and *A. cepa* (Kirov et al. 2017).

5.5.2 Retrotransposons

Retrotransposons are the commonest class of eukaryotic transposable elements. They are distinct from other transposons by their ability to transpose via an RNA intermediate, which is converted into extrachromosomal DNA by reverse transcription before reinsertion. Tyl-copia group retrotransposons are present throughout the plant kingdom as highly heterogeneous populations of high copy number elements (Kumar 1996), and there are 100,000-200,000 copies within the *A. cepa* diploid genome (Pearce et al. 1996). FISH to metaphase chromosomes reveals that Tyl-copia retrotransposons are distributed throughout the euchromatin of all chromosomes of *A. cepa* but are enriched in the terminal heterochromatic regions, which contain tandem arrays of satellite sequences (Pearce et al. 1996; Pich and Schubert 1998).

5.6 Centromere

The centromere is essential for the proper segregation and inheritance of genetic information. Visually, the centromere appears on metaphase chromosomes as a primary constriction. The DNA sequence underlying the centromere is not evolutionarily conserved and, in most species, is composed of megabases of rapidly evolving tandem repeats (Melters et al. 2013). In all flowering plants investigated so far, the centromere is generally composed of large arrays of centromeric satellite repeats and centromeric retrotransposons. The abundance and the arrangement of these repeats vary substantially, both within and among species (Jiang et al.2003; Nagaki et al. 2004, 2009; Nagaki and Murata 2005). The satellite sequences that occur within the centromeres of most eukaryotes are usually species-specific (Houben and Schubert 2003). Among *Allium* species, centromeric DNA sequences were identified only for *A. fistulosum* by Chromatin Immuno Precipitation (ChIP) and Tail-PCR (Nagaki et al 2012). Three clones with inserts of CHIP isolated centromeric sequences produced FISH signals on the centromere position of all 16 chromosomes of *A. fistulosum* (Fig. 5.1). These clones were sequenced and their sequences are available in GenBank: Afi11 (AB735740), Afi19 (AB735741), and Afi56 (AB735743). The authors revealed two more clones that produced the centromeric FISH signals but not on all 16 chromosomes. For Afi54 (AB735742), 12 strong and four weak signals, and for Afi61 (AB735744), one strong signal pair and 14 weak signals on the centromeres were observed. The DNA sequence of these five clones did not show similarity to annotated genomic sequences in NCBI databases.

The centromeric and pericentromeric regions of plant chromosomes are colonized by Ty3/gypsy retrotransposons. Centromeric retrotransposons belong to a number of lineages of the chromo-virus family of Ty3/gypsy LTR (long terminal repeat) retrotransposones (Neuman et al. 2011). Centromeric retrotransposons are found between the centromeric satellite repeats (Cheng et al. 2002) and they also can take the major role in the centromeric structure (Li et al. 2013). The presence of Ty3/gypsy-like retrotransposons in the centromeric region of *Allium cepa* and *Allium fistulosum* was reported by Kiseleva et al. (2014). The putative copy number of Ty3/gypsy centromeric retrotransposons constituted about 26,000 for *A. cepa* and about 7000 for *A. fistulosum*. In silico identification of centromeric retrotransposons has also been performed, followed by their clustering. Using the NCBI Entrez in the NCBI database, a total of 10,725 GSSs (genome survey sequences) of *A. cepa* were identified. These sequences were used as query data for the RepeatExplorer server. FISH with PCR product obtained with designed primers on reverse transcriptase of Beetle1 and CRM Ty3/gypsy elements showed strong hybridization signals in the centromeric regions of *A. cepa*. FISH on the chromosomes of *A. fistulosum* showed hybridization signals of different intensity in the centromeric region, as well as in other chromosomal regions. The estimation of a retrotransposon insertion time in the onion genomes suggests a high activity of some of LTR retrotransposons in their recent history (Vitte et al. 2013). It is still unclear why some retrotransposons insert selectively into the centromere region. A possible explanation of this phenomenon may be that the centromeric region of a chromosome provides a safe environment for retrotransposons reducing the chance of their elimination from the genome via recombination, which is suppressed in this region.

5.7 Telomere

The ends of eukaryotic chromosomes are capped by a special structure called the telomere. The telomere protects the termini of eukaryotic chromosomes from degradation by nucleases, illegitimate fusion, and progressive shortening as a result of incomplete replication of linear DNA molecules at their 5′-ends. The DNA component of the telomere is typically formed by long arrays of a G-rich tandemly repeated short minisatellite sequence that, depending on the organism, extend for tens of base pairs to as much as 150 kilobase pairs. In contrast to the centromere, the telomere is highly conserved DNA sequence in large groups of organisms, e.g., TTAGGG in vertebrates (Cheng et al. 1989; Meyne et al. 1989), TTTAGGG in plants (Richards and Ausubel 1988) and TTAGG in insects (Okazaki et al.1993). The strong conservation of telomeric repeats is likely a result of the interaction between telomeric DNA and telomere-specific binding proteins (Watson and Riha 2010). Surprisingly, telomere composed of (TTTAGGG)n DNA repeats typically for most of the plants was absent in the *Alliaceae* family. For decades, scientists have been trying to find out how *Alliaceae* stabilize their chromosome ends in the absence of

TTTAGGG sequences. Candidates for alternative telomeric sequences included ribosomal RNA genes or subtelomeric satellite sequences which could be spread by homologous recombination were suggested (Pich et al. 1996b). A possible involvement of Ty1-copia retrotransposons was also suggested for Allium (Kumar et al. 1997), but the subsequent study did not support this idea (Pich and Schubert 1998). To clarify the enigma of the *Allium* telomere maintenance, extensive studies of telomere evolutionary variability have been performed in Asparagales plants. In a number of families in this order, starting from the divergence of the *Iridaceae* family, the typical plant-type telomeric sequence (TTTAGGG)n had been partly or fully replaced by the *vertebrate*-type sequence (TTAGGG)n synthesized by telomerase (Sykorova et al. 2003, 2006a, b). However, the genus *Allium* was an exception where neither of the plant-type repeats or their known variants could be detected at chromosome termini, and a corresponding telomerase activity was absented as well (Fajkus et al. 2005; Sykorova et al. 2006a). Finally, in 2015 the group of professor Jiří Fajkus reported that *Allium* telomeres are unmasked. The unusual telomeric sequence (CTCGGTTATGGG)n is synthesized by telomerase (Fajkus et al. 2016). Due to recent advances in bioinformatics on transcriptomic and genomic data in combination with conventional approaches of molecular biology and molecular cytogenetics, the researchers succeeded in finding the unusual telomeric sequence of *Allium* and demonstrate its synthesis by telomerase.

5.8 Integration of Recombination, Cytogenetic and Sequence (Contigs) Maps

Since King et al. (1998) published the first genetic (recombination) map of *Allium* species, AFLP, SSRa, and EST-derived SNP markers have been employed to increase its marker density (McCallum et al. 2012). The densified genetic maps considerably improved the efficiency of the breeding process and expanded our knowledge on the onion genetics. The genetic maps display the linear order of genes or markers and the recombination frequencies between them. The distance between markers is expressed in centi-Morgans (cM) and 1 cM is equal to 1% of recombination. While genetic maps are important in biology, they do not show the real physical distance between gene/markers due to unequal distribution of recombination frequencies along the chromosomes. One cM on a genetic map can be equivalent to a few kilobases as well as to millions of base pairs of physical distance (Khrustaleva et al. 2005; Sun et al. 2013; Si et al. 2015; Romanov et al. 2015).

Cytogenetic maps allow us to determine the approximate distance between genetic markers, but not the exact distance (number of base pairs). Cytogenetic maps show the positions of genetically mapped markers on chromosomes, relative to centromeres, telomeres, heterochromatin, and euchromatin. It is important to note that cytogenetic mapping is an essential tool for the ordering of markers in region where there is the absence of recombination. The discrepancy between marker positions on the genetic map and their actual location relative to each other on the physical chromosome in the region of the suppressed recombination often occur (Szinay et al. 2008). Multiple approaches have been used to develop the integrated genetic and cytogenetic maps. For instance, in wheat, 436 deletion lines were constructed to relate the deletions as landmarks to recombination (Sandhu et al. 2001). However, this approach cannot be used for diploid species as onion because diploid organisms do not tolerate large deletions. To overcome this problem, Khrustaleva et al. (2005) applied a novel strategy to relate markers with their position on a physical chromosome. For the development of an integrated map, the authors used *Allium* trihybrid population, which originated from a cross between *Allium cepa* and (*A. roylei* x *A. fistulosum*). This population represents an ideal source for integrated mapping, because in each chromosome pair, one homoeologous chromosome originates from the interspecific hybrid between *A. roylei* and *A. fistulosum* and another nonrecombinant homoeologous chromosome originates from *A. cepa* (Fig. 5.1i). The recombination sites on the recombinant chromosome between *A. roylei* and *A. fistulosum* were visualized via GISH. The AFLP profiles of individual genotypes were compared with the corresponding recombinant chromosome. Simultaneously for *A. roylei* and *A. fistulosum,* integrated physical and recombination maps of chromosome 5 and 8 were constructed. The integration of genetic and chromosome maps demonstrated how genetic and physical distance between markers varied depending on the marker position on physical chromosome: the base pair/centimorgan estimates were 1.4 Mb/cM in the hotspot recombination and 74.3 Mb/cM in the region of low recombination.

However, with an above mentioned approach for integrated mapping, regions lacking recombination will be not covered. The most direct and effective way to construct cytogenetic maps for organisms with large and complex genomes like onion is to directly localize single-copy genes on chromosomes. With this approach, the LFS gene (Masamura et al. 2012) (Fig. 5.1f, 2b), the EST-based markers (Romanov et al. 2015), SNPs, and RFLP tightly linked to the male fertility restoration gene (Khrustaleva et al. 2016) (Fig. 5.1g, 2a) were mapped on physical chromosome and integrated into genetic maps.

To conclude, the reader may be skeptical about integrated mapping because now with the next generation sequencing, the whole-genome sequencing of any organism became fast and robust. However, scaffold order arrangement and whole-genome assembly remain challenge. Recent publications showed massive discrepancies between in silico assembled version of the genome and nuclear genome. Mostly misassembled genome sequences occurred in the "cold" spot recombination regions (Yang et al. 2014; Karafiátová et al. 2013; Shearer et al. 2014). Therefore, the order and orientation of sequenced scaffolds in pseudomolecules should be corrected using independent physical methods, such as FISH and optical mapping. Optical mapping has been used to improve de novo plant genome assemblies (Tang et al. 2015). This method allows to visualize the locations of the restriction sites or sequence motifs under light microscopes.

5.9 Application of Molecular Cytogenetics in the Onion Interspecific Breeding

Interspecies hybridization plays an important role in bulb onion breeding since its gene pool appeared to be rather depleted within its long history of more than 5000 years of human cultivation (Jones 1983). An ancestral species, *A. cepa*, was lost together with the depletion of multiple valuable properties. Closely related species might be used as donors of economically valuable properties in bulb onion breeding. Effective identification of alien chromosomes is essential for monitoring the alien genetic material. Molecular markers, however, can only reveal the regions from which they have been derived. A large number of markers that represent different chromosomal regions would have to be used to analyze a complete chromosome. Furthermore, the presence of a marker (or a syntenic group of markers) often does not distinguish whether one copy or multiple copies of a particular chromosome are present in the plant (Dong et al. 1999). Molecular cytogenetic identification of alien chromatin in breeding lines includes determination of both genomic origin and chromosomal specificity of a chromosome or chromosomal segment. GISH provides a direct visual identification of parental genomes in interspecific hybrids and their backcross progenies. GISH is a modification of fluorescent in situ hybridization, when labeled total genomic DNA of one parental species is used as a probe with unlabeled genomic DNA from another species at a higher concentration, which serves as a blocking DNA, hybridizing with the sequences in common with both genomes.

Allium fistulosum is a rich source of desirable traits which are very beneficial for the breeding of new onion (*A. cepa*) cultivars. The first attempt of a gene transfer from *A. fistulosum* to *A. cepa* using interspecific hybridization was performed in 1935 by Emsweller and Jones. The F_2 plants proved to be largely sterile, although occasionally seed set occurred upon selfing, while they succeeded to obtain several sterile BC_1 plants. Levan (1936) could also obtain F_2

plants between *A. cepa* and *A. fistulosum*, and only those with spontaneous chromosomal duplication were shown to be fertile. Maeda (1937) revealed that a number of the F_2 derivatives and backcross progenies with *A. fistulosum* were mostly sterile and diploids. It was reported that the F_1BC_3 plants with *A. cepa* cytoplasm showed a certain level of fertility and a similar morphology as *A. cepa* (Hou and Peffley 2000; Peffley and Hou 2000). Pathak and colleagues (2001) succeeded in producing the F_3 progenies resistant to stemphylium leaf blight with the high level of pollen fertility (40–80%) and seed set (20–60%). GISH analysis of F_2 and advanced generation of interspecific hybrids between *A. cepa* and *A. fistulosum* that were relatively resistant to downy mildew was reported by Budylin and coleagues (Budylin et al. 2014). The GISH analysis of its advanced generation revealed that the F_5 and BC1F5plants which produced few seeds was amphidiploid with 32 chromosomes. In the same way, it was concluded that a sterile BC_1F_5 plant was triploid possessing eight *A. fistulosum* and 16 *A. cepa* chromosomes and that a partially fertile BC_1F_5 plant was amphidiploid with 4 recombinant chromosomes. Because colchicine treatment was not used in any generation of hybrids authors suggested the presence of a spontaneous polyploidization. The GISH analysis of recombinant chromosomes allowed to suggest the formation of 2*n* gametes as a result of second division restitution (SDR) since both sister chromatids of recombinant chromosomes remained in the same gamete. Levan (1936) previously demonstrated polyploid F_2 plants as a result of a crossing between *A. cepa* and *A. fistulosum* without the artificial chromosome duplication, and explained this phenomenon by 2*n*-gametes formation in both macrosporogenesis and microsporogenesis. Consequently, it seems that viable F_2 zygotes with triploid or tetraploid chromosome set might be formed. Contrary to the studies of Budylin et al. (2014) and Levan (1936), other scientists reported the diploid F_1, F_2, and BC_1 plants of hybrids between *A. cepa* and *A. fistulosum* (Emsweller and Jones 1935; Maeda 1937; Peffley and Hou 2000; Hou and Peffley 2000). Levan (1941) tried to explain inconsistencies between his data and the results reported by Emsweller and Jones (1935), and carried out a comparative analysis of meiosis in his hybrids and hybrids reported by them. Levan observed significant differences in chromosomal behavior during meiosis between his hybrids and the hybrids obtained by Emsweller and Jones (1935) and suggested that the origin of parental cultivars played an important role in the polyploidization. The hybrids of Emsweller and Jones (1935) and Maeda (1937) were obtained from crossings between "Yellow Danvers" of *A. cepa* and the Japanese cultivars of *A. fistulosum* ("Nebuka" in the first case and "Hidanegin" in the second). Later Peffley and Hou (2000) and Hou and Peffley (2000) obtained diploid BC_1 and F_1BC_3 plants, respectively, as a result of interspecific hybridization by using "Ishikura" as *A. fistulosum* parent. In the crossing experiments researches of Levan (1936) and Budylin et al (2014), the European cultivars of *A. fistulosum* were used as a pollinator. Taking all these into consideration, no fertile breeding line possessing a single gene from *A. fistulosum* in the diploid background of *A. cepa* had been obtained despite numerous reports on interspecific $F_{1\text{-}5}$ hybrids and its backcross progenies.

To circumvent the aforementioned problems, an idea was born to use *A. roylei* as an intermediate species to transfer genes from *A. fistulosum* to *A. cepa* (de Vries et al. 1992). It was shown that *A. roylei* crossed readily with both *A. cepa* (van der Meer and de Vries1990) and *A. fistulosum* (McCollum 1982). The appropriateness of this strategy was fully confirmed by the GISH analyses of several generations in the bridge-cross (*A. cepa* x (*A roylei* x *A. fistulosum*)) (Khrustaleva and Kik 1998, 2000). The first and second bridge-cross generations were fertile and the recombination between chromosomes of *A. cepa* and *A. fistulosum* occurred frequently. The power of GISH analyses in breeding process was clearly demonstrated in producing onion breeding lines resistant to downy mildew (Scholten et al. 2007). The GISH analysis of the advanced generations showed that an *A. roylei* fragment bearing the gene locus *Pd1* was located in the distal end of the long arm of chromosome 3 as reconfirmed by the genetic mapping of van Heusden et al. (2000a). Moreover, the GISH analysis could not detect the large *A. roylei* segments on both homologous chromosome 3 in 14 downy mildew resistant plants of $F_1BC_5S_3$ progeny 2348, a progeny that segregated for resistance, and six plants from $F_1BC_5S_3$ progeny 3591, a progeny that consisted of only resistant plants. This observation allowed to hypothesize the presence of the recessive lethal factor proximally located to the downy mildew resistant gene within the large *A. roylei* segment. The further developing of molecular marker closely linked to downy mildew resistant gene validated this hypothesis. Thus, the breeding process was complicated by a lethal factor that seems to be expressed only in an *A. cepa* background. By crossing overs between homoeologus chromosomes, the downy mildew resistance locus and the lethal factor were separated and the remaining small segment of *A. roylei* harboring *Pd1* could be made homozygous without any problems.

5.10 Summary and Perspectives for the Future

To conclude it can be said that the FISH technology together with technical progress in genome sequencing and bioinformatics must be very useful to study the molecular organization of repeat and single-copy sequences along the chromosomes. Summarizing past achievements in the molecular cytogenetic study of alliums; (1) A number of markers (SNP, RFLP, EST, AFLP), genes (45S rDNA, 5S rDNA, LFS, and Pd1) were mapped on the respective physical chromosomes, (2) Several species-specific markers for identification of individual chromosomes were developed, (3) Centromeric DNA sequences for *A. fistulosum* and centromeric Ty3/gypsy retrotransposons for *A. cepa* and *A. fistulosum* were found and (4) The unusual telomeric sequence (CTCGGTTATGGG)n of alliums was discovered (Figs. 5.3 and 5.4).

For the future, the development of FISH markers and construction of high-density cytogenetic maps will accelerate the ongoing genome sequencing projects of *A. cepa* and *A. fistulosum*. An integrated approach including different sequencing strategies is needed due to problems of whole-genome assembly. The approaches will be comprised of long-size insert libraries, long-read sequencing (e.g., PacBio sequencing), and the Hi-C data on the basis of scaffolding as well as genetic map, independent tools such as cytogenetic mapping and optical mapping (Korbel and Lee 2013; Cao et al. 2016; Chaney et al. 2016), etc. To anchor assembled pseudochromosomes to each arm of the physical chromosomes of *A. cepa* and *A. fistulosum* as well as to determine their north–south orientation, at least 32 probes, two probes per pseudochromosome should be developed. Such a cytogenetic map will contribute the progress of *Allium* breeding via effective map-based cloning and accurate genome assembly. The development of robust cytogenetic markers will extend our knowledge of *Allium* genome organization and evolution and will fill the gaps between genome sequencing

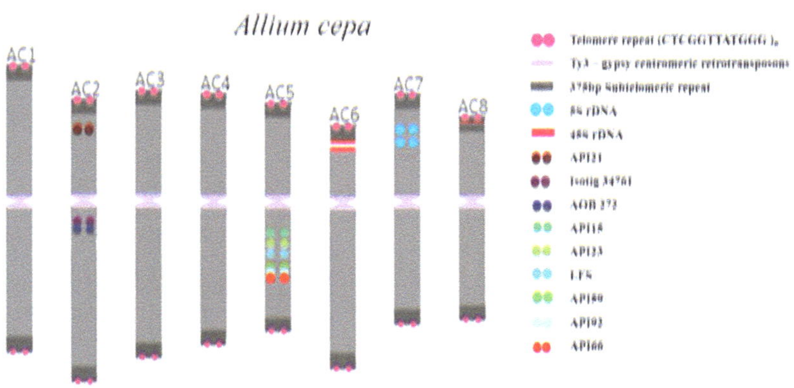

Fig. 5.3 Idiograms of *A. cepa* chromosomes with indicated position of FISH-mapped repetitive and unique DNA sequences

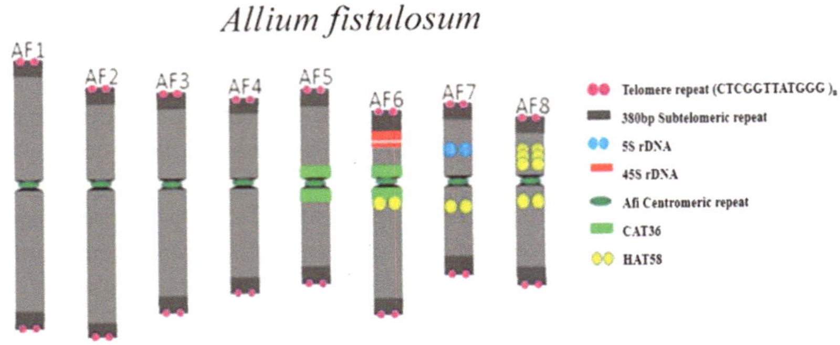

Fig. 5.4 Idiograms of *A. fistulosum* chromosomes with indicated position of FISH-mapped repetitive and unique DNA sequences

and sub-chromosomal measurement data. Furthermore, the integration of linkage and cytogenetic maps will extend our scanty knowledge of recombination rates and patterns in higher plants and, probably, will shed light on several issues through in-depth studies; (1) Why do the recombination sites localize adjacent to the centromeres of *A. fistulosum* chromosomes?, (2) Why do the recombination sites distribute randomly on *A. cepa* chromosomes and (3) Why is the recombination frequency of plants higher than that of animals. In 1913, Alfred Sturtevant constructed the first genetic map on the basis of Morgan's theories of crossing-over. James Watson and Francis Crick had solved the three-dimensional structure of DNA in 1953. Today, we need to solve a big puzzle of the organization and function of whole-genome. Molecular cytogenetics will play an essential role as a bridge between genomics, classical genetics, and other biological disciplines.

Acknowledgements I am grateful for the unpublished images provided by Ilya Kirov (Shemyakin-Ovchinnikov Institute of Bioorganic Chemistry, Russian Academy of Sciences), Dmitry Romanov and Nataliya Kudryavtseva (Russian State Agrarian University-Moscow Timiryazev Agricultural Academy), and to my diploma student Alexey Ermolaev for the technical assistance. This work was supported by the Russian Science Foundation, Grant No. 16-16-10031.

References

Albert PS, Gao Z, Danilova TV, Birchler JA (2010) Diversity of chromosomal karyotypes in maize and its relatives. Cytogenet Genome Res 129:6–16. https://doi.org/10.1159/000314342

Alberts B, Bray D, Lewis J, Raff M, Roberts K, Watson JD (1989) Molecular biology of the cell, 2nd edn. Garland Publishing, New York, p 1218

Albini SM, Jones GH (1988) Synaptonemal complex spreading in *Allium cepa* and *Allium fistulosum*. II. Pachytene observations: the SC karyotype and the correspondence of late recombination nodules and chiasmata. Genome 30:399–410. https://doi.org/10.1139/g88-069

Anderson LK, Stack SM, Fox MH, Zhang C (1985) The relationship between genome size and synaptonemal complex length in higher plants. Expl Cell Res 156:367–377. https://doi.org/10.1016/0014-4827(85)90544-0

Bang H, Kim S, Park SO, Yoo KS, Patil BS (2013) Development of a codominant CAPS marker linked to the Ms locus controlling fertility restoration in onion (*Allium cepa* L.). Sci Hortic 153:42–49. https://doi.org/10.1016/j.scienta.2013.01.020

Barnes SR, James AM, Jamieson G (1985) The organization, nucleotide sequence, and chromosomal distribution of a satellite DNA from *Allium cepa*. Chromosoma 92:185–192. https://doi.org/10.1007/bf00348692

Budylin MV, Kan LY, Romanov VS, Khrustaleva LI (2014) GISH study of advanced generation of the hybrids between *Allium cepa* L. and *Allium fistulosum* L. with relative resistance to downy mildew. Russ J Genet 50:387–394. https://doi.org/10.1134/S1022795414040036

Cao HX, Vu GTH, Wang W, Appenroth KJ, Messing J, Schubert I (2016) The map-based genome sequence of *Spirodela polyrhiza* aligned with its chromosomes, a reference for karyotype evolution. New Phytol 209:354–363. https://doi.org/10.1111/nph.13592

Chaney L, Sharp AR, Evans CR, Udall JA (2016) Genome mapping in plant comparative genomics. Trends Plant Sci 21:770–780. https://doi.org/10.1016/j.tplants.2016.05.004

Cheng JF, Smith CL, Cantor CR (1989) Isolation and characterization of a human telomere. Nucleic Acids Res 17:6109–6127. https://doi.org/10.1093/nar/17.15.6109

Cheng Z, Dong F, Langdon T, Ouyang S, Buell CR, Gu M, Blattner FR, Jiang J (2002) Functional rice centromeres are marked by a satellite repeat and a centromere-specific retrotransposon. Plant Cell 14:1691–1704. https://doi.org/10.1105/tpc.003079

de Vries JN (1990) Onion chromosome nomenclature and homoeology relationships—workshop report. Euphytica 49:1–3. https://doi.org/10.1007/bf00024124

de Vries JN, Jongerius MC (1998) Interstitial C-bands on the chromosomes of *Allium* species from the section *Cepa*. In: Riggs TJ, Astley D, Brewster JL, Dawson DR, Maude RB (eds) Proceedings of the 4th *Allium* symposium. Institute of Horticultural Research, Wellesbourne, UK, pp 71–78

de Vries JN, Wietsma WA, Appels M (1992) Direct and *Allium roylei*-mediated transfer of *A. fistulosum* genes to onion. Allium Improve Lett 2:9–10

Do GS, Seo BB (2000) Phyiogenetic relationships among allium subg. rhizirideum species based on the molecular variation of 5S rRNA genes. Korean Journal of Biological Sciences 4:77–85. https://doi.org/10.1080/12265071.2000.9647527

Do GS, Seo BB, Yamamoto M, Suzuki G, Mukai Y (2001) Identification and chromosomal location of tandemly repeated DNA sequences in *Allium cepa*. Genes Genet Syst 76:53–60. https://doi.org/10.1266/ggs.76.53

Dong F, Novy RG, Helgeson JP, Jiang J (1999) Cytological characterization of potato-*Solanum etuburosum* somatic hybrids and their backcross progenies by genomic in situ hybridization. Genome 42:987–992. https://doi.org/10.1139/g99-037

Emsweller SL, Jones HA (1935) An interspecifc hybrid in *Allium*. Hilgardia 9:265–273

Fajkus J, Sykorova E, Leitch AR (2005) Telomeres in evolution and evolution of telomeres. Chromosome Res 13:469–479. https://doi.org/10.1007/s10577-005-0997-2

Fajkus P, Peška V, Sitová Z, Fulnečková J, Dvořáčková M, Gogela R, Sýkorová E, Hapala J, Fajkus J (2016) *Allium* telomeres unmasked: the unusual telomeric sequence (CTCGGTTATGGG)n is synthesized by telomerase. Plant J 85:337–347. https://doi.org/10.1111/tpj.13115

Fesenko IA., Khrustaleva LI, Karlov GI (2002) Organization of the 378-bp satellite repeat in terminal heterochromatin of Allium fistulosum. Russ J Genet 38(7):745–753. https://doi.org/10.1023/A:1016379319030

Fredotović Z, Šamanić I, Weiss-Schneeweiss H, Kamenjarin J, Jang T-S, Puizina J (2014) Triparental origin of triploid onion, *Allium* × *cornutum* (Clementi ex Visiani, 1842), as evidenced by molecular, phylogenetic and cytogenetic analyses. BMC Plant Biol 14:24. https://doi.org/10.1186/1471-2229-14-24

Gerlach WL, Dyer TA (1980) Sequence organization of the repeating units in the nucleus of wheat which contain 5S rRNA genes. Nucleic Acids Res 8 (21):4851–4865. PMC55228

Gökçe AF, Havey MJ (2002) Linkage equilibrium among tightly linked RFLPs and the *Ms* locus in open-pollinated onion populations. J Amer Soc Hort Sci 127:944–946

Gökçe AF, McCallum J, Sato Y, Havey MJ (2002) Molecular tagging of the *Ms* locus in onion. J Amer Soc Hort Sci 127:576–582

Guerra M, Kenton A, Bennett MD (1996) rDNA sites in mitotic and polytene chromosomes of *Vigna unguiculata* (L.) Walp. and *Phaseolus coccineus* L. revealed by in situ hybridization. Ann Bot 78(2):157–161. https://doi.org/10.1006/anbo.1996.0108

Havey MJ (2002) Genome organization in *Allium*. In: Rabinowitch HD, Currah L (eds) *Allium* crop science: recent advances. CABI Publishing, UK, pp 59–79

Havey MJ (2013) Single nucleotide polymorphisms in linkage disequilibrium with the male-fertility restoration (*Ms*) locus of onion. J Amer Soc Hort Sci 138:306–309

Henikoff S, Ahmad K, Malik HS (2001) The centromere paradox: stable inheritance with rapidly evolving DNA. Science 293(5532):1098–1102. https://doi.org/10.1126/science.1062939

Hizume M (1994) Allodiploid nature of *Allium wakegi* Araki revealed by genomic *in situ* hybridization and localization of 5S and 18S rDNAs. Jpn J Genet 69:407–415. https://doi.org/10.1266/jjg.69.407

Hoa QV, Yoshimatsu Y, Khrustaleva LI, Yamauchi N, Shigyo M (2012) Alien genes introgression and development of alien monosomic addition lines from a threatened species, *Allium roylei* Stearn, to *Allium cepa* L. Theor Appl Genet 124:1241–1257. https://doi.org/10.1007/s00122-011-1783-5

Hou A, Peffley EB (2000) Recombinant chromosomes of advanced backcross plants between *Allium cepa* L. and *A. fistulosum* L. revealed by *in situ* hybridization. Theor Appl Genet 100:1190–1196. https://doi.org/10.1007/s001220051423

Hou Y, Miao J, Liu B, Yang Y, Zhang Y, Tahara Y, Meng Q, He Q, Kitano H, Wu X (2012) The expression of pectin methylesterase in onion flower buds is associated with the dominant male-fertility restoration allele. Plant Breed 131:211–216. https://doi.org/10.1111/j.1439-0523.2011.01907.x

Houben A, Schubert I (2003) DNA and proteins of plant centromeres. Curr Opin Plant Biol 6:554–560. https://doi.org/10.1016/j.pbi.2003.09.007

Irifune K, Hirai K, Zheng J, Tanaka R, Morikawa H (1995) Nucleotide sequence of a highly repeated DNA sequence and its chromosomal localization in *Allium fistulosum*. Theor Appl Genet 90:312–316. https://doi.org/10.1007/BF00221970

Jiang J, Gill BS (2006) Current status and the future of flourescence in situ hybridization (FISH) in plant genome research. Genome 49:1057–1068. https://doi.org/10.1139/g06-076

Jiang J, Gill BS, Wang G-L, Ronald PC, Ward DC (1995) Metaphase and interphase fluorescence in situ hybridization mapping of the rice genome with bacterial artificial chromosomes. Proc Natl Acad Sci USA 92:4487–4491

Jiang J, Birchler JA, Parrott WA, Dawe RK (2003) A molecular view of plant centromeres. Trends Plant Sci 8:570–575. https://doi.org/10.1016/j.tplants.2003.10.011

Jones RN (1983) Cytogenetic evolution in the genus *Allium*. In: Swaminathan MS, Gupta PK, Sinha U (eds) Cytogenetics of crop plants. Macmillan, New York, pp 516–554

Jones RN (1990) Cytogenics. In: Rabinowitch HD, Brewster JL (eds) Onions and Allied crops, vol I. Botany, physiology and genetics. CRC Press, Florida, pp 199–214

Jones HA, Davis GN (1944) Inbreeding and heterosis and their relation to the development of new varieties of onions. US Department of Agriculture Technical Bulletins No 874, Washington, p 28

Jones RN, Rees H (1968) Nuclear DNA variation in allium. Heredity 23:591–605 10.1038/hdy.1968.76

Kalkman ER (1984) Analysis of the C-banded karyotype of *Allium cepa* L. Standard system of nomenclature and polymorphism. Genetica 65:141–148. https://doi.org/10.1007/BF00135278

Karafiátová M, Bartoš J, Kopecký D, Ma L, Sato K, Houben A, Stein N, Doležel J (2013) Mapping nonrecombining regions in barley using multicolor FISH. Chromosome Res 21:739–751. https://doi.org/10.1007/s10577-013-9380-x

Khrustaleva LI, Kik C (1998) Cytogenetical studies in the bridge cross Allium cepa x (A. fistulosum x A. roylei). Theor Appl Genet 96:8-14. https://doi.org/10.1007/s001220050702

Khrustaleva LI, Kik C (2000) Introgression of Allium fistulosum into A. cepa mediated by A. roylei. Theor Appl Genet 100:17–26. https://doi.org/10.1007/s001220050003

Khrustaleva LI, Kik C (2001) Localization of single copy T-DNA insertion in transgenic shallots (*Allium cepa* L.) by using ultrasensitive FISH with tyramide signal amplification. Plant J 25:699–707. https://doi.org/10.1046/j.1365-313x.2001.00995.x

Khrustaleva LI, de Melo PE, van Heusden AW, Kik C (2005) The integration of recombination and physical maps in a large-genome monocots using haploid genome analysis in a trihybrid *Allium* population. Genetics 169:1673–1685. https://doi.org/10.1534/genetics.104.038687

Khrustaleva L, Jiang J, Havey MJ (2016) High-resolution tyramide-FISH mapping of markers tightly linked to the male-fertility restoration (*Ms*) locus of onion. Theor Appl Genet 129(3):535–545. https://doi.org/10.1007/s00122-015-2646-2

Kim S (2014) A codominant molecular marker in linkage disequilibrium with a restorer-of-fertility gene (*Ms*) and its application in reevaluation of inheritance of fertility restoration in onions. Mol Breed 34:769–778. https://doi.org/10.1007/s11032-014-0073-8

King JJ, Bradeen JM, Bark O, McCallum JA, Havey MJ (1998) A low-density genetic map of onion reveals a role for tandem duplication in the evolution of an extremely large diploid genome. Theor Appl Genet 96:52–62. https://doi.org/10.1007/s001220050708

Kirov I, Divashuk M, Van Laere K, Soloviev A, Khrustaleva L (2014) An easy "SteamDrop" method for high quality plant chromosome preparation. Mol Cytogenet 7:21. https://doi.org/10.1186/1755-8166-7-21

Kirov IV, Kiseleva AV, Van Laere K, Van Roy N, Khrustaleva LI (2017) Tandem repeats of *Allium fistulosum* associated with major chromosomal landmarks. Mol Genet Genomics 292:453. https://doi.org/10.1007/s00438-016-1286-9

Kiseleva AV, Kirov IV, Khrustaleva LI (2014) Chromosomal organization of centromeric Ty3/gypsy retrotransposons in Allium cepa L. and Allium fistulosum L. Russ J Genet 50:586–592. https://doi.org/10.1134/S102279541404005X

Konishi T, Funamoto T, Smirnov SV, Damdinsuren O (2011) Chromosome studies in seven species of *Allium* (*Alliaceae* sensu stricto) in Mongolia and Russian Altai. Chromosome Bot 6:53–60. https://doi.org/10.3199/iscb.6.53

Koo D-H, Jiang J (2009) Supper-stretched pachytene chromosomes for fluorescence *in situ* hybridization mapping and immunodetection of DNA methylation. Plant J 59:509–516. https://doi.org/10.1111/j.1365-313x.2009.03881.x

Koo D-H, Hong CP, Batley J, Chung YS, Edwards D, Bang J-W, Hur Y, Lim YP (2011) Rapid divergence of repetitive DNAs in *Brassica* relatives. Genomics 97:173–185. https://doi.org/10.1016/j.ygeno.2010.12.002

Korbel JO, Lee C (2013) Genome assembly and haplotyping with Hi-C. Nat Biotechnol 31:1099–1101

Kornneef M, Fransz P, de Jong H (2003) Cytogenetic tools for *Arabidopsis thaliana*. Chromosome Res 11:183–194. https://doi.org/10.1023/a:1022827624082

Kumar A (1996) The adventures of the Ty1-copia group of retrotransposons in plants. Trends Genet 12:41–43. https://doi.org/10.1016/0168-9525(96)81393-X

Kumar A, Pearce SR, McLean K, Harrison G, Heslop-Harrison JS, Waugh R, Flavell AJ (1997) The Ty1-copia group of retrotransposons in plants: genomic organisation, evolution, and use as molecular markers. Genetica 100:205–217. https://doi.org/10.1023/a:1018393931948

Leitch IJ, Heslop-Harrison JS (1992) Physical mapping of the 18S–5.8S–26S rRNA genes in barley by in situ hybridization. Genome 35:1013–1018. https://doi.org/10.1139/g92-155

Lee SH, Seo BB (1997) Chromosomal localization of 5S and 18S-26S rRNA genes using fluorescence in situ hybridization in Allium wakegi. Korean J Genet 19:19–26

Levan A (1936) Die cytologie von *Allium cepa* x *fistulosum*. Hereditas 21:195–214. https://doi.org/10.1111/j.1601-5223.1936.tb03199.x

Levan A (1941) The cytology of the species hybrid *Allium cepa* × *fistulosum* and its polyploidy derivates. Hereditas 27:253–272. https://doi.org/10.1111/j.1601-5223.1941.tb03260.x

Li B, Choulet F, Heng Y, Hao W, Paux E, Liu Z, Yue W, Jin W, Feuillet C, Zhang X (2013) Wheat centromeric retrotransposons: the new ones take a major role in centromeric structure. Plant J 73:952–965. https://doi.org/10.1111/tpj.12086

Long EO, David ID (1980) Repeated genes in eukaryotes. Ann Rev Biochem 49:727–764. https://doi.org/10.1146/annurev.bi.49.070180.003455

Maeda T (1937) Chiasma studies in *Allium fistulosum* L., *Allium cepa* L. and their F_1, F_2 and backcross hybrids. Japn J Genet 13:146–169. https://doi.org/10.1266/jjg.13.146

Mancia FH, Sohn SH, Ahn YK (2015) Distribution of various types of repetitive DNAs in *Allium cepa* L. based on dual color FISH. Hortic Environ Biotechnol 56:793. https://doi.org/10.1007/s13580-015-1100-3

Martin WJ, McCallum J, Shigyo M, Jakse J, Kuhl JC, Yamane N, Meeghan P-J, Ali FG, Kenneth CS, Christopher DT, Michael JH (2005) Genetic mapping of expressed sequences in onion and in silico comparisons show scant collinearity with rice. Mol Genet Genomics 274:197–204. https://doi.org/10.1007/s00438-005-0007-6

Masamura N, McCallum J, Khrustaleva L, Kenel F, Pither-Joyce M, Shono J, Suzuki G, Mukai Y, Yamauchi N, Shigyo M (2012) Chromosomal organization and sequence diversity of genes encoding lachrymatory factor synthase in Allium cepa L. 3G: Genes Genomes Genet 2:643–651. https://doi.org/10.1534/g3.112.002592

Masuzaki Sh-I, Shigyo M, Yamauchi N (2006) Complete assignment of structural genes involved in flavonoids biosysnthesis influencing bulb color to individual chromosomes of shallot (*Allium cepa* L.). Genes Genet Syst 81:255–263. https://doi.org/10.1266/ggs.81.255

McCallum J, Baldwin S, Shigyo M, Deng Y, van Heusden S, Pither-Joyce M, Kenel F (2012) AlliumMap-A comparative genomics resource for cultivated *Allium* vegetables. BMC Genom 13:168. https://doi.org/10.1186/1471-2164-13-168

McCollum GD (1982) Experimental hybrids between *Allium fistulosum* and *A. roylei*. Bot Gaz 143:238–242. https://doi.org/10.1086/337295

Melters DP, Bradnam KR, Young HA, Telis N, May MR, Ruby JG, Sebra R, Peluso P, Eid J, Rank D, Garcia JF, DeRisi JL, Smith T, Tobias C, Jeffrey R-I, Korf I, Chan SWL (2013) Comparative analysis of tandem repeats from hundreds of species reveals unique insights into centromere evolution. Genome Biol 14: R10. https://doi.org/10.1186/gb-2013-14-1-r10

Meyne J, Ratliff RL, Moyzis RK (1989) Conservation of the human telomere sequence $(TTAGGG)_n$ among vertebrates. Proc Nat Acad Sci, USA 86:7049–7053

Nagaki K, Murata M (2005) Characterization of CENH3 and centromere associated DNA sequences in sugarcane. Chromosome Res 13:195–203. https://doi.org/10.1007/s10577-005-0847-2

Nagaki K, Cheng Z, Ouyang S, Talbert PB, Kim M, Jones KM, Henikoff S, Buell CR, Jiang J (2004) Sequencing of a rice centromere uncovers active genes. Nat Genet 36:138–145. https://doi.org/10.1038/ng1289

Nagaki K, Kashihara K, Murata M (2009) A centromeric DNA sequence colocalized with a centromere-specific histone H3 in tobacco. Chromosoma 118:249–257. https://doi.org/10.1007/s00412-008-0193-1

Nagaki K, Yamamoto M, Yamaji N, Mukai Y, Murata M (2012) Chromosome dynamics visualized with an anti-centromeric histone H3 antibody in *Allium*. PLoS One 7(12):e513–515. https://doi.org/10.1371/journal.pone.0051315

Neumann P, Navratilova A, Koblizkova A, Kejnovský E, Hřibová E, Hobza R, Widmer A, Doležel J, Macas J (2011) Plant centromeric retrotransposons: a structural and cytogenetic perspective. Mob DNA 2:4. https://doi.org/10.1186/1759-8753-2-4

Ohri D (1998) Genome size variation and plant systematics. Ann Bot 82:75–83. https://doi.org/10.1006/anbo.1998.0765

Okazaki S, Tsuchida K, Maeka HWA, Ishikawa H, Fujiwara H (1993) Identification of a pentanucleotide telomeric sequence, (TTAGG)n, in the silkworm *Bombyx mori* and in other insects. Mol Cell Biol 13:1424–1432. https://doi.org/10.1128/MCB.13.3.1424

Park J, Bang H, Cho DY, Yoon MK, Patil BS, Kim S (2013) Construction of high-resolution linkage map of the *Ms* locus, a restorer-of-fertility gene in onion (*Allium cepa* L.). Euphytica 192:267–278. https://doi.org/10.1007/s10681-012-0851-5

Pathak CS, Black LL, Cheng SJ, Wang TC, Ko SS (2001). Breeding onions for stemphylium leaf blight resistance. Acta Hortic. 555:77-81. https://doi.org/10.17660/ActaHortic.2001.555.7

Pearce SR, Pich U, Harrison G, Flavell AJ, Heslop-Harrison JS, Schubert I, Kumar A (1996) The Ty1-copia group retrotransposons of *Allium cepa* are distributed throughout the chromosomes but are enriched in the terminal heterochromatin. Chromosome Res 4:357–364. https://doi.org/10.1007/bf02257271

Peffley EB, Currah L (1988) The chromosomal locations of enzyme-coding genes *Adh-1* and *Pgm-1* in *Allium fistulosum* L. Theor Appl Genet 75:945–949. https://doi.org/10.1007/BF00258058

Peffley EB, Hou A (2000) Bulb-type onion introgressants possessing *Allium fistulosum* L. genes recovered from interspecific hybrid backcrosses between *A. cepa* L. and *A. fistulosum* L. Theor Appl Genet 100:528–534. https://doi.org/10.1007/s001220050069

Pich U, Schubert I (1998) Terminal heterochromatin and alternative telomeric sequences in *Allium cepa*. Chromosome Res 6:315–321. https://doi.org/10.1023/A:1009227009121

Pich U, Fritsch R, Shubert I (1996a) Closely related *Allium* species (*Alliaceae*) share a very similar satellite sequence. Plant Syst Evol 202:255–264. https://doi.org/10.1007/BF00983386

Pich U, Fuchs J, Schubert I (1996b) How do *Alliaceae* stabilize their chromosome ends in the absence of TTTAGGG sequences? Chromosome Res 4:207–213. https://doi.org/10.1007/bf02254961

Putter M, Van de Vooren JG (1988) Identification of *Allium cepa* L. cultivars by means of statistical analysis of C-banded chromosomes. Euphytica 39:153–160. https://doi.org/10.1007/BF00039868

Raap A, van de Corput M, Vervenne R, van Gijlswijk R, Tanke H, Wiegant J (1995) Ultra-sensitive FISH using peroxidase-mediated deposition of biotin- or fluorochrome-tyramides. Hum Mol Genet 4:529–534. https://doi.org/10.1093/hmg/4.4.529

Richards EJ, Ausubel FM (1988) Isolation of a higher eukaryotic telomere from *Arabidopsis thaliana*. Cell 53:127–136. https://doi.org/10.1016/0092-8674(88)90494-1

Ricroch A, Peffley EB, Baker RJ (1992) Chromosomal location of rDNA in Allium: in situ hybridization using biotin- and fluorescein-labelled probe. Theor Appl Genet 83(4):413–418. https://doi.org/10.1007/BF00226528

Ricroch A, Yockteng R, Brown SC, Nadot S (2005) Evolution of genome size across some cultivated *Allium* species. Genome 48:511–520. https://doi.org/10.1139/g05-017

Roa F, Guerra M (2012) Distribution of 45S rDNA sites in chromosomes of plants: Structural and evolutionary implications. BMC Evol Biol 12:225. https://doi.org/10.1186/1471-2148-12-225

Romanov D, Divashuk M, Havey MJ, Khrustaleva L (2015) Tyramide-FISH mapping of single genes for development of an integrated recombination and cytogenetic map of chromosome 5 of *Allium cepa* L. Genome 58:111–119. https://doi.org/10.1139/gen-2015-0019

Sandhu D, Champoux JA, Bondareva SN, Gill KS (2001) Identification and physical location of useful genes and markers to a major gene-rich region on wheat group 1S chromosomes. Genetics 157:1735–1747

Scholten OE, van Heusden AW, Khrustaleva LI, Burger-Meijer K, Mank RA, Antonise RGC, Harrewijn JL, Van Haecke W, Oost EH, Peters RJ, Kik C (2007) The long and winding road leading to the successful introgression of downy mildew resistance into onion. Euphytica 156:345–353. https://doi.org/10.1007/s10681-007-9383-9

Shearer LA, Anderson LK, de Jong H, Smit S, Goicoechea JL, Roe BA, Hua A, Giovannoni JJ, Stack SM (2014) Fluorescence *in situ* hybridization and optical mapping to correct scaffold arrangement in the tomato genome. G3: Genes Genomes Genet 4:1395–1405. https://doi.org/10.1534/g3.114.011197

Shibata F, Hizume M (2002) Evolution of 5S rDNA units and their chromosomal localization in Allium cepa and Allium schoenoprasum revealed by microdissection and FISH. Theor Appl Genet 105(2–3):167–172. https://doi.org/10.1007/s00122-002-0950-0

Shigyo M, Tashiro Y, Miyazaki S (1994) Chromosomal locations of glutamate oxaloacetate transaminase gene loci in Japanese bunching onion (*Allium fistulosum* L.) and shallot (*A. cepa* L. Aggregatum group). Jpn J Genet 69:417–424. https://doi.org/10.1266/jjg.69.417

Shigyo M, Tashiro Y, Isshiki S, Miyazaki S (1995a) Chromosomal location of five isozyme gene loci (*Lap-1, Got-1, 6-Pgdh-2, Adh-1* and *Gdh-1*) in shallot *Allium cepa* L. Aggregatum group. Jpn J Genet 70:399–407. https://doi.org/10.1266/jjg.70.399

Shigyo M, Tashiro Y, Isshiki S, Miyazaki S (1995b) Chromosomal location of isocitrate dehydrogenase and phosphoglucoisomerase gene loci in shallot *Allium cepa* L. Aggregatum group. Jpn J Genet 70:627–632. https://doi.org/10.1266/jjg.70.627

Shigyo M, Tashiro Y, Isshiki S, Miyazaki S (1996) Establishment of a series of nonosomic alien additional lines of Japanese bunching onion (*Allium fistulosum* L.) with extra chromosomes from shallot (*A. cepa* L. Aggregatum group). Genes Genet Syst 71:363–371. https://doi.org/10.1266/ggs.71.363

Shigyo M, Miyazaki T, Isshiki S, Tashiro Y (1997) Assignment of randomly amplified polymorphic DNA markers to all chromosomes of shallot (*Allium cepa* L. Aggregatum group). Genes Genet Syst 72:249–252. https://doi.org/10.1266/ggs.72.249

Shigyo M, Imamura K, Iino M, Yamashita K, Tashiro Y (1998) Identification of alien chromosomes in a series of *Allium fistulosum—A. cepa* monosomic addition lines by means of genomic *in situ* hybridization. Genes Genet Syst 73:311–315. https://doi.org/10.1266/ggs.73.311

Si W, Yuan Y, Huang J, Zhang X, Zhang Y, Zhang Y, Tian D, Wang C, Yang Y, Yang S (2015) Widely distributed hot and cold spots in meiotic recombination as shown by the sequencing of rice F_2 plants. New Phytol 206:1491–1502. https://doi.org/10.1111/nph.13319

Son J-H, Park K-Ch, Lee S-I, Jeon E-J, Kim H-H, Kim N-S (2012) Sequence variation and comparison of the 5S rRNA sequences in *Allium* species and their chromosomal distribution in four *Allium* species. J Plant Biol 55:15–25. https://doi.org/10.1007/s12374-011-9185-4

Stack SM, Comings DE (1979) The chromosomes and DNA of *Allium cepa*. Chromosoma 70:161–181. https://doi.org/10.1007/BF00288404

Sun J, Zhang Z, Zong X, Huang S, Li Z, Han Y (2013) A high-resolution cucumber cytogenetic map integrated with the genome assembly. BMC Genom 14(1):461. https://doi.org/10.1186/1471-2164-14-461

Suzuki G, Ura A, Saito N, Do G, So B, Yamamoto M, Mukai Y (2001) BAC FISH analysis in Allium cepa. Genes Genet Syst (Japan) 76:251–255. https://doi.org/10.1266/ggs.76.251

Sykorova E, Lim KY, Kunicka Z, Chase MW, Bennett MD, Fajkus J, Leitch AR (2003) Telomere variability in the monocotyledonous plant order *Asparagales*. P Roy Soc B-Biol Sci 270:1893–1904. https://doi.org/10.1098/rspb.2003.2446

Sykorova E, Fajkus J, Meznikova M, Lim KY, Neplechova K, Blattner FR, Chase MW, Leitch AR (2006a)

Minisatellite telomeres occur in the family *Alliaceae* but are lost in *Allium*. Am J Bot 93:814–823. https://doi.org/10.3732/ajb.93.6.814

Sykorova E, Leitch AR, Fajkus J (2006b) Asparagales telomerases which synthesize the human type of telomeres. Plant Mol Biol 60:633–646. https://doi.org/10.1007/s11103-005-5091-9

Szinay D, Chang S-B, Khrustaleva LI, Peters S, Schijlen E, Stiekema W, Roeland CHJ, van Ham RCHJ, de Jong H, Klein Lankhorst R (2008) High-resolution chromosome mapping of BACs using multi-colour FISH and pooled-BAC FISH as a backbone for sequencing tomato chromosome 6. Plant J 56:627–637. https://doi.org/10.1111/j.1365-313x.2008.03626.x

Tang H, Lyons E, Town CD (2015) Optical mapping in plant comparative genomics. GigaScience 4(1):3. https://doi.org/10.1186/s13742-015-0044-y

Tsukazaki H, Yamashita K, Yaguchi S, Masuzaki S, Fukuoka H, Yonemaru J, Kanamori H, Kono I, Hang T, Shigyo M, Kojima A, Wako T (2008) Construction of SSR-based chromosome map in bunching onion (*Allium fistulosum*). Theor Appl Genet 117:1213–1223. https://doi.org/10.1007/s00122-008-0849-5

Tsukazaki H, Yamashita K, Yaguchi S, Yamashita K, Hagihara T, Shigyo M, Kojima A, Wako T (2010) Direct determination of the chromosomal location of bunching onion and bulb onion markers using bunching onion-shallot monosomic additions and allotriploid-bunching onion single alien deletions. Theor Appl Genet 122:501–510. https://doi.org/10.1007/s00122-010-1464-9

van der Meer QP, de Vries JN (1990) An interspecific cross between *Allium roylei* Stearn and *Allium cepa* L., and its backcross to *A. cepa*. Euphytica 47:29–31. https://doi.org/10.1007/BF00040359

van Heusden A, Shigyo M, Tashiro Y, Vrielink-van RG, Kik C (2000) AFLP linkage group assignment to the chromosomes of *Allium cepa* L. via monosomic addition lines. Theor Appl Genet 100:480–486. https://doi.org/10.1007/s001220050062

van Heusden AW, van Ooijen JW, Vrielink-van GR, Verbeek WHJ, Wietsma WA, Kik C (2000a) A genetic map of an interspecific cross in *Allium* based on amplified fragment length polymorphism (AFLP) markers. Theor Appl Genet 100:118–126. https://doi.org/10.1007/s001220050017

Vanzela ALL, Cuadrado A, Guerra M (2003) Localization of 45S rDNA and telomeric sites on holocentric chromosomes of *Rhynchospora tenuis* Link (Cyperaceae). Genet Mol Biol 26:199–201. https://doi.org/10.1590/S1415-47572003000200014

Vitte C, Estep MC, Leebens-Mack J, Bennetzen JL (2013) Young, intact and nested retrotransposons are abundant in the onion and asparagus genomes. Ann Bot 112:881–889. https://doi.org/10.1093/aob/mct155

Watson MJ, Riha K (2010) Comparative biology of telomeres: Where plants stand. FEBS Lett 584(17):3752–3759. https://doi.org/10.1016/j.febslet.2010.06.017

Yang YY, Huo YM, Miao J, Liu BJ, Kong SP, Gao LM, Liu C, Wang ZB, Tahara Y, Kitano H, Wu X (2013) Identification of two SCAR markers co-segregated with the dominant *Ms* and recessive *ms* alleles in onion (*Allium cepa* L.). Euphytica 190:267–277

Yang L, Koo DH, Li D, Zhang T, Jiang J, Luan F, Renner SS, Hénaff E, Sanseverino W, Garcia-Mas J, Casacuberta J, Senalik DA, Simon PW, Chen J, Weng Y (2014) Next-generation sequencing, FISH mapping and synteny-based modeling reveal mechanisms of decreasing dysploidy in Cucumis. Plant J 77:16–30. https://doi.org/10.1111/tpj.12355

6 Cytoplasmic Genome

Mai Tsujimura and Toru Terachi

Abstract
The chloroplasts and mitochondria are the organelles essential for normal growth and development of plants. They have their original genomes, and genes encoded in these organelles are expressed in concordance with factors originating from the nuclear genome. Researchers have analyzed cytoplasmic genomes to understand this complicated mechanism. The sizes of the chloroplast genome and the mitochondrial genome are relatively small, with the former being approximately 150 kbp and the latter being several hundred kbp to several Mbps. The widespread use of next-generation sequencing (NGS) in recent years has enabled sequencing of the entire genome of chloroplasts and mitochondria with relative ease and affordability. For example, several chloroplast and mitochondrial genomes have been sequenced for onions. Onions carry several different types of cytoplasms characterized by distinct cytoplasmic genomes. In particular, male sterility induced by certain types of cytoplasms is a trait that is crucial to F_1 hybrid breeding, and male sterility phenotypes are determined by their compatibility with the nuclear genome. This chapter explains the cytoplasmic form of these onion cultivars and lays out the characteristics of the cytoplasmic genome of each one of them.

M. Tsujimura · T. Terachi (✉)
Faculty of Life Sciences, Kyoto Sangyo University, Motoyama, Kamigamo, Kita-ku, Kyoto 603-8555, Japan
e-mail: terachi@cc.kyoto-su.ac.jp

6.1 Introduction

Cytoplasmic genomes are present in the cytoplasm and are distinct from nuclear genomes. In plants, these include chloroplast and mitochondrial genomes.

Mitochondria originated from an endosymbiosis between a host cell and an alpha-proteobacterium 1.45 billion years ago. Chloroplasts, on the other hand, originated from the endosymbiosis of a cyanobacterium 1.5 billion years ago. The host cell took control over the symbiotic fungi over the years, and eventually, mitochondria and chloroplasts came to play their own parts within the cell as an organ that synthesizes adenosine triphosphate (ATP) and an organelle responsible for carbon fixation, respectively. The genome of each bacteria disappeared as the host's nuclear genome acquired its genes. Although hydrogenosomes present in certain anaerobic protists (e.g., *Trichomonas*) have entirely lost their original genomes following an endosymbiotic process, mitochondria and chloroplasts still retain their original genomes with functional genes, which are crucial for

M. Shigyo et al. (eds.), *The Allium Genomes*, Compendium of Plant Genomes,
https://doi.org/10.1007/978-3-319-95825-5_6

proper functioning of the organelles and the successful passage of hereditary information to the next generation (Dyall et al. 2004; Keeling 2010; Dunning Hotopp 2011).

Particular attention should be drawn to the distinct characteristics of the chloroplast genome, the plant mitochondrial genome, and the animal mitochondrial genome. The chloroplast genomes of most plant species are composed of two inverted repeats that separate a large single-copy (LSC) region from a small single-copy (SSC) region. The plant mitochondrial genomes, on the other hand, are believed to take on the shape of a circle, but the size and sometimes even the structure of them vary with species. The mammalian mitochondrial genomes are well conserved, with almost identical structure and encoded genes.

The genes encoded in the nuclear genome are essential for these organelles to function normally. Of all genes that are associated with mitochondria, over 1000 are transferred to mitochondria. These proteins are indispensable for normal photosynthesis and respiration. Furthermore, genes originating from the nuclear genome are deeply associated with the expression of genes encoded in each organelle genome. A chloroplast contains genes that use a plastid-encoded RNA polymerase (PEP) and a nucleus-encoded RNA polymerase (NEP). Although PEP and NEP use different promoters, nearly all genes have both PEP and NEP promoters like in most cases of bacterial expression, and these promoters are transcribed as polycistronic RNAs. Although mitochondria are also expressed by the bacterial RNA polymerase, they do not carry their own polymerase and therefore are dependent on a nucleus-encoded RNA polymerase (Borner et al. 2015; Hammani and Giege 2014). Transcripts are subject to post-transcriptional modifications, such as splicing, RNA editing, and definition at their 3′ end and 5′ end.

The presence of signal transduction that enables interactive control and regulation of genes encoded in the nucleus and organelles has also been identified, which plays a crucial role in stress and development responses. Although not much is known about the metabolic pathways and biosynthesis of these genes, several molecules have been identified to carry signals. These molecular components are referred to as anterograde (nuclear to organelle) and retrograde (organelle to nuclear) signals, which affect signal transduction pathways and switch gene expression on or off. Intermediates in the metabolic pathways of reactive oxygen species (ROS), calcium ion, and tricarboxylic acid (TCA) have been demonstrated to be involved in these mechanisms, and the alternative oxidase (AOX) pathway, in particular, has been studied extensively as a model for mitochondrial retrograde regulation (MRR) (Ng et al. 2014; Rhoads 2011).

In view of these facts, it is of extreme importance to focus attention on cytoplasms during the analysis of growth, development, and functionality of plants, and it is also essential to consider nucleus-cytoplasm compatibility. With agricultural crops, cytoplasmic male sterility/fertility restoration (CMS/Rf) systems are often used for F_1 hybrid breeding. The use of a line exhibiting CMS as the female parent and a line with restorer genes as pollen parent allows for F_1 hybrid seed production without the need for emasculation. These systems are especially useful for breeding agricultural crops not only because hybrids often exhibit hybrid vigor but also because they help protect information regarding genes carrying superior traits. With a typical CMS-Rf system, CMS-inducing genes are present in mitochondria, whereas the genes that restore male fertility are present in the nuclear genome (Hanson and Bentolila 2004; Bohra et al. 2016).

Furthermore, comparing cytoplasmic genomes between different species or genera helps to identify a lineage relationship between those species or genera. The estimation of a lineage relationship by using the nuclear genome can be quite challenging with species that have large nuclear genomes or polyploidy genomes. Cytoplasmic genomes, on the other hand, are extremely small in size, and because they are transmitted from mother to offspring, investigating cytoplasmic genomes can reveal maternal-line ancestry.

The widespread use of NGS in recent years has enabled sequencing of the entire genome of chloroplasts and mitochondria with relative ease and affordability. As a result, organelle genome data has been steadily accumulated from diverse plant species, In fact, chloroplast and mitochondrial genomes have been completely sequenced for several onion species. This chapter lays out the characteristics of the cytoplasmic genome of these onions.

6.2 Cytoplasm Species and Their Origin in Onion

Onion, one of the most important crops in the world, is a diploid species ($2n = 2x = 16$) that has been cultivated for more than 5000 years (Khosa et al. 2016). Onions are cultivated in 147 countries in the world, with the world's onion production reaching 111 metric tons in 2014 (FAOSTAT 2014).

No wild form of onions has ever been found, and only cultivated varieties exist today. *Allium vavilovii*, which can be crossed with *Allium cepa*, grows naturally in the Tien-shan mountains and the Pamir–Altai mountains, where onions are believed to have originated (Kik 2002). *Allium* and its close relatives are classified into four grades, and it has been shown that *A. cepa* seems to originate from *A. vavilovii* or from hybridization between *Allium galanthum* and *Allium fistulosum* (Gurushidze et al. 2007).

The presence of CMS in onions has long been known, and onion was, in fact, the first crop to be used for F_1 hybrid breeding. Since the discovery of CMS in onion by Jones and Emsweller (1936), a considerable increase in crop yields has been achieved through heterosis breeding (Khosa et al. 2016). To date, two types of male-sterility-inducing cytoplasm, namely the CMS-S and CMS-T types, have been widely used for commercial breeding.

The CMS-S type shows normal development until the tetrad stage of pollen development, which is followed by degeneration of the protoplasm in the tetrads, resulting in empty pollen grain (Holford et al. 1991a). CMS-T, on the other hand, has an abnormal pollen meiosis (Kik 2002).

S-type cytoplasmic male sterility is restored to fertility by a single nuclear locus (Jones and Clarke 1943; Jones and Davis 1944), whereas male sterility due to T cytoplasm is restored by three independent loci: *aa*, *bb*, and *cc* (Schweisguth 1973). The Normal-type cytoplasm, on the other hand, does not exhibit sterility. These three types of cytoplasms are identified by the restriction fragment patterns of chloroplast and mitochondrial DNAs.

Normal-type and CMS-T cytoplasms resemble one another and are, therefore, categorized into the M group, which is distinguished from the CMS-S-type cytoplasm (de Courcel et al. 1989; Holford et al. 1991b). In 2000, Havey showed that the CMS-S type originates from *Allium x proliferum* and that the CMS-T type and the Normal-type originate from *A. vavilovii* (Havey 2000).

Not much progress has been made in the improvement of *Allium* by use of wild-type subspecies in introgression breeding. This is attributable to the fact that onions are biennials that have a 2-year life cycle and that there are a limited number of onion species available for breeding. A previous study has demonstrated that while *A. vavilovii* can be crossed with *A. cepa*, crossing the triploid ovum *A. x proliferum* with *A. cepa* can be quite challenging (Kik 2002).

Introgression of *A. fistulosum* or *A. galanthum* with *A. cepa* produces sterility. Although CMS-induced by *A. galanthum* has been shown to produce empty pollen grains that are phenotypically similar to the pollens produced by the CMS-S type, CMS-S restorer genes do not restore fertility to *galanthum*-CMS (Havey 1999). A single locus has been found to restore male fertility in *A. fistulosum* (Japanese bunching onion) in which CMS has been induced by *A. galanthum* (Yamashita et al. 2005).

6.3 Chloroplast Genome

The chloroplast genomes are generally composed of two inverted repeats that separate an LSC from an SSC. This structure is common to all plants.

The same is true for the following five chloroplast genomes identified so far for onions: CMS-S type and Normal type identified by von Kohn et al. (2013) and CMS-S type, CMS-T type, and Normal-type identified by Kim et al. (2015).

In the analysis performed by von Kohn et al. (2013), 28 single nucleotide polymorphisms (SNPs), 2 restriction fragment length polymorphisms (RFLPs), and 1 insertion–deletion (InDel) were found in the genetic code between the CMS-S type and the Normal type. Furthermore, an InDel of 45 bp was present in the accD gene, with the CMS-S type showing the same sequence as tobacco, orchid, and rice. InDels of 99, 59, and 50 bp were also present in the *trnL-trnT, psbM-petN, and* rps16-*trnQ* intergenic regions, respectively, with CMS-S having the shorter sequence. An InDel of 90 bp was also found in the *rps4-trnT* intergenic region, with S having one repeat of 45 bp and N having three tandem repeats of 45 bp.

In the analysis by Kim et al. (2015), on the other hand, a SNP call for all the regions revealed 323 SNPs and 141 InDels between the CMS-S type and the Normal type. Furthermore, SNPs at four locations and InDels at two locations were found between the Normal type and the CMS-T type.

Between the two registered types of CMS-S, differences were found in SNPs at three locations and an InDel at one location (Table 6.1). Between the two Normal types and the CMS-T type, differences were found in SNPs at three locations and InDels at two locations (Table 6.2). None of the differences found in those SNPs and InDels were specific to the Normal type or the CMS-T type. Mutations seen in not only the S type but also between the Normal types or between N and T appeared to be interspecies variation rather than mutations specific to Normal-type and CMS-T type. These findings suggest that while chloroplast genomes are useful for distinguishing between the CMS-S type and the Normal-type, they may be of little value when it comes to distinguishing between the Normal-type and the CMS-T type.

6.4 Mitochondrial Genome

As described in the previous section, it is important to distinguish between different types of onion cytoplasms because those that induce CMS are used with the Normal-type to develop F_1 hybrids. Like chloroplast genomes, mitochondrial DNAs are also widely used for classification of onion cytoplasms, and as demonstrated by Havey, these cytoplasms have been classified into the CMS-S type, the Normal-type, and the CMS-T type, the latter two of which belong to the M group (Havey 2000). These three types of cytoplasms (i.e., CMS-S type, CMS-T type and Normal type) were first identified through an RFLP analysis of mitochondrial DNAs, which revealed three types of mitochondrial genomes (de Courcel et al. 1989).

A specific gene called *orf725* in mitochondria has been shown to induce CMS (Kim et al. 2009). This gene, which is a *cox1* gene with an additional sequence of 576 bp attached to its 3′ end, is found in both the CMS-S type and the CMS-T type. This gene has been shown to be transcribed in both CMS lines, and due to a stop codon resulting from the editing at a position of 30 bases inward from the 3′ end, this gene in the CMS lines turns out to be 184aa longer than the normal-type *cox1* gene. This additional sequence

Table 6.1 Positions of SNPs and InDels between two CMS-S chloroplast genomes

Position(KF728079; CMS-S type)	Type	KF728079; CMS-S type	KM088014; CMS-S type
32,557	Substitution	T	A
32,559	Substitution	A	T
141,791	Substitution	G	A
153,338–153,355	InDel	ACTCAAATTCGAGTGGAA	–

Table 6.2 Positions of SNPs and InDels among three chloroplast genomes classified as M group

Position(KF728080; N type)	Type	KF728080; N type	KM088013; N type	KM088015; CMS-T type
29,169–29,211	InDel	TTTCTATTCTATATGTACATC TACTATATACATATTGTAATTT	–	TTTCTATTCTATATGTACATC TACTATATACATATTGTAATTT
46,534–46,550	InDel	TAATATTAATAATATAT	–	–
50,776	Substitution	C	C	A
73,827	Substitution	A	G	A
119,803	Substitution	T	A	T
121,628	Substitution	A	G	G
122,659	Substitution	GAA	GAA	TTC

shows a high level of homology with *orfA501*, which is a gene specific to the CMS cytoplasm of chive (*A. schoenoprasum* L.) (Engelke and Tatlioglu 2002), and according to protein structure prediction, it contains transmembrane domains at two positions.

Although *orf725* is transcribed in both CMS lines, the normal-type *cox1* is not transcribed to the CMS-S type. However, in the CMS-T type, the normal-type *cox1* that is the same in size as that of the Normal type is transcribed (Kim et al. 2009). Furthermore, as mentioned earlier, the restorer genes in CMS-S and CMS-T are located at different loci, and the growth process of sterile pollen varies between these two cytoplasms. For this reason, this gene has long been considered a potential inducer of CMS; however, it has yet to be identified whether it is common to both the CMS-S type and the CMS-T type. On the other hand, a separation experiment carried out by using a line having a cytoplasm resembling CMS-T not only revealed an F_1 of 3:1 but also successfully identified the presence or absence of restorer genes by using the same locus markers as those of the restorer genes of the CMS-S type (Kim 2014). Thus, future research must reassess CMS-T. In addition, it is not yet known what systems are involved in the induction of male sterility and how it is restored to fertility, which would be an interesting topic to explore.

The plant mitochondrial genomes are far more complicated than the plant chloroplast genomes. Their sizes vary with species, ranging from several mega-bases to several dozen kilo-bases. Although the structure of plant mitochondrial genomes is often expressed as a single master circle, it has also been suggested that they may exist as multiple circles or linear molecules. Furthermore, mitochondrial genomes are characterized by the many repeats they have, through which numerous recombinant molecules exist.

Kim et al. were the first to identify the CMS-S type of onion mitochondrial genome in 2016 (Kim et al. 2016), followed by the identification of other species of the CMS-S type (Tsujimura et al. 2018) and then the Normal type by our group.

Figure 6.1 shows the RFLPs of the mitochondrial genomes we identified. The

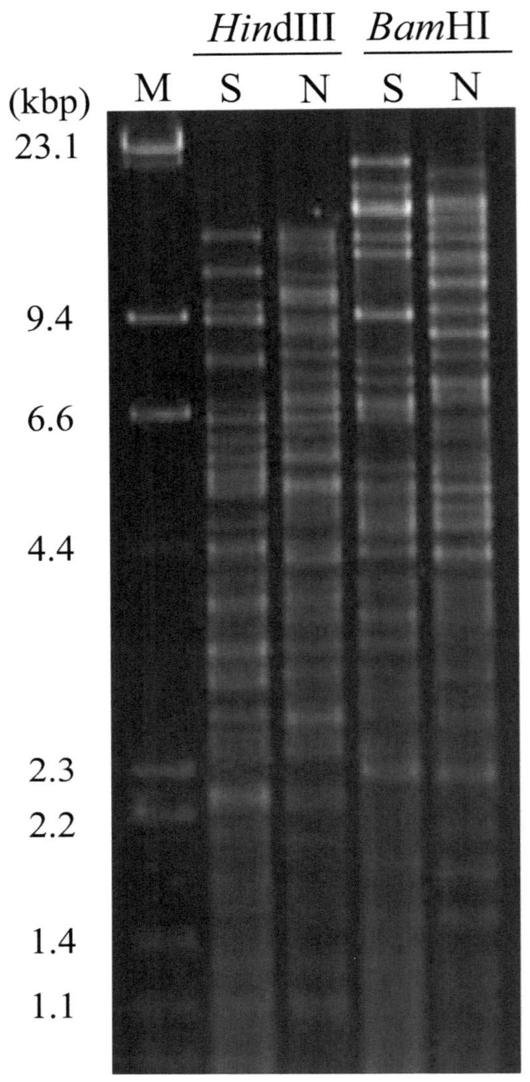

Fig. 6.1 Restriction fragment patterns of onion mitochondrial genome. M: molecular marker, S: CMS-S-type mtDNA, and N: Normal-type mtDNA

comparison of our RFLPs with the RFLP patterns presented by Courcel et al. shows that their RFLP patterns resemble those of our CMS-S type and Normal type (de Courcel et al. 1989). Furthermore, it is evident that the CMS-S type and the Normal-type are different species.

Interestingly, as it turned out, the CMS-S type we identified seemed to exist as three circles. Our initial NGS result found two circles, i.e., a 170-kbp (MC1) circle, and a 140-kbp circle. MC1 contains a direct repeat of 3.4 kbp, and when polymerase chain reaction (PCR) was carried out by using primers that had been designed to flank this repeat region, amplification and recombination were observed with all primer combinations. This PCR result suggests that the direct repeat yields two circles. Furthermore, another recombination occurred through a direct repeat of 3.4 kbp in MC1, possibly giving rise to two circles (Fig. 6.2). These findings suggest that the CMS-S type could be mapped as a total of three circles.

The Normal-type mitochondrial genomes, on the other hand, was illustrated as a single circle, according to the NGS result (Fig. 6.3). The total genome length was approximately 310 kbp for the CMS-S type and 537 kbp for the Normal-type, the latter of which turned out to be nearly 1.7 times longer than that of the "Momiji-3" onion cultivar. The size difference between these two species can be explained by the number of repeat regions. In fact, the percentage of the entire genome occupied by repeat regions was 1.5% for CMS-S-type and 30.7% for the Normal-type. Although it is possible that these repeat regions in the Normal-type induce recombination that could give rise to multiple circles like in the case of the CMS-S type, no band patterns were detected with pulsed-field gel electrophoresis (PFGE).

Table 6.3 summarizes the differences in the gene regions into which protein is coded by the CMS-S type and the Normal-type. Of the base substitutions and InDels found a total of 14 positions, 12 were accompanied by amino acid substitution, with a particularly significant difference, including a frameshift, found in *cox1*, *cox3*, and *nad6* (Fig. 6.4). Furthermore, we also obtained information on transcripts for each line by using RNA sequencing (RNAseq) and estimated the transcribed regions with respect to the coding regions.

In the case of *cox1*, the sequence after the stop codon differed between the CMS-S type and the Normal-type as previously reported, and with the CMS-S type, *cox1* was encoded as *orf725*.

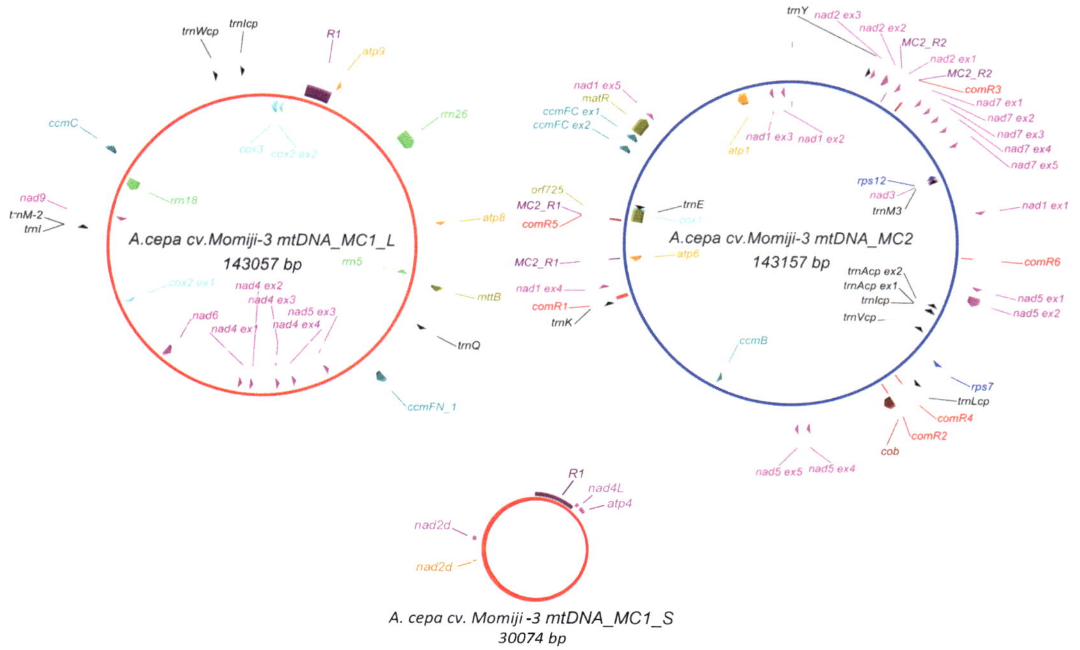

Fig. 6.2 Mitochondrial genome structure of *A. cepa* cv. "Momiji-3". Genes encoding proteins (red), rRNA (green), and tRNA (black) are shown outside (forward direction) and inside (reverse complementary direction) the circle. A direct repeat sequence of 3.4 kbp (R1) is also indicated as a box (purple) outside the MC1 circles

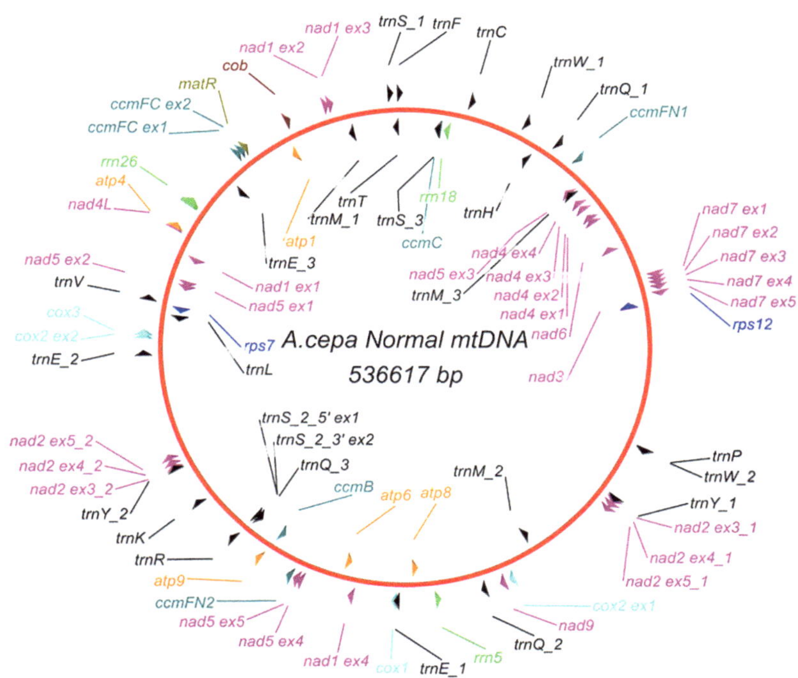

Fig. 6.3 Mitochondrial genome structure of *A. cepa* "Normal type". Genes encoding proteins (red), rRNA (green), and tRNA (black) are shown outside (forward direction) and inside (reverse complementary direction) the circle

Table 6.3 Nucleotide differences in the coding region of protein-coding genes between "Momiji-3" and "Normal type"

Gene	Position (Momiji-3)	Momiji-3	Normal	Description of differences
atp1	81	A	C	Synonymous substitution
atp6	92	A	C	Non-synonymous substitution
ccmFC	1168–1170	GGA	TCC	Non-synonymous substitution
ccmFN1	13–14	GA	TC	Non-synonymous substitution
	519–524	TGGAAG	–	InDels without causing frameshift
cox1 (*orf725*)	1595	C	T	Non-synonymous substitution
	1604–2151			Low homology within this region
cox3	52–58	GTATC	–	InDels causing frame shift
nad1	819–820	GA	TC	Non-synonymous substitution
nad4	297	T	G	Non-synonymous substitution
nad5	1314	A	C	Synonymous substitution
nad6	720–729	GACCT	–	InDels causing frameshift
rps7	268–269	GA	TC	Non-synonymous substitution
	360	A	C	Non-synonymous substitution

The C-terminus of the CMS-S type was longer than that of the Normal-type by 184aa, with each transcribed region covering the coding region. In the case of *cox3*, due to frameshift caused by an InDel of 5 bp at the 5′ side, the N-terminus of the Normal type was shorter than that of the CMS-S type by 43aa. The transcribed region started from the same position in both genomes. In the case of *nad6*, the C-terminus of the Normal type was shorter than that of the CMS-S type by 75aa due to an InDel. In both genomes, however, the transcribed region ended within the coding region. Even with *Arabidopsis*, transcription termination in *nad6* has been shown to occur upstream of the stop codon, suggesting the presence of a unique transcription termination system (Raczynska et al. 2006; Forner et al. 2007). Because the position of transcription termination is almost identical between the CMS-S type and the Normal-type, a similar transcription system is probably at work in onions.

Western blot analysis of *orf725*, a suspected inducer of CMS, using COX1 antibody demonstrated differences between the CMS-S type and the Normal-type (Fig. 6.5). No difference was found, however, between the CMS-S lines that have the *Ms* restorer gene and those that do not have the *Ms* restorer gene. This implies that the presence of a restorer gene does not affect the translated products of *orf725*. Two hypotheses can be inferred from this finding. First, genes other than *orf725* may actually be responsible for inducing CMS. Genes such as the above-mentioned *cox3* and *nad6* may show differences even at the translation level. Alternatively, there could be an unknown open reading frame (ORF) that may exhibit unexpected effects. Secondly, a restorer gene may be playing a part in fixing a defect caused during the *orf725* assembly process. *cox1* is a subunit comprising respiratory chain complex IV. It could be that, even when no functional protein is produced during the process of complex formation, the helping hand from a restorer gene ensures that a functional complex is produced. Further analysis is needed, which may include the investigation of the activity of respiratory chain complexes, to verify these hypotheses.

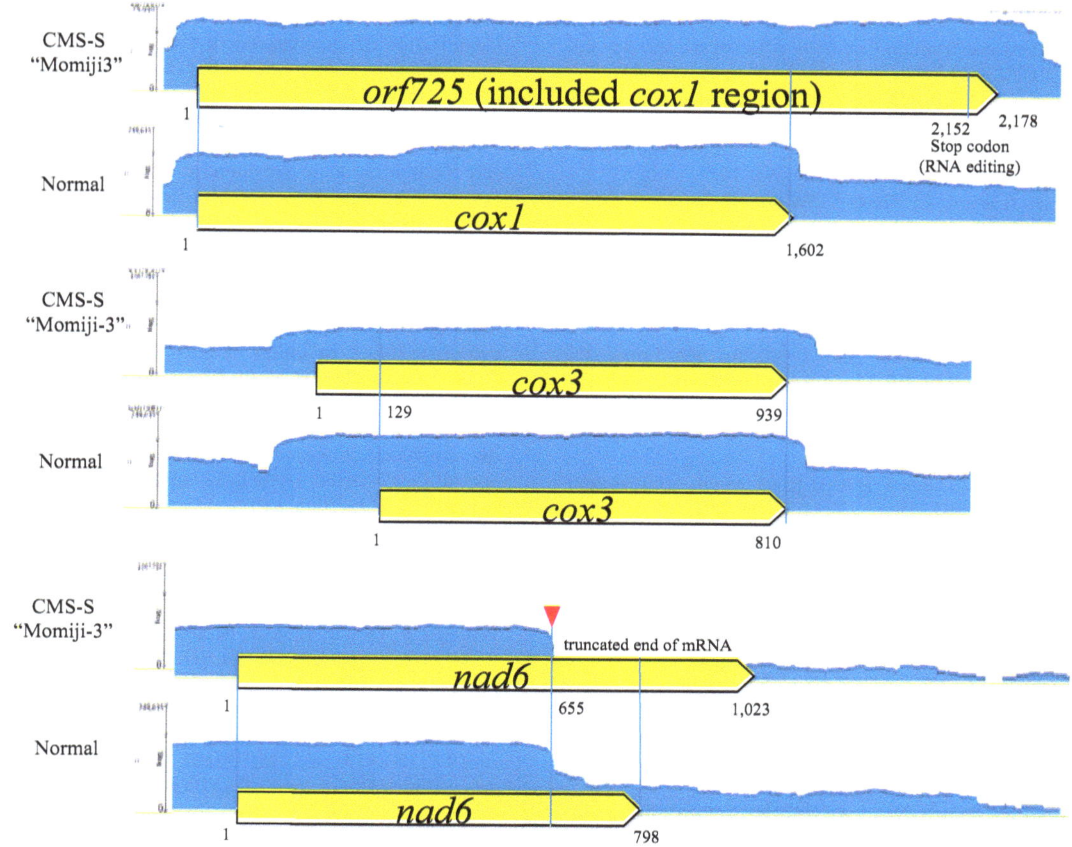

Fig. 6.4 Details of coding region differences in the genes *cox1*, *cox3*, and *nad6*. Block allows show protein-coding regions and graphs on the back of block allow indicate transcript abundance calculated by the coverage of sequence reads in RNA experiments

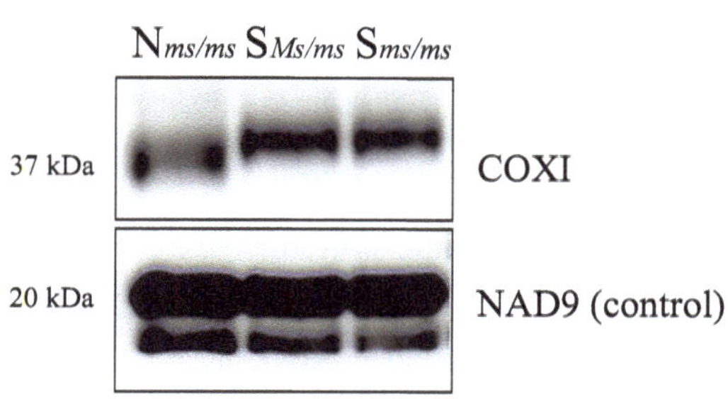

Fig. 6.5 Western blot analysis of COX1 and NAD9. N*ms/ms*; Normal-type with *ms/ms*, CMS-S type with *Ms/ms, and* CMS-S type with *ms/ms*

References

Bohra A, Jha UC, Adhimoolam P, Bisht D, Singh NP (2016) Cytoplasmic male sterility (CMS) in hybrid breeding in field crops. Plant Cell Rep 35:967–993

Borner T, Aleynikova AY, Zubo YO, Kusnetsov VV (2015) Chloroplast RNA polymerases: role in chloroplast biogenesis. Biochim Biophys Acta 1847:761–769

de Courcel AGL, Vedel F, Boussac JM (1989) DNA polymorphism in *Allium cepa* cytoplasms and its implications concerning the origin of onions. Theor Appl Genet 77:793–798

Dunning Hotopp JC (2011) Horizontal gene transfer between bacteria and animals. Trends Genet 27:157–163

Dyall SD, Brown MT, Johnson PJ (2004) Ancient invasions: from endosymbionts to organelles. Science 304:253–257

Engelke T, Tatlioglu T (2002) A PCR-marker for the CMS_1 inducing cytoplasm in chives derived from recombination events affecting the mitochondrial gene *atp9*. Theor Appl Genet 104:698–702

Forner J, Weber B, Thuss S, Wildum S, Binder S (2007) Mapping of mitochondrial mRNA termini in *Arabidopsis thaliana*: t-elements contribute to 5′ and 3′ end formation. Nucleic Acids Res 35:3676–3692

FAOSTAT, 2014. Agriculture Organization of the United Nations

Gurushidze M, Mashayekhi S, Blattner FR, Friesen N, Fritsch RM (2007) Phylogenetic relationships of wild and cultivated species of *Allium* section *Cepa* inferred by nuclear rDNA ITS sequence analysis. Plant Syst Evol 269(3):259–269

Hammani K, Giege P (2014) RNA metabolism in plant mitochondria. Trends Plant Sci 19:380–389

Hanson MR, Bentolila S (2004) Interactions of mitochondrial and nuclear genes that affect male gametophyte development. Plant Cell 16(Suppl):S154–S169. https://doi.org/10.1105/tpc.015966

Havey MJ (1999) Seed yield, floral morphology, and lack of male-fertile restoration of male sterile onion (*Allium cepa*) population possessing the cytoplasm of *Allium galanthum*. J Am Soc Hort Sci 124:626–629

Havey MJ (2000) Diversity among male-sterility-inducing and male-fertile cytoplasms of onion. Theor Appl Genet 101:778–782

Holford P, Croft J, Newbury HJ (1991a) Structural studies of microsporogenesis in fertile and male-sterile onions (*Allium cepa* L.) containing the cms-S cytoplasm. Theor Appl Genet 82:745–755

Holford P, Croft JH, Newbury HJ (1991b) Differences between, and possible origins of, the cytoplasms found in fertile and male-sterile onions (*Allium cepa* L.). Theor Appl Genet 82:737–744

Jones HA, Clarke AE (1943) Inheritance of male sterility in the onion and the production of hybrid seed. Proc Am Soc Hortic Sci 43:189–194

Jones HA, Davis G (1944) Inbreeding and heterosis and their relation to the development of new varieties of onions. USDA Tech Bull 874:1–28

Jones HA, Emsweller SL (1936) A male-sterile onion. Proc Am Soc Hortic Sci 34:582–585

Keeling PJ (2010) The endosymbiotic origin, diversification and fate of plastids. Philos Trans R Soc Lond B Biol Sci 365:729–748. https://doi.org/10.1098/rstb.2009.0103

Khosa JS, McCallum J, Dhatt AS, Macknight RC (2016) Enhancing onion breeding using molecular tools. Plant Breed 135:9–20. https://doi.org/10.1111/pbr.12330

Kik C (2002) Exploitation of wild relatives for the breeding of cultivated *Allium* species. CABI Publishing, pp 81–100. https://doi.org/10.1079/9780851995106.0081

Kim B, Kim K, Yang TJ, Kim S (2016) Completion of the mitochondrial genome sequence of onion (*Allium cepa* L.) containing the CMS-S male-sterile cytoplasm and identification of an independent event of the $ccmF_N$ gene split. Curr Genet 62:873–885. https://doi.org/10.1007/s00294-016-0595-1

Kim S (2014) A codominant molecular marker in linkage disequilibrium with a restorer-of-fertility gene (*Ms*) and its application in reevaluation of inheritance of fertility restoration in onions. Mol Breed 34:769–778. https://doi.org/10.1007/s11032-014-0073-8

Kim S, Lee ET, Cho DY, Han T, Bang H, Patil BS, Ahn YK, Yoon M-K (2009) Identification of a novel chimeric gene, *orf725*, and its use in development of a molecular marker for distinguishing among three cytoplasm types in onion (*Allium cepa* L.). Theor Appl Genet 118:433–441. https://doi.org/10.1007/s00122-008-0909-x

Kim S, Park JY, Yang T (2015) Comparative analysis of the complete chloroplast genome sequences of a normal male-fertile cytoplasm and two different cytoplasms conferring cytoplasmic male sterility in onion (*Allium cepa* L.). J Hortic Sci Biotechnol 90:459–468. https://doi.org/10.1080/14620316.2015.11513210

Ng S, De Clercq I, Van Aken O, Law SR, Ivanova A, Willems P, Giraud E, Van Breusegem F, Whelan J (2014) Anterograde and retrograde regulation of nuclear genes encoding mitochondrial proteins during growth, development, and stress. Mol Plant 7:1075–1093. https://doi.org/10.1093/mp/ssu037

Raczynska KD, Le Ret M, Rurek M, Bonnard G, Augustyniak H, Gualberto JM (2006) Plant mitochondrial genes can be expressed from mRNAs lacking stop codons. FEBS Lett 580:5641–5646. https://doi.org/10.1016/j.febslet.2006.09.010

Rhoads DM (2011) Plant mitochondrial retrograde regulation. In: Kempken F (ed) Plant mitochondria. Springer, New York, pp. 411–437. https://doi.org/10.1007/978-0-387-89781-3_16

Schweisguth B (1973) Étude d'un nouveau type de stérilité male chez l'oignon. Allium cepa L. Ann. Amélior Plant 23:221–233

Tsujimura M, Kaneko T, Sakamoto T, Kimura S, Shigyo M, Yamagishi H, Terachi T (2018) Multichromosomal structure of the onion mitochondrial genome and a transcript analysis. Mitochondrion (in press). https://doi.org/10.1016/j.mito.2018.05.001

von Kohn C, Kielkowska A, Havey MJ (2013) Sequencing and annotation of the chloroplast DNAs and identification of polymorphisms distinguishing normal male-fertile and male-sterile cytoplasms of onion. Genome 56:737–742. https://doi.org/10.1139/gen-2013-0182

Yamashita K-I, Takatori Y, Tashiro Y (2005) Chromosomal location of a pollen fertility-restoring gene, *Rf*, for CMS in Japanese bunching onion (*Allium fistulosum* L.) possessing the cytoplasm of *A. galanthum* Kar. et Kir. revealed by genomic *in situ* hybridization. Theor Appl Genet 111:15–22. https://doi.org/10.1007/s00122-005-1941-8

Repetitive Sequences

7

Maki Yamamoto

Abstract

Repetitive sequences compose a major component of genome constitution in the genus *Allium*. They contribute to genomic amplification by reciprocal translocation of repetitive sequences within the chromosome, and thus thought to enlarge the genome. The *Allium* genome mainly consists of middle-repetitive sequences. Only edible *Allium* species, such as onion and Welsh onion have been studied for their repetitive DNA characterization based on sequence analysis of isolated DNA. The repetitive sequences of onion have specific feature as compared with the other plants, and the long repetitive sequences exist in the telomeric heterochromatin regions and the nucleolus organizer regions (NORs). Most of the satellite DNA sequences are associated with telomeric, subtelomeric, and centromeric regions. Ty1-copia retrotransposon and En/Spm-transposable element-like sequence were found out in repetitive sequences of onion. Lately, BAC DNA library and genome-wide analysis using next-generation sequencing and flow cytometry have contributed to finer aspects of genome analysis.

The genus *Allium* sports telomeric sequences different from the *Arabidopsis*-type sequences, but novel information related to telomeric sequences goes on increasing. The retrotransposon element-like sequences seemed not to be included in centromeric sequences of the *Allium* genus as much as other plants. However, Ty3/gypsy-like retrotransposon was found out recently. Since the repetitive sequence of *Allium* genome is characteristic to the genus *Allium*, therefore, further studies are needed for its exhaustive utilization in other species of the genus for their characterization and speciation.

7.1 Introduction

The genus *Allium* has relatively few, but very large chromosomes, which directly reflect the huge amount of nuclear DNA. The basic chromosome number of the genus *Allium* is eight, and most of the *Allium* species are diploid ($2n = 2x = 16$). The cell nucleus of onion, *Allium cepa* L ($2n = 2x = 16$) contains 16.75–17.26 pg of DNA per 1C nucleus as deduced by flow cytometry, that corresponds to the genome size of about 1.6×10^4 Mbp (Doležel and Greilhuber 2010). The genome size of Welsh onion, *Allium fistulosum* ($2n = 2x = 16$) is estimated to be 11.7 pg/1C or 1.2×10^4 Mbp (Ricroch et al. 2005).

M. Yamamoto (✉)
Faculty of Education, Kansai University of Welfare Sciences, Kashiwara, Osaka 582-0026, Japan
e-mail: myamamoto@tamateyama.ac.jp

M. Shigyo et al. (eds.), *The Allium Genomes*, Compendium of Plant Genomes,
https://doi.org/10.1007/978-3-319-95825-5_7

The size of onion genome is biggest in cultivated crops. For example, its genome size is 6, 16, and 107 times greater than those of maize, tomato, and *Arabidopsis*, respectively, and equals that of hexaploid wheat. And each chromosome of onion has the DNA amount of 75% of the 1C nuclear genome of maize (Havey 2002).

The onion genome consists of middle-repetitive sequences occurring in short-period interspersions among low-copy regions (Jakse et al. 2006; Havey 2002). Analysis of the satellite DNA in bulb-onion shows that except for 375 bp telomeric DNA that amounts to 4% of the genome (Barnes et al. 1985), 80% is constituted by repetitive sequences. Only, 6% of the genomic DNA is single copy, and the nature of the rest of the 10% remains to be unknown (Stack and Comings 1979). These repetitive sequences are conserved beyond species. It has been observed that chromosomes of different sizes form eight bivalent chromosomes at the meiotic stage of hybrids of onion and Welsh onion (Narayan 1988). There is no seed fertility in the hybrid gametocyte. The origin of *Allium* genus is old, but phylogenetic evolutionary research is lagging behind. It is believed that species differentiation may have occurred a long time ago because it is quite difficult to realize artificial hybrids of the existing interspecific *Allium* species. For example, GISH experiment performed using total genomic DNA from each species as probes in *Allium* hybrids (onion × Welsh onion, Chinese chives × Welsh onion), revealed distinct genomic territories and clear chromosome identities of the two genomes with no sign of intergenomic introgression (Hizume 1994; Yamamoto 2015) (Fig. 7.1).

In order to seek an answer to the presence of large chromosome and huge genome size, there could be two possibilities, (1) reciprocal translocation by transposon or retrotransposon, and/or (2) tandem duplication within the genome by crossing-over via repetitive sequence. Jones and Rees (1968) and Ranjekar et al. (1978) have proposed genome duplication by reciprocal translocation within the chromosome. A genetic map based on restriction fragment length polymorphisms (RFLP) supports intrachromosomal duplication as a mechanism contributing to the large onion genome (King et al. 1998). Duplicated RFLPs showing linkage could be due to retroviral-like sequences in intrachromosomal duplications, not whole genome, duplications because this duplication frequency is less than that reported for paleopolyploids but higher than for diploid species.

It is thought that paleopolyploid of *Allium* was established as autopolyploid or allopolyploid several million years ago. During the course of the evolutionary process of their diploidization, the duplicated genes were purged out from the

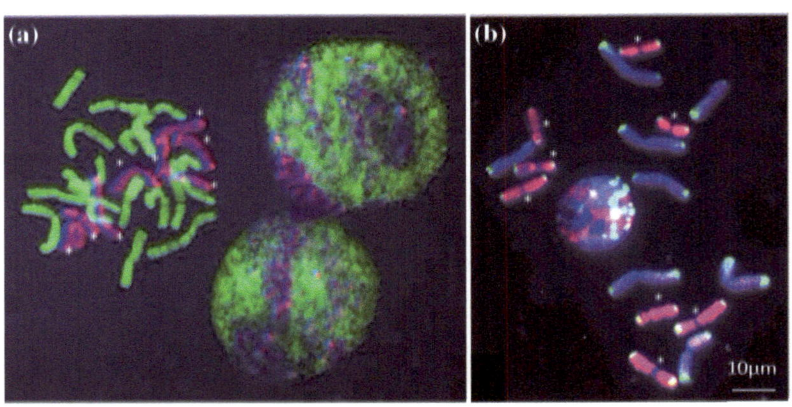

Fig. 7.1 GISH analysis of mitotic chromosomes of *Allium* hybrids, using genomic DNA probes. **a** Negi-Nira (*A. tuberosum* × *A. fistulosum*) and **b** *A. wakegi* (*A. cepa* × *A. fistulosum*). * indicates *A. fistulosum* chromosomes. Reproduced from Yamamoto (2015)

genome. However, evidence to explain the duplication of the onion genome has not been elucidated yet. With regard to this problem, it has been analyzed how the DNA of *Allium* changed with evolution by a survey of nuclear DNA amount by flow cytometry and genome GC content (Ricroch et al. 2005). The GC content of onion and green onions is about 39%, which is lower than any angiosperms (Ricroch et al. 2005), corresponding to the GC content of Arabidopsis rather than rice (Kuhl et al. 2004). There is no correlation between GC content and phylogenetic correlation of *Allium* plants and genome size.

Phylogenetic analysis using internal transcribed spacers (ITS), covering 30 species of cultivated species and wild species revealed that primitive species form bulbs, and those maintaining bulbaceous in the process of evolution have become larger genome size, but the species changed to rhizomes has become smaller genome size (Fig. 7.2) (Ricroch et al. 2005). As an explanation, the species that produces rhizomes has evolved from the corm shape of the ancestral species to the subspecies *Rhizirideum*, followed by the voluntary disappearance of the genome at least twice within this subspecies. The phylogenetic positioning of onions has diverged from species that evolved once from ancestral rhizome species to Welsh onion, and recently it has become bulbaceous, and its genome size is also increasing. In fact, the genome size of onion is 28% larger than that of Welsh onion (Jones and Rees 1968; Labani and Elkington 1987). This difference (5.4 pg/1C) is comparable to the nuclear genome of barley (*Hordeum vulgare*), pepper (*Capsicum annuum*), and radish (*Raphanus sativus*) (Bennett and Smith 1976).

7.2 Repetitive Sequences in General

Repetitive DNA sequences are the main component of eucaryote genomes and account for up to 90% of the genome size. They can be called satellite, minisatellite, and microsatellite DNA. Satellite DNA sequences are considered to be a fast-evolving component of eukaryotic genomes. The tandem repeated sequences are composed of the single unit that was highly conserved. The single unit of the satellite DNA is 150–400 base pairs (bp) in length. Repetitive sequences may be species- or genus-specific and maybe centromeric or subtelomeric in nature. Repetitive sequences accumulate variations in sequences and copy number, accrued by the process of the evolution. Therefore, those are important tools

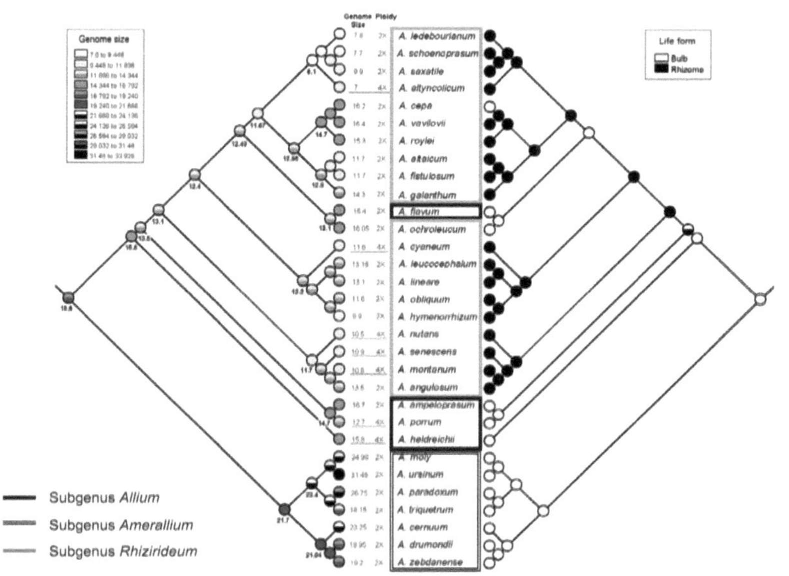

Fig. 7.2 Comparative phylogenetic analysis of between genome size (pg) and ploidy level in *Allium* species (Ricroch et al. 2005)

for taxonomic and phylogenetic studies and are known as "tuning knobs" in the evolution (Kashi et al. 1997; Kashi and King 2006). Repetitive sequences have cytoplasmic, cellular, and developmental effects and play a role in chromosomal recombination and the genome evolution (Mehrotra and Goyal 2014).

7.2.1 Feature of Repetitive Sequences in *A. cepa*

Repetitive sequences consisting of approximately 95% in the genome of *A. cepa*, have an essential role as a component of a huge genome. The most constituent of the genome is middle-repetitive sequences (Stack and Comings 1979). The caesium chloride (CsCl) density gradient method shows that the onion genome contains 4% satellite DNA. The satellite has a repeat length of 375 bp (Barnes et al. 1985). The experiment of FISH was carried out in the genus *Allium* using the 375 bp repetitive sequences as a probe. The 375 bp repetitive sequences probe were detected at the end of chromosomes and the interstitial regions at the section *Cepa* including onion and the section *Rhizirideum* (*A. roylei*), but not detected at the section *Allium* and the section *Schoenoprasum* (Stevenson et al. 1999). Pich and Schubert (1998) reported that this repetitive sequence is distributed over all chromosome ends except for the end of the short arm of chromosome 6 in onion. At the end of the short arm of chromosome 8, a 375 bp repetitive sequence probe is colocated with rDNA, and only rDNA is located at the end of the short arm of chromosome 6. Furthermore, Ty1-copia retrotransposon and En/Spm-transposable element-like sequence were found out in repetitive sequences of onion (Jakse et al. 2008). The Cot analysis of *A. cepa* DNA indicates that onion genome consists of two major middle-repetitive sequences (Fractions I and II) and a single copy sequence (Fraction III). Fraction I DNA is dispersed throughout the onion genome. Compared with other plant species of Monocots, *Ornithogalum virens* (Liliaceae) and *Secale cereale* (Poaceae), Fraction I is characteristic in onion and long repeated sequences are present in telomeric heterochromatin and nucleolus organizer regions (NORs) (Stack and Comings 1979). As other satellite DNAs, p*Ac*074 having a repetitive sequence of 314 bp (Do et al. 2001) and 370 bp sequences of ACSAT1, ACSAT2, and ACSAT3 have been reported (Barnes et al. 1985). The p*Ac*074 repetitive DNA probe was detected on the ends of all chromosomes except for the short arm of chromosome 6 by FISH experiment on the onion chromosomes and was simultaneously located with a 45S rDNA probe on the subtelomeric region in the short arm of chromosome 8 (Fig. 7.3) (Do et al. 2001).

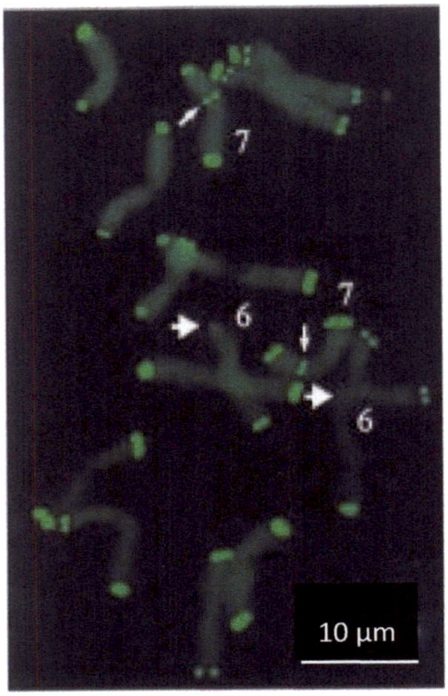

Fig. 7.3 Detection of the p*Ac*074 repetitive DNA probe on the ends of all *A. cepa* chromosomes except for the short arm of chromosome 6 by FISH experiment. Reproduced from Do et al. (2001)

7.2.2 Distribution of Repetitive Sequences in *Allium* Genome Using BAC Clones

A BAC DNA library of onion has been prepared for molecular, cytological analysis (Suzuki et al. 2001). It consists of 48,000 clones covering 30%

of the whole genome of onion (1C = 15,000 Mb) (average insert size: 100 kb). To cover most (95%) of the onion genome, 450,000 clones are required with an average insert size of 100 kb. BAC-FISH analysis of the constituent sequences of this BAC DNA library revealed that more than 80% contained repetitive sequences, 15% were telomere and centromere, and 5% were chromosome-specific sequences (Suzuki et al. 2001). Furthermore, using the onion BAC DNA, when the Southern hybridization analysis with the *A. fistlosum* genome of the same *Cepa* section was carried out, for example, one 108 kb BAC clone had one gene, but more than 50% of the sequence was a transposon (Jakse et al. 2008). We isolated 86 kb BAC DNA that seems to contain onion alliinase gene and performed 35 kb sequencing analysis. As a result, the alliinase like gene (*ALL1*) was found and embedded in the transposon-related sequences and highly repetitive sequences in the onion genome (Do et al. 2003). The 33.2% G+C content of the 35 kb *ALL1* region is similar to that of dicot plants and lower than that of monocot cereal plants, although onion is classified into monocots. Sequence analysis of the 18,484 EST of onion showed that among 11,008 unique sequences, 6,609 sequences (60%) were known proteins, but 0.8% (145 sequences) out of the total (18,484 sequences) were TE (transposable element), of which 25% (36 sequences) are class I DNA elements (mutator, Ac/Ds, En/Spm) and the remaining 75% are class II DNA elements (copia, gypsy, LTR retrotransposons) (Kuhl et al. 2004). Thus, the onion genome is composed of various repetitive sequences.

7.2.3 Feature of Repetitive Sequences in *A. fistulosum*

The size of *A. fistulosum* genome (1C = 12,275, Narayan 1988) is 1.2×10^4 Mb (Ricroch et al. 2005), which is 28% smaller than the onion genome. Highly repeated DNA sequences with a repeating unit of approximately 380 and 400 bp were found in *A. fistulosum*. Three independent clones containing a unit of 380 bp were isolated from *Eco*RV digests of the total genomic DNA of *A. fistulosum*. These units showed more than 94% sequence homology, and copy number was estimated to be about 2.8×10^6 per haploid genome (Irifune et al. 1995). A 400 bp new tandem repeat (CL26) was obtained from the genome DNA by the next-generation sequencing data (Sheikh et al. 2016) and occupied 0.2% of *A. fistulosum* genome. A total of 1100 BAC clones were constructed from the *A. fistulosum* genomic DNA. A single BAC clone possessing the CL26 tandem repeats were screened using the CL26 sequence. It was reported that CL26 was localized in the subtelomeric region of short arms of chromosome 6 and 8 in onion of the same section *Cepa*, but was not detected in other species of the genus *Allium* or the section *Schoenoprasum* (Sheikh et al. 2016).

It is known that rDNA is localized at the end of the short arm of chromosome 6 and 8 of onion, and it substitutes for telomere (Pich et al. 1996). CL26 clearly shows that sequences showing high homology with sequences in the IGC spacer region of onion 45S rDNA exist by analysis using BLASTN. CL26 is located in the subtelomeric region of the short arm of chromosome 6 in *A. fistulosum*, but no rDNA is there. rDNA is co-localized with a repetitive sequence of 378 bp in the subtelomeric region of the short arm of chromosome 8 of *A. fistulosum* (Fesenko et al. 2002).

7.3 Telomeric Sequences of *Allium*

Among the tandem sequences in genomes, the telomeric sequence of the genus *Allium* is different from other plant species. In general, the telomere sequence of higher plants is conserved in the *Arabidopsis*-type sequence ((TTTAGGG) n, in *Arabidopsis*) (Richards and Ausubel 1988), and in other organisms such as vertebrate animals and insects, (TTAGGG)n in Humans (Cheng et al. 1989; Meyne et al. 1989) and (TTAGG)n in *Bombyx* (Okazaki et al. 1993) are conserved. However, as exceptional cases, it was reported that a typical sequence as described above lacked in the insect, *Diptera* (Biessmann et al. 1990; Nielsen and Edstrom 1993; Biessmann et al. 1996) and in the plant, *A. cepa* (Fushs et al. 1995; Pich et al. 1996). In *Allium*, FISH experiments using a probe of *Arabidopsis*-type sequence, PCR analysis, and Southern hybridization analysis revealed that there was no *Arabidopsis*-type repeat sequence in onion telomere and whole genome at all (Fushs et al. 1995; Pich et al. 1996).

7.3.1 Characteristics of Telomeric DNA Sequences in the Genus *Allium*

Among the Asparagales, it is known that telomeres in *Iridaceae* and *Amaryllidaceae* are mutated to human-type (TTAGGG)n (Sykorova et al. 2003c; Monkheang et al. 2016). Furthermore, as far as the genus *Allium* of *Allioideae* belonging to *Amaryllidaceae*, *Allium* telomere sequence has been replaced by other sequences from human telomere in the course of its evolution (Sykorova et al. 2003c). The telomere sequence of higher plants is generally Arabidopsis-type (TTTAGGG) n, but in dicotyledonous plants, it has been reported that the sequence of TTTAGGG does not exist in a telomere of the genus *Cestrum* of *Solanaceae* (Sykorova et al. 2003a, b). Thus, it has become clear that telomere differentiation occurs at the genus or family level.

Detailed analysis of *Alliaceae* telomeres was attempted by Sykorova et al. (2003c, 2006). Slot-blot hybridization was performed using the genomic DNA of *Alliaceae*. Among various telomeric sequences used as probes, *Alliaceae* genomic DNA hybridized to Arabidopsis-type sequence (ATSB; TTTAGGG), human-type sequence (HUSB; TTAGGG), and/or Tetrahymena-type sequence (TTSB; TTGGGG). Species in *Allium* are divided into clades; clade1 has only small amounts of TTSB; clade2 has strong signals of HUSB; clade3 shows a heterogeneous pattern with these telomeric probes. FISH experiments were conducted using telomere sequences of various groups as probes. In plants, other than *Allium* among *Alliaceae*, a HUSB probe was detected in the telomeric region of chromosomes. In *Allium*, however, telomere sequences were detected in the interposed portion of the chromosome, not in the telomeric region. These results clearly showed that the consensus telomeric repeats are not a sequence of *Allium* telomeres.

7.3.2 Molecular Cytogenetic Analysis of Repetitive Telomeric Sequences in the Genus *Allium*

In order to clarify the telomere sequence of onion, several types of repetitive satellite sequences constituting the onion genome were analyzed as candidates for the telomeric or subtelomeric sequence. Although the 375 bp sequence (Barnes et al. 1985) was detected only at the end of some chromosomes (Stevenson et al. 1999; Pich et al. 1996), the existence of 314 bp sequence newly obtained using the random primer set (OPA07 primer; 5'-GAAACGGGTG-3') was confirmed by FISH experiment that the sequence exists on all chromosome ends excluding the rDNA region at the terminal of the short arm of chromosome 6 of onion (Do et al. 2001). The 314 bp sequence was also sequenced and found to be 81, 81, and 78% homologous to each 370 bp sequence of ACSAT1, ACSAT2, and ACSAT3 (Barnes et al. 1985), respectively (Do et al. 2001) (Fig. 7.4).

Suzuki et al. (2001) conducted BAC-FISH by selecting 91 clones from 48,000 clones of Onion BAC library and found three clones detected in all the terminal regions of the onion chromosome (Table 7.1). The FISH analysis was performed on onion chromosome and extended DNA fiber using clones of 400 bp sequence contained in two BAC clones (BAC 1-31 and BAC 2-17), p*Ac*074, BAC 1-31 and BAC 2-17 (Fig. 7.5) (Yamamoto 2002; Yamamoto and Mukai 2007). Signals on chromosomes of both BAC DNAs are not exactly the same, and since sequence homology is expected to be low also from DNA fiber FISH results, these are thought to be a repetitive sequence specific to the vicinity

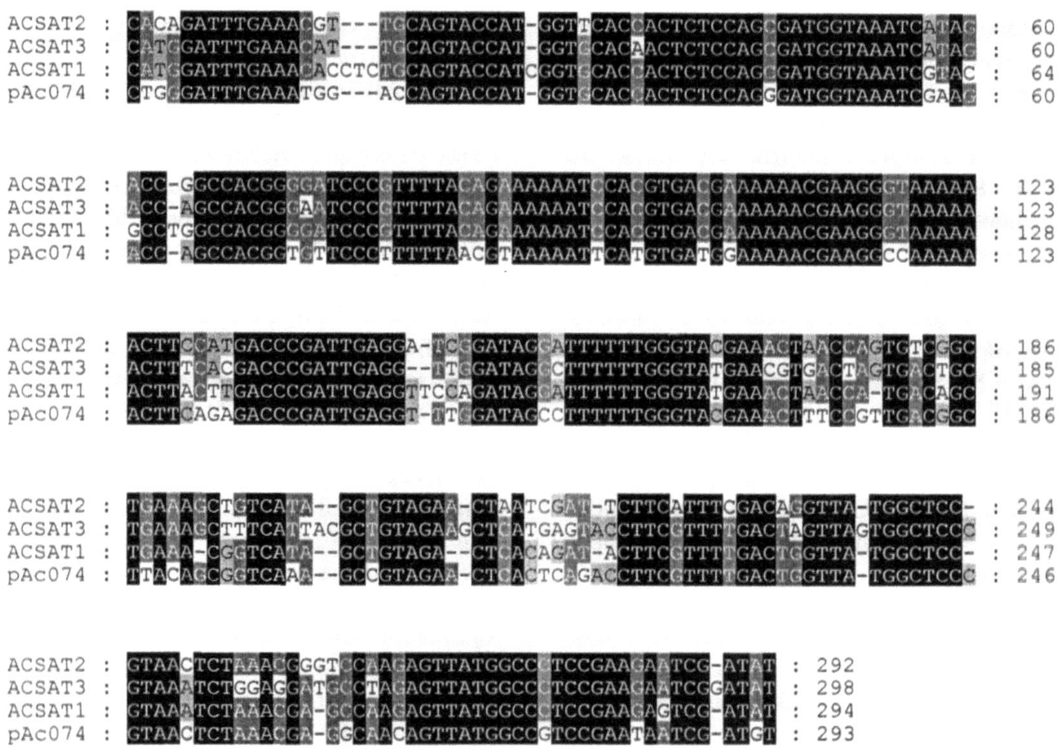

Fig. 7.4 The alignment of nucleotide sequences of the clone p*Ac*074 and the satellite DNA sequences (ACSAT1, ACSAT2, and ACSAT3) of *A. cepa* (Do et al. 2001)

Table 7.1 FISH signals of 91 onion BAC clones (Reproduced from Suzuki et al. 2001)

Detected region	Number of clones
Centromere	9
Telomere	3
Entire chromosomes	68
Specific regions	4
No signal	7
Total	91

of telomere rather than an onion telomere sequence. When simultaneous FISH of 1-31 and 400 bp is performed on the same chromosome, the 400 bp sequence is distributed more in the terminal part of the chromosome. Both p*Ac*074 and 400 bp sequences were subtelomeric sequences of onion, and when FISH was performed on the chromosomes of hybrid plants in *Allium*, these were detectable at the end of the chromosome of *A. fistulosum* (the section *Cepa*) as well as onion, but were not detected at all on the *A. tuberosum* chromosome (the section *Rhizirideum*). Therefore, the structure of repetitive sequences of the subtelomeric region is quite different depending on the section (Yamamoto 2004).

In situ hybridization, with the 380 bp repeating unit (pAfi100) as a probe showed that the repeated sequence of *A. fistulosum* is closely associated with the C-heterochromatin located in the terminal regions of all 16 chromosomes at mitotic metaphase (Irifune et al. 1995). From FISH results, although a 378 bp sequence was detected at the end of all *A. fistlosum*

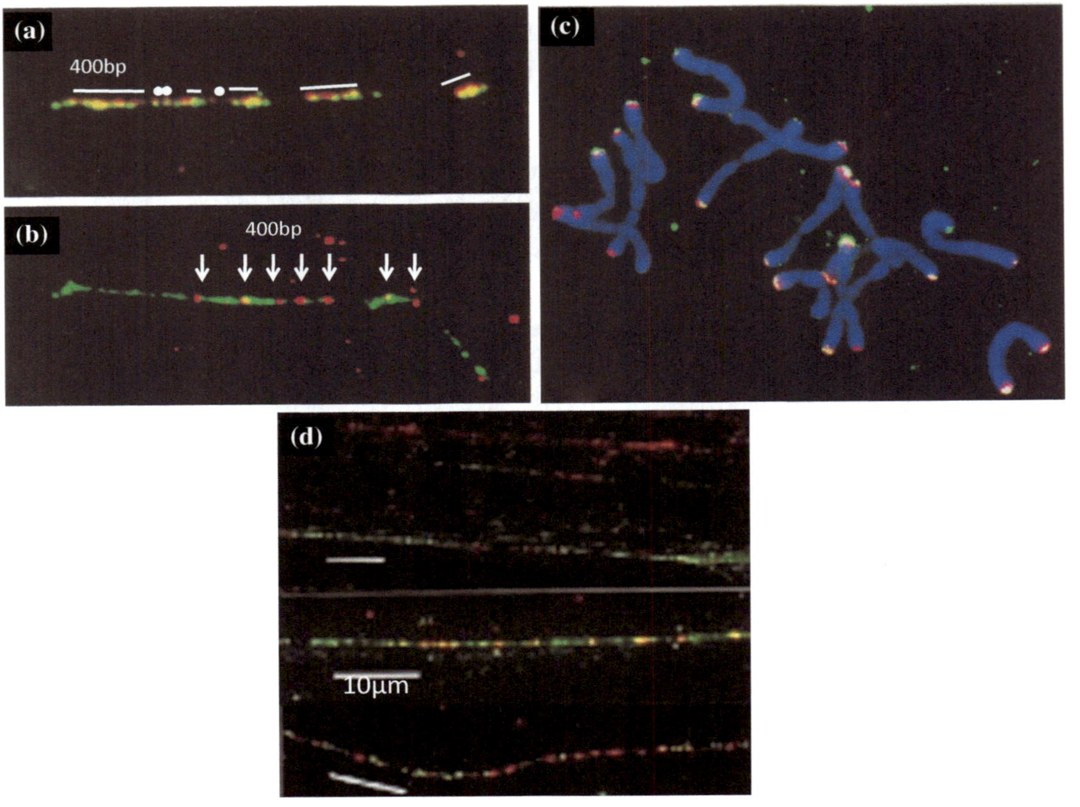

Fig. 7.5 FISH mapping on two BAC DNAs (**a** BAC2-17, **b** BAC1-31) using 400 bp repetitive DNA probe, and on onion chromosomes (**c**) using 400bp probe (green color) and BAC 1-31 probe (red color), and extended DNA fibers (**d**) using two BAC DNA as probes. (d) BAC 1-31 and BAC 2-17 probes were detected with different color. Reproduced from Yamamoto (2002), Yamamoto and Mukai (2007)

chromosomes (Fesenko et al. 2002; Sheikh et al. 2016), CL26 was located more proximal on the short arm of chromosome 6 as compared to the 378 bp subtelomeric repeat location (Sheikh et al. 2016).

7.3.3 Search for *Allium* Telomere Sequence by Next-Generation Sequencing (NGS) and Telomerase Repeat Amplification Protocol (TRAP) Methods

On the onion telomeric sequence, a search for candidate sequences was attempted by next-generation sequencing (NGS) (Fajkus et al. 2016). Repetitive sequences of fragments of up to 15 bp were searched using the TRFi (Tandem Repeats Finder) of NGS, and 17 patterns were obtained. Among them, searching for a tandem sequence cleaved at BAL 31 of the nuclease resulted in only a 12 bp minisatellite (CTCGGTTATGGG). As for the BAL 31 treatment, it is clear that the digested sequence is present at the terminal end of the chromosome because 45S rDNA is digested and 5S rDNA is not digested, with BAL 31 treatment of 45S rDNA sequence and 5S rDNA sequence as a control. FISH experiments of two species (*A. ursinum* and *A. cepa*) with different clade revealed that this candidate sequence was detected in the telomeric region of all chromosomes. Whether this sequence is synthesized with telomerase was confirmed by the TRAP assay (Kim et al. 1994; Mieczyslaw et al. 1995). Extract of the telomerase protein from the young sprout tissue on day 5 of germination of *A. cepa* and its reaction with the CTCGGTTACGGG sequence as a telomerase substrate sequence (TS primer) add to the telomere. Since the extension product by telomerase was amplified by PCR using TS primer and CX primer and a ladder of seven bases could be confirmed by electrophoresis, it was revealed that this candidate sequence is an onion telomere sequence synthesized by onion telomerase (Fajkus et al. 2016).

Elucidating the telomere sequence of the genus *Allium* by analyzing single mutations in the RNA template region of telomerase would have value in understanding the process of realizing these unique sequences from Arabidopsis-type through human-type. Such information would help understand the evolution and differentiation of the genus *Allium*. The data set of onion's BAC library (Suzuki et al. 2001) as well as the onion genome projects "Sequon" (http://www.oniongenome.net/) and "SCINET of Allium" (http://alliumgenetics.org/) (McCallum et al. 2012) are expected to become a useful tool.

7.4 Centromere-Specific Repetitive Sequences

Functional proteins constituting kinetochores first found CENP-A in humans (Moroi et al. 1980; Earnshaw and Rothfield 1985). CENP-A is well conserved across species from yeast to human. It has been reported that centromere-specific protein (CENH3) in plants is also well conserved. For example, antibodies of Welsh onion CENH3 have an affinity for any centromeres of onion, Chinese chives and garlic as well (Fig. 7.6) (Nagaki et al. 2012).

The centromere DNA sequence that binds to CENH3 is composed of highly repetitive sequences similar to the subtelomeric sequences and consists of a region in which a sequence of hundreds of base pairs is repeated about 10,000 times, so the centromere sequence occupies a large proportion in the genome. Thus, although the size of centromeric DNA closely resembles among species, it is also known that the DNA sequence itself is specific to species. There are reports on garlic and onion in the genus *Allium* (Nagaki et al. 2012). CENH3-specific DNA sequences have already been reported for monocotyledonous species: barley (Sanei et al. 2011), maize (Zhong et al. 2002), rice (Nagaki et al. 2004), sugarcane (Nagaki and Murata 2005), and wheat (Li et al. 2012), and for dicotyledonous species: carrot (Dunemann et al. 2014), cabbage (Wang et al. 2011), common bean (Iwata et al. 2013), cotton (Luo et al. 2012), pea (Neumann et al. 2012), potato (Gong et al. 2012), soybean (Tek et al. 2010), tobacco (Nagaki et al. 2009).

7.4.1 Analysis of Centromeric Sequences in *Allium*

Analysis of centromere DNA sequence is difficult because of the following reasons: (1) it must search from repetitive sequences in the genome; (2) use of database homologs cannot be expected due to different sequences depending on species, and (3) cloning is difficult because of highly repetitive DNA sequences. In particular, since the species with large genome size such as *Allium* contains a large amount and kinds of repetitive sequences, it is more difficult to analyze. Nagaki et al. (2012) succeeded in identifying centromeric DNA sequences of Welsh onion by ChIP analysis using the anti-AfiCENH3 antibody. These DNA sequences were cloned, and five clones (Afi11, Afi19, Afi54, Afi56, Afi61) were confirmed to be localized in the centromere of chromosomes by FISH experiment (Fig. 7.7) (Nagaki et al. 2012; Yamamoto 2014, 2015).

7.4.2 Feature of Repetitive Sequences in Centromere Region of *Allium* Chromosomes

With respect to five clones (Afi11, Afi19, Afi54, Afi56, Afi61), Tail-PCR experiment revealed the orientation of these clones and their peripheral

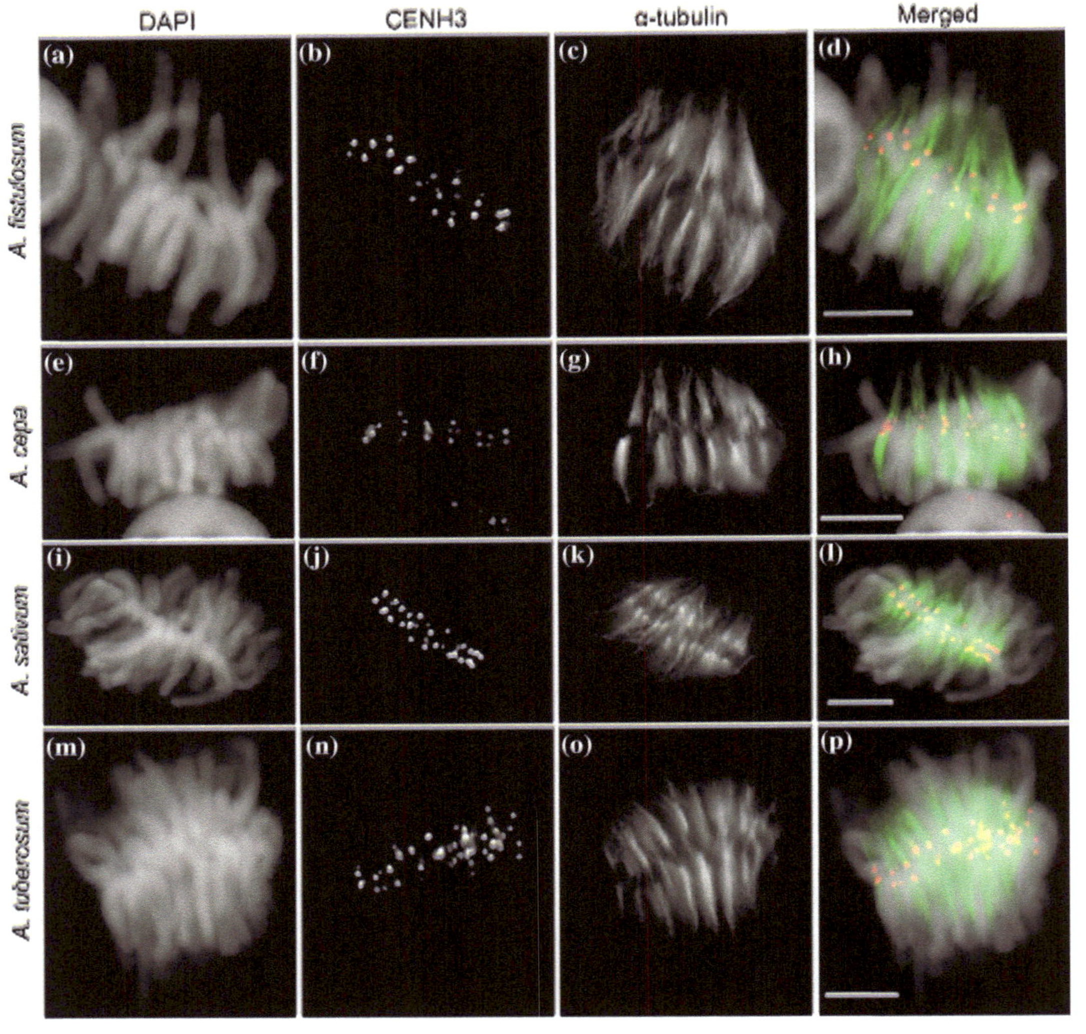

Fig. 7.6 Immunostaining of chromosomes of *Allium* species using an anti-AfiCENH3 antibody (Nagaki et al. 2012)

structure (Fig. 7.8) (Nagaki et al. 2012). In this investigation, the existence of a transposon-like sequence has not been confirmed, and it was suggested that the centromere-specific DNA sequence of the genus *Allium* has a different characteristic from a general centromere DNA sequence in which a large number of transposon sequences are present. However, this approach has limitations to obtain sufficient clones to grasp the whole centromeric-specific sequence. In the future, it will be expected to explore in the onion's BAC library or to incorporate analysis by the next-generation sequencing.

Recently, the presence of Ty3/gypsy-like retrotransposon has been reported near the centromere sequence of onion and Welsh onion, as opposed to the prediction that no transposon-like sequence is present in such centromere sequence of Welsh onion (Kiseleva et al. 2014). Analysis from onion data of NCBI database showed that Ty3/gypsy-like retrotransposon sequence in onion and Welsh onion was present in 26,000 copies and 7,000 copies, respectively. For this retrotransposon-like sequence, primers were prepared from the known sequence of "Genome Survey Sequence database" (ET 645811), and a

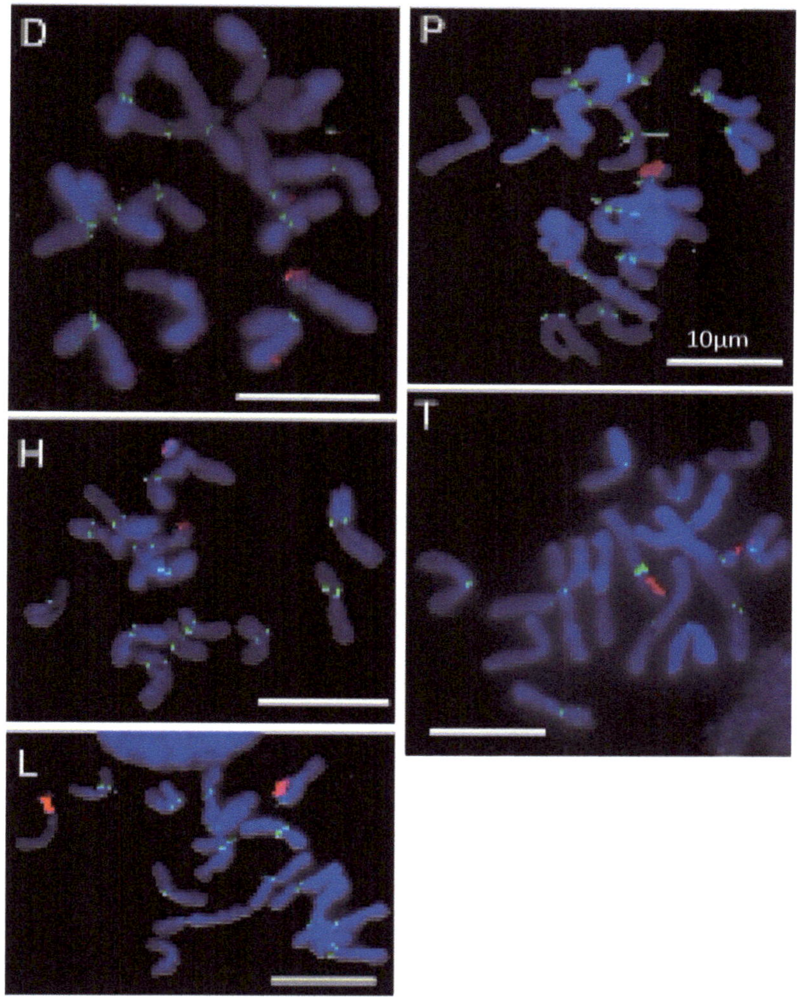

Fig. 7.7 FISH using the CENH3 ChIP clones, (D) Afi11, (H) Afi19, (L) Afi54, (P) Afi56, (T) Afi61, on each centromeric regions of all chromosomes in *A. fistulosum*. Reproduced from Nagaki et al. (2012)

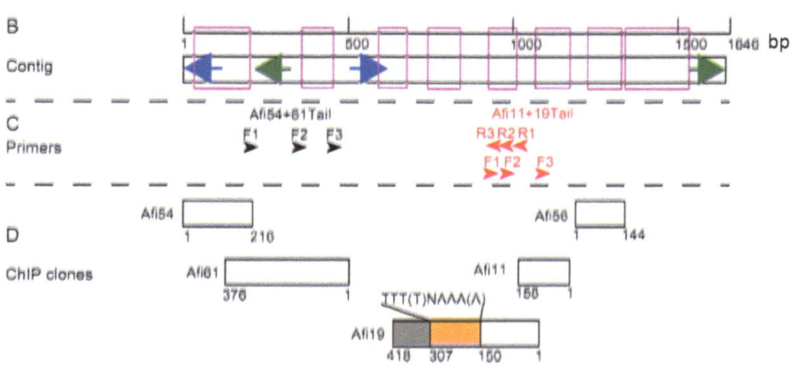

Fig. 7.8 Analysis of centromeric DNA sequences of *A. fistulosum*. Reproduced from Nagaki et al. (2012)

probe of 300 bp was prepared from each genome by PCR and used for the FISH experiment. The signal was mainly detected on the centromere of the onion chromosome and was not necessarily centromere-specific in Welsh onion (Kiseleva et al. 2014). Furthermore, two repetitive sequences (HAT 58 and HAT 36) isolated from Welsh onion genome are derived from the sequence in the peripherals of the centromere of chromosome 5, 6, 7, and 8 by FISH experiments (Kirov et al. 2017).

References

Barnes SR, James AM, Jamieson G (1985) The organisation, nucleotide sequence, and chromosomal distribution of satellite DNA from *Allium cepa*. Chromosoma 92:185–192

Bennett MD, Smith JB (1976) Nuclear DNA amounts in angiosperms. Philos Trans R Soc Lond B Biol Sci 274:227–274

Biessmann H, Donath J, Walter MF (1996) Molecular characterization of the *Anopheles gambiae* 2L telomeric region via an integrated transgene. Insect Mol Biol 5:11–20

Biessmann H, Mason JM, Ferry K, d'Hulst M, Valgeirsdottir K, Traverse KL, Pardue M-L (1990) Addition of telomere-associated HeT DNA sequences "heals" broken chromosome ends in *Drosophila*. Cell 61:663–673. https://doi.org/10.1016/0092-8674(90)90478-W

Cheng JF, Smith CL, Cantor CR (1989) Isolation and characterization of a human telomere. Nucleic Acids Res 17:6109–6127

Do GS, Seo BB, Yamamoto M, Suzuki G, Mukai Y (2001) Identification and chromosomal location of tandemly repeated DNA sequences in *Allium cepa*. Genes Genet Syst 76:53–60. https://doi.org/10.1266/ggs.76.53

Do GS, Suzuki G, Mukai Y (2003) Genomic organization of a novel root alliinase gene, *ALL1*, in onion. Gene 325:17–24. https://doi.org/10.1016/j.gene.2003.09.033

Doležel J, Greilhuber J (2010) Nuclear genomic size: are we getting closer? Cytometry A 77:635–642. https://doi.org/10.1002/cyto.a.20915

Dunemann F, Schrader O, Budahn H, Houben A (2014) Characterization of centromeric histone H3 (CENH3) variants in cultivated and wild carrots (*Daucus* sp.). PLoS ONE 9:e98504. https://doi.org/10.1371/journal.pone.0098504

Earnshaw WC, Rothfield N (1985) Identification of a family of human centromere proteins using autoimmune sera from patients with scleroderma. Chromosome 91:313–321

Fajkus P, Peska V, Sitova Z, Fulnečková J, Dvořáčková M, Gogela R, Sýkorová E, Hapala J, Fajkus J (2016) *Allium* telomeres unmasked: the unusual telomeric sequence (CTCGGTTATGGG)n is synthesized by telomerase. Plant J 85:337–347. https://doi.org/10.1111/tpj.13115

Fesenko IA, Khrustaleva LI, Karlov GI (2002) Organization of the 378-bp satellite repeat in terminal heterochromatin of *Allium fistulosum*. Genetika 38:894–903

Fushs J, Brandes A, Schubert I (1995) Telomere sequence localization and karyotype evolution in higher plants. Pl Syst Evol 196:227–241

Gong Z, Wu Y, Koblizkova A, Torres GA, Wang K, Iovene M, Neumann P, Zhang W, Novák P, Buell CR, Macas J, Jiang J (2012) Repeatless and repeat-based centromeres in potato: implications for centromere evolution. Plant Cell 24:3559–3574

Havey MJ (2002) Genome organization in *Allium*. In: Rabinowitch HD, Currah L (eds) *Allium* crop science: recent advances, Wallingford, Oxfordshire. pp 59–79

Hizume M (1994) Allodiploid nature of *Allium wakegi* Araki revealed by genomic *in situ* hybridization and localized of 5S and 18S rDNAs. Jpn J Genet 69:407–415

Irifune K, Hirai K, Zheng J, Tanaka R, Morikawa H (1995) Nucleotide sequence of a highly repeated DNA sequence and its chromosomal localization in *Allium fistulosum*. Theor Appl Genet 90:312–316. https://doi.org/10.1007/BF00221970

Iwata A, Tek AL, Richard MMS, Abernathy B, Fonsêca A, Schmutz J, Chen NWG, Thareau V, Magdelenat G, Li Y, Murata M, Pedrosa-Harand A, Geffroy V, Nagaki K, Jackson SA (2013) Identification and characterization of functional centromeres of the common bean. Plant J 76:47–60

Jakse J, Meyer JDF, Suzuki G, McCallum J, Cheung F, Town CD, Havey MJ (2008) Pilot sequencing of onion genomic DNA reveals fragments of transposable elements, low gene densities, and significant gene enrichment after methyl filtration. Mol Genet Genomics 280:287–292

Jakse J, Telgmann A, Jung C, Khar A, Melgar S, Cheung F, Town CD, Havey MJ (2006) Comparative sequence and genetic analyses of asparagus BACs reveal no microsynteny with onion or rice. Theor Appl Genet 114:31–39. https://doi.org/10.1007/s00112-006-0407-y

Jones RN, Rees H (1968) Nuclear DNA variation in *Allium*. Heredity 23:591–605

Kashi Y, King DG (2006) Simple sequence repeats as advantageous mutators in evolution. Trends Genet 22:253–259

Kashi Y, King DG, Soller M (1997) Simple sequence repeats as a source of quantitative genetic variation. Trends Genet 13:74–78

Kim NW, Piatyszek MA, Prowse KR, Harley CB, West MD, Ho PL, Coviello GM, Wright WE, Weinrich SL, Shay JW (1994) Specific association of human telomerase activity with immortal cells and cancer. Science 266:2011–2015

King JJ, Bradeen JM, Bark O, McCallum JA, Havey MJ (1998) A low-density genetic map of onion reveals a role for tandem duplication in the evolution of an extremely large diploid genome. Theor Appl Genet 96:52–62

Kirov IV, Kiseleva AV, Van Leare K, Van Roy N, Khrustaleva LI (2017) Tandem repeats of *Allium fistulosum* associated with major chromosomal landmarks. Mol Genet Genomics 292:453–464. https://doi.org/10.1007/s00438-016-1286-9

Kiseleva AV, Kirov IV, Khrustaleva LI (2014) Chromosomal organization of centromeric Ty3/gypsy retrotransposons in *Allium cepa* L. and *Allium fistulosum* L. Genetika 50:670–676. https://doi.org/10.1134/S102279541404005X

Kuhl JC, Cheung F, Yuan Q, Martin W, Zewdie Y, McCallum J, Catanach A, Rutherford P, Sink KC, Jenderek M, Prince JP, Town CD, Havey MJ (2004) A unique set of 11,008 onion expressed sequence tags reveals expressed sequence and genomic differences between the monocot orders Asparagales and Poales. Plant Cell 16:114–125

Labani RM, Elkington TT (1987) Nuclear DNA variation in the genus *Allium* L. (Liliaceae). Heredity 59: 119–128

Li B, Choulet F, Heng Y (2012) Wheat centromeric retrotransposons: the new ones take a major role in centromeric structure. Plant J 73:952–965

Luo S, Mach J, Abramson B, Ramirez R, Schurr R, Barone P, Copenhaver G, Folkerts O (2012) The cotton centromere contains a Ty3-gypsy-like LTR retroelement. PLoS ONE 7:e35261. https://doi.org/10.1371/journal.pone.0035261

McCallum J, Baldwin S, Shigyo M, Deng Y, van Heusden S, Pither-Joyce M, Kenel F (2012) AlliumMap-A comparative genomics resource for cultivated *Allium* vegetables. BMC Genom 13:168. https://doi.org/10.1186/1471-2164-13-168

Mehrotra S, Goyal V (2014) Repetitive sequences in plant nuclear DNA: types, distribution, evolution and function. Genomics Proteomics Bioinform 12:164–171. https://doi.org/10.1016/j.gpb.2014.07.003

Meyne J, Ratliff RL, Moyzis RK (1989) Conservation of the human telomere sequence (TTAGGG)n among vertebrates. Proc Natl Acad Sci USA 86:7049–7053

Mieczyslaw AP, Kim NW, Scott LW, Hiyama K, Hiyama E, Wright WE, Shay JW (1995) Detection of telomerase activity in human cells and tumors by a telomeric repeat amplification protocol (TRAP). Methods Cell Sci 17:1–15

Monkheang P, Chaveerach A, Sudmoon R, Tanee T (2016) Karyotypic features including organizations of the 5S, 45S rDNA loci and telomeres of *Scadoxus multiflorus* (*Amaryllidaceae*). Comp Cytogent 10:637–646. https://doi.org/10.3897/CompCytogen.v10i4.9958

Moroi Y, Peebles C, Fritzler MJ, Steigerwald J, Tan EM (1980) Autoantibody to centromere (kinetochore) in scleroderma sera. Proc Natl Acad Sci USA 77: 1627–1631

Nagaki K, Cheng Z, Ouyang S, Talbert PB, Kim M, Jones KM, Henikoff S, Buell CR, Jiang J (2004) Sequencing of a rice centromere uncovers active genes. Nat Genet 36:138–145

Nagaki K, Kashihara K, Murata M (2009) A centromeric DNA sequence colocalized with a centromere-specific histone H3 in tobacco. Chromosoma 118:249–257

Nagaki K, Murata M (2005) Characterization of CENH3 and centromere-associated DNA sequences in sugarcane. Chromosom Res 13:195–203

Nagaki K, Yamamoto M, Yamaji N, Mukai Y, Murata M (2012) Chromosome dynamics visualized with an anti-centromeric histone H3 antibody in *Allium*. PLoS ONE 7(12):e51315. https://doi.org/10.1371/journal.pone.0051315

Narayan RKJ (1988) Constraints upon the organization and evolution of chromosomes in *Allium*. Theor Appl Genet 75:319–329. https://doi.org/10.1007/BF00303971

Neumann P, Navratilova A, Schroeder-Reiter E, Koblížková A, Steinbauerová V, Chocholová E, Novák P, Wanner G, Macas J (2012) Stretching the rules: monocentric chromosomes with multiple centromere domains. PLoS Genet 8:e1002777. https://doi.org/10.1371/journal.pgen.1002777

Nielsen L, Edstrom JE (1993) Complex telomere-associated repeat units in members of the genus *Chironomus* evolve from sequences similar to simple telomeric repeats. Mol Cell Biol 13:1583–1589

Okazaki S, Tsuchida K, Maekawa H, Ishikawa H, Fujiwara H (1993) Identification of a pentanucleotide telomeric sequence, (TTAGG)n, in the silkworm *Bombyx mori* and in other insects. Mol Cell Biol 13:1424–1432

Pich U, Fuchs J, Schubert I (1996) How do Alliaceae stabilize their chromosome ends in the absence of TTTAGGG sequences? Chromosome Res 4:207–213

Pich U, Schubert I (1998) Terminal heterochromatin and alternative telomeric sequences in *Allium cepa*. Chromosom Res 6:315–321

Ranjekar PK, Pallotta D, Lafontaine JG (1978) Analysis of plant genomes. V. Comparative study of molecular properties of DNAs of seven *Allium* species. Biochem Genet 16:957–970

Richards EJ, Ausubel FM (1988) Isolation of a higher eukaryotic telomere from Arabidopsis thaliana. Cell 53(1):127–136

Ricroch A, Yockteng R, Brown SC, Nadot S (2005) Evolution of genome size across some cultivated *Allium* species. Genome 48:511–520. https://doi.org/10.1139/G05-017

Sanei M, Pickering R, Kumke K, Nasuda S, Houben A (2011) Loss of centromeric histone H3 (CENH3) from centromeres precedes uniparental chromosome elimination in interspecific barley hybrids. Proc Natl Acad Sci USA 108:498–505

Sheikh BGMA, Kirov IV, Khrustaleva LI (2016) A new chromosome specific subtelomeric tandem repeat in *Allium fistulosum* (L.). Izvestiya of Timiryazev Agricultural Academy 4:17–26

Stack SM, Comings DE (1979) The chromosomes and DNA of *Allium cepa*. Chromosoma 70:161–181. https://doi.org/10.1007/BF00288404

Stevenson M, Armstrong SJ, Jones GH, Ford-Lloyd BV (1999) Distribution of a 375 bp repeat sequence in *Allium* (Alliaceae) as revealed by FISH. Pl Syst Evol 217:31–42

Suzuki G, Saito N, Do GS, Seo BB, Yamamoto M, Mukai Y (2001) BAC FISH analysis in *Allium cepa*. Genes Genet Syst 76:251–255

Sykorova E, Fajkus J, Meznikova M, Lim KY, Neplechová K, Blattner FR, Chase MW, Leitch AR (2006) Minisatellite telomeres occur in the family *Alliaceae* but are lost in *Allium*. Am J Bot 93: 814–823

Sykorova E, Lim KY, Chase MW, Knapp S, Leitch IJ, Leitch AR, Fajkus J (2003a) The absence of Arabidopsis-type telomeres in *Cestrum* and closely related genera *Vestia* and *Sessea* (*Solanaceae*): first evidence from eudicots. Plant J 34:283–291

Sykorova E, Lim KY, Fajkus J, Leitch AR (2003b) The signature of the *Cestrum* genome suggests an evolutionary response to the loss of (TTTAGGG)(n) telomeres. Chromosoma 112:164–172

Sykorova E, Lim KY, Kunicka Z, Chase MW, Bennett MD, Fajkus J, Leitch AR (2003c) Telomere variability in the monocotyledonous plant order Asparagales. Proc R Soc Lond B 270:1893–1904. https://doi.org/10.1098/rspb.2003.2446

Tek AL, Kashihara K, Murata M, Nagaki K (2010) Functional centromeres in soybean include two distinct tandem repeats and a retrotransposon. Chromosom Res 18:337–347

Wang G, He Q, Liu F, Cheng Z, Talbert PB, Jin W (2011) Characterization of CENH3 proteins and centromere-associated DNA sequences in diploid and allotetraploid *Brassica* species. Chromosoma 120:353–365

Yamamoto M (2002) FISH analysis of the telomeric region in *Allium cepa* L. Bull Kansai Women's Coll 12:69–76

Yamamoto M (2004) Dynamics of functional domains in meiotic chromosomes of *Allium* by FISH. Bull Kansai Women's Coll 13:7–84

Yamamoto M (2014) Meiotic abnormality and chromosome dynamics in an *Allium* hybrid. Bull Kansai Univ Welfare Sci 18:51–62

Yamamoto M (2015) Molecular cytogenetic perspectives of crossing barriers and peculiar reproduction ability in *Allium* hybrids. Nucleus 58:165–172. https://doi.org/10.1007/s13237-015-0153-0

Yamamoto M, Mukai Y (2007) Extended DNA fiber FISH in plants: visible messages from cell nuclei. Nucleus 50:437–448

Zhong CX, Marshall JB, Topp C, Mroczek R, Kato A, Nagaki K, Birchler JA, Jiang J, Dawe RK (2002) Centromeric retroelements and satellites interact with maize kinetochore protein CENH3. Plant Cell 14:2825–2836

Simple Sequence Repeat

8

Hikaru Tsukazaki

Abstract

DNA markers are powerful tools for molecular breeding or checking the genetic homogeneity of F_1 hybrid seeds. For the simple and fast approach to detect DNA polymorphisms, simple sequence repeats (SSRs), which are tandem repeats of short sequence motifs (mainly 1–6 bases), are powerful markers because of their codominant mode of inheritance, higher reliability, and abundance in genomes. In *Allium*, GT/CA repeats are abundant than GA/CT repeats, and AAG/CTT motif is the most frequent in the trinucleotide SSRs. NGS technologies including public sequence data will be a mainstream for identifying new SSR markers.

8.1 Introduction

For the simple and fast approach to detect DNA polymorphisms, Random Amplified Polymorphic DNA (RAPD), Amplified fragment length polymorphism (AFLP), and Restriction Fragment Length Polymorphism (RFLP) analyses have been applied for genetic study in alliums. Two polymerase chain reaction (PCR)-based, simple and fast approaches for detecting DNA polymorphisms, i.e., RAPD and AFLP analyses, were previously applied to genetic studies of *Allium* (Wilkie et al. 1993; Bradeen and Havey 1995; van Heusden et al. 2000; Ohara et al. 2005; Ipek et al. 2005). However, these markers are usually expressed as dominant characters. In contrast, King et al. (1998) reported a linkage map of bulb onion based on RAPDs and RFLPs, the latter of which is expressed as a codominant character and has high specificity over RAPDs and AFLPs. Despite these advantages, RFLP analysis is both time consuming and labor intensive (Crockett et al. 2000). Thus, it is difficult for other populations or species to apply these markers and compare genome components.

Simple sequence repeats (SSRs), on the other hand, are powerful molecular markers because of their codominant mode of inheritance, higher reliability, and abundance in genomes over RAPD (Jones et al. 1997). SSRs are tandem repeats of short sequence motifs (mainly 1–6 bases) that occur ubiquitously in eukaryotic genomes (Weising et al. 1998; Zane et al. 2002; Kalia et al. 2011) (Fig. 8.1). Most of the SSRs are nuclear SSRs, however, they are also distributed in mitochondria and chloroplasts. SSRs are highly useful as molecular markers due to several reasons; (1) widely dispersed throughout the genome, (2) highly variable among varieties mainly due to

H. Tsukazaki (✉)
Tohoku Agricultural Research Center (TARC), National Agriculture and Food Research Organization (NARO), 4 Akahira, Shimokuriyagawa, Morioka, Iwate 020-0198, Japan
e-mail: tsuka@affrc.go.jp

M. Shigyo et al. (eds.), *The Allium Genomes*, Compendium of Plant Genomes,
https://doi.org/10.1007/978-3-319-95825-5_8

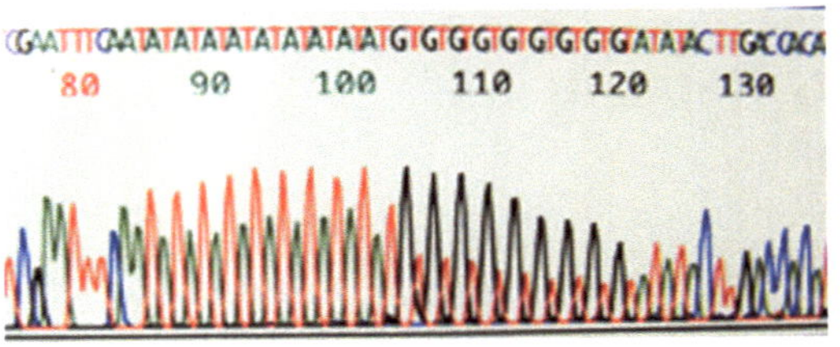

Fig. 8.1 An example of simple sequence repeat (SSR) motif (repeats of $(AT)_{10}$ and $(GT)_9$)

variable numbers of repeats, (3) most SSRs are probably nonfunctional, thus selectively neutral. The SSR mutation rates (10^{-2}–10^{-6} events per locus per generation) are very high as compared with the rates of point mutation at coding loci (Li et al. 2002). High degree of allelic variation revealed by SSR markers results from variation in number of repeat motifs at a locus caused by replication slippage and/or unequal crossing-over during meiosis (Kalia et al. 2011). Thus, SSRs are used for constructing linkage maps, marker-assisted selection, phylogenetic studies, genomic synteny analysis, and cultivar identification (Jones et al. 1997; Bredemeijer et al. 1998). In this chapter, I have reviewed the recent research about SSRs in *Allium*.

8.2 Genomic SSRs (gSSRs)

Studies on the isolation of genomic SSRs in the genus *Allium* are listed in Table 8.1. Fischer and Bachmann (1998) reported for the first time isolation of 30 genomic SSR (gSSR) markers from bulb onion (*A. cepa* L.). However, these markers were recalcitrant for wider use, due to their complicated programs to amplify PCR fragments (Fischer and Bachmann 2000) and the problems with interpreting complex banding patterns.

Song et al. (2004) isolated 52 SSR clones from bunching onion (*A. fistulosum* L.). They screened with a mixture of $(GA)_{15}$ and $(GT)_{15}$ oligonucleotide probes from approximately 180,000 clones of a size-fractionated genomic library. Although relatively simple, this method needs to construct a genomic library and then screen clones by repeated rounds of filter hybridization using an oligonucleotide probe.

SSR-enrichment, on the other hand, has also been developed to increase the proportion of clones in a library containing the microsatellite motif of interest. Several strategies for microsatellite enrichment have been reported (Edwards et al. 1996; Fischer and Bachmann 1998; Zane et al. 2002; Nunome et al. 2006) and this technology is widely used for isolation of SSR clones in *Allium* (Table 8.1).

Tsukazaki et al. (2007) isolated 1796 SSR clones from SSR-enriched DNA libraries of bunching onion. To increase the efficiency of isolation of SSR-containing clones and to decrease redundancy of effort in developing new markers, they constructed 10 SSR-enriched libraries. A total of 10,922 clones from each SSR-enriched DNA library were randomly selected and sequenced. Of them, 3752 (34.4%) were found to contain more than five SSR motif repeats and 1796 SSR-containing uniclones were isolated. Some of these SSR loci were polymorphic among bunching onion inbred lines (Fig. 8.2). Ma et al. (2009) and Cunha et al. (2012) independently isolated SSRs from an SSR-enrichment library of garlic (*A. sativum* L.). Also, Zhang et al. (2014) isolated SSRs from an

Table 8.1 Development of genomic SSRs in *Alliums*

Species	Methods	Number of clones	References
A. cepa	SSR-enrichment	30	Fischer and Bachmann (1998)
	NGS sequencing	814	Baldwin et al. (2012a)
A. fistulosum	Genomic libraries	52	Song et al. (2004)
A. sativum	SSR-enrichment	1,796	Tsukazaki et al. (2007)
	SSR-enrichment	139	Ma et al. (2009)
A. mongolicum	SSR-enrichment	16	Cunha et al. (2012)
	SSR-enrichment	38	Zhang et al. (2014)

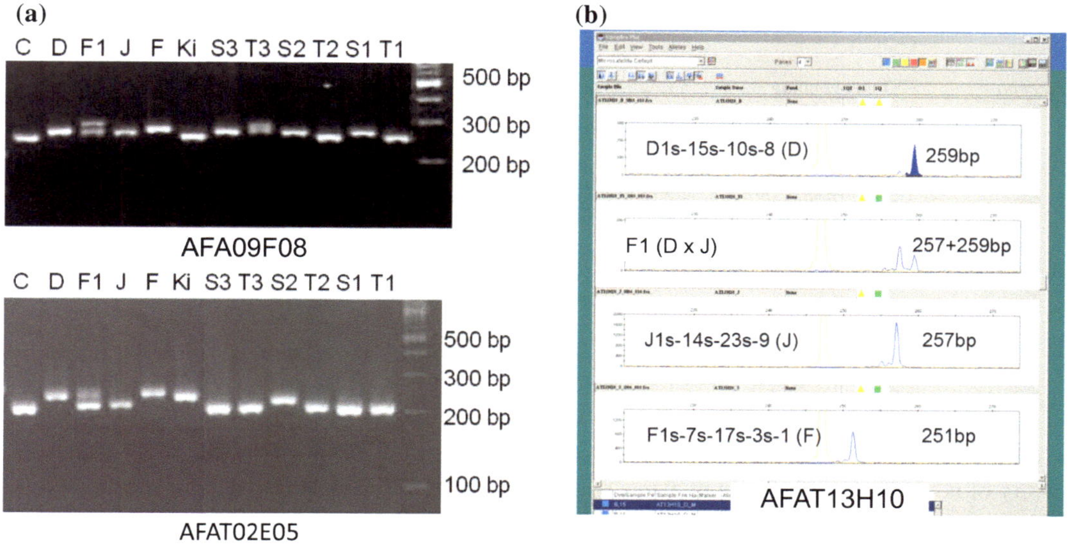

Fig. 8.2 DNA polymorphisms among inbred lines of bunching onion. PCR fragments were separated by 3% agarose gel (**a**) or DNA sequencer (**b**)

SSR-enrichment library of *A. mongolicum* (referred to as wild onion or mountain onion).

Recently, next-generation sequencing (NGS) technologies have enabled large amounts of sequence data to be acquired. By using NGS technologies, the development of genomic SSRs markers was facilitated by genomic sequencing, which has been shown to be highly efficient in uncharacterized genomes without SSR-enrichment (Abdelkrim et al. 2009; Castoe et al. 2010). In *Allium*, Baldwin et al. (2012a) identified 921 SSR motifs from bulb onion genomic DNA using skim sequencing by GS-FLX (Roche). It was estimated that the trimmed sequence data (6.63 Mbps) represents only 0.02% of the onion genome, 3% of all the reads contained an SSR motif.

8.3 Genic-SSRs

Although SSRs constitute a large fraction of noncoding DNA, a lot of SSRs are located in transcribed regions of the genome. It has been reported that noncoding (genomic) SSRs are preferentially located in heterochromatic regions (Frary et al. 2005; Ohyama et al. 2009), in contrast, most genic-SSRs are found in euchromatin. Genic-SSRs have an advantage over genomic SSRs because their variations often directly affect the corresponding gene products and even cause phenotypic changes (Li et al. 2004). Thus, genic-SSR markers can be used as "perfect" markers.

Studies on the isolation of genic SSRs in the genus *Allium* are listed in Table 8.2. In the years preceding 2010, only medium-scale sequencing of expressed sequence tags (ESTs) was conducted in bulb onion (Kuhl et al. 2004) and garlic (Kim et al. 2009).

Kuhl et al. (2004) isolated RNA samples from immature bulbs, callus, and roots of bulb onion and synthesized a normalized cDNA library. A total of 20,000 clones were randomly selected and sequenced, and 11,008 unique sequences (unigenes) were obtained. From these ESTs, they identified 336 SSRs and the frequency of SSR sequence represented 1 SSR/25 kb sequence.

In garlic, Kim et al. (2009) obtained 7720 nonredundant garlic ESTs, however, they did not describe EST-SSRs. Subsequently, Ipek et al. (2015) found 132 SSR motifs in 125 ESTs by screening of garlic ESTs reported by Kim et al. (2009).

Table 8.2 Development of genic SSRs in *Alliums*

Species	Methods	Number of clones	References
A. cepa	EST sequencing	336	Kuhl et al. (2004)
	Transcriptome shotgun assemblies	4,437	Han et al. (2016)
A. fistulosum	Transcriptome shotgun assemblies	1,374	Liu et al. (2014)
	Transcriptome shotgun assemblies	2,074	Tsukazaki et al. (2015)
	Transcriptome shotgun assemblies	1,892	Sun et al. (2016)
A. sativum	EST sequencing	125	Ipek et al. (2015) EST sequences was obtained from garlic EST database (Kim et al. 2009)
A. tuberosum	Transcriptome shotgun assemblies	2,125	Zhou et al. (2015)

Transcriptome sequencing utilizing NGS is a powerful and cost-efficient approach for the discovery of novel genes, especially in plants where full-genome sequencing is not economically feasible. In *Allium*, Sun et al. (2012) were the first to obtain 127,933 unigenes from garlic by transcriptome sequencing utilizing NGS. Although they did not describe transcriptome shotgun assembly (TSA)-derived SSRs (TSA-SSRs), the data is available for identifying SSR motifs.

Recently, several thousands of TSA-SSRs were newly identified in bulb onion (Han et al. 2016), bunching onion (Liu et al. 2014; Tsukazaki et al. 2015; Sun et al. 2016), and Chinese chive (*A. tuberosum* Rottl. ex Spr., also called as *A. ramosum* L.) (Zhou et al. 2015) by transcriptome sequencing utilizing NGS (Table 8.2). For example, Liu et al. (2014) isolated RNA samples from mature leaves of waxy and glossy cultivars of bunching onion, and RNA-Seq was conducted using Hiseq 2000 (Illumina). A total of 42,881 unigenes were obtained by assembling from 47.3 M read pairs containing 9.55 Gbps, and 1558 SSRs were identified.

Tsukazaki et al. (2015) applied two NGS technologies: the GS-FLX and HiSeq 2000 systems. For 454 GS-FLX sequencing, total RNAs were isolated from nine tissue types (two-week-old seedlings, leaf, roots, basal meristem, immature flower bract, mature bract (about one week before anthesis), opened flowers, immature fruit, and sliced pseudostem), synthesized a full-length normalized cDNA library, and then sequenced. A total of 42,511 contigs were obtained from the sequence analysis of 1.35 M high-quality reads. For sequencing using the HiSeq 2000 system, eight samples from three bunching onion inbred lines were used to construct cDNA libraries. Each library was constructed using 91 bps of the paired-end sequence method. A total of 121,354 unigenes were obtained from the analysis of 25.9–27.7 M clean reads containing 2.33–2.50 Gbps. Finally, a TSA set of 54,903 unigenes was obtained by the integration of both the data sets. From these unigenes, 2396 di- to pentanucleotide SSRs among 2074 unigenes were found (3.8%). The estimated frequency of SSR motifs detection was 0.14 SSR/kbp. In the case of bulb onion, the frequency of SSRs was similar to the results obtained for ESTs (2.8%; Kuhl et al. 2004) and genomic skim sequencing (3%; Baldwin et al. 2012a).

8.4 Comparison of gSSR and cDNA-Derived SSRs

8.4.1 Core Sequences

In general, plants are rich in AT/TA repeats followed by GA/CT repeats, whereas GT/CA repeat is the most common in mammals (Lagercrantz et al. 1993; Kalia et al. 2011). AT/TA repeats are difficult to isolate from genomic libraries because they are palindromic (Powell et al. 1996). Thus, $(GA/CT)_n$ and $(GT/CA)_n$ probes were mainly used for plaque screening or SSR-enrichment to isolate gSSRs.

In bulb onion, Fischer and Bachmann (1998, 2000) developed 30 bulb onion SSR markers, the majority of which had GT/CA repeats (21 contained GT/CA repeats, while 3 contained GA/CT_n). However, they did not describe the $(GT/CA)_n$: $(GT/CA)_n$ ratio, since they did not $(GA/CT)_n$ probe for SSR-enrichment. In bunching onion, Song et al. (2004) revealed that GT/CA repeats are surprisingly more frequent than GA repeats (49 contained GT/CA cores, while only one contained GA/CT_n). Tsukazaki et al. (2007) also showed that 74.1% of the bunching onion gSSRs had GT/CA cores, while only 17.5% had GA/CT cores (Table 8.3). These results confirm that GT/CA repeats are far more abundant than GA/CT repeats in the bunching onion genome. In garlic, the GT/TG repeats are also more frequently identified (27.0%) than GA/CT repeats (11.5%) (Ma et al. 2009), nevertheless, they did not use $(GA/CT)_n$ as a probe for SSR-enrichment.

On the other hand, Baldwin et al. (2012a) revealed that the 65% of all repeats were AT cores by 6.6 Mbps of genomic skim sequence. This strategy is not necessary for SSR-enrichment, it will be valuable for identifying new SSRs because SSRs with AT cores are not fully identified in *Allium*.

As for the genic-SSRs, the frequency of trinucleotide is higher than that of gSSRs. Kuhl et al. (2004) revealed that trinucleotide SSRs (60%) were more than dinucleotide (35%) of the total detected EST-SSRs in bulb onion. Similar result (33.9% of trinucleotides and 13.9% of dinucleotides) was obtained by Liu et al. (2014), (48.3 and 20.9%) Zhou et al. (2015), (50.9 and 22.0%) Sun et al. (2016), and (62.8 and 28.2%) Han et al. (2016). Whereas, there are some reports that dinucleotide cores are more abundant than trinucleotide in cDNA-derived SSRs by Ipek et al. (2015) (60% of dinucleotides and 26%

Table 8.3 The comparison of core motifs in bunching onion SSRs

	gSSRs (Tsukazaki et al. 2007)				genic SSRs (Tsukazaki et al. 2015)			
Motif	No.	Frequency (%)	Frequency (%)	Average repeats	No.	Frequency (%)	Frequency (%)	Average repeats
2bp			95.7	10.4			63.6	5.6
GT	1,331	74.1		10.5	368	15.4		5.7
GA	314	17.5		10.4	530	22.1		5.7
AT	151	4.1		7.9	614	25.6		5.5
GC					12	0.5		5.3
3bp			4.3	7.3			35.2	5.6
AAG	51	2.8		7.0	264	11.0		5.7
CCT	2	0.1		6.0	116	4.8		5.7
AGC	24	1.3		7.8	84	3.5		5.7
AAT					71	3.0		5.7
ATC					70	2.9		5.5
ATG					66	2.8		5.5
ACG					57	2.4		5.5
GGT	1	0.1		6.0	52	2.2		5.5
AAC					48	2.0		5.6
CCG					16	0.7		5.5
4bp					24		1.0	5.5
5bp					4		0.2	5.0
Total	1,796		100.0	10.2	2,396		100.0	5.6

of trinucleotides) and Tsukazaki et al. (2015) (63.6 and 35.2%), respectively.

Among the dinucleotide genic-SSRs, the frequency of GA/CT as a core nucleotide (22.1%) was higher than GT/CA (15.4%) in bunching onion transcriptome sequences (Tsukazaki et al. 2015) (Table 8.3). Similar results were obtained for bulb onion ESTs (GA/CT 10.1%, GT/CA 9.2%) Kuhl et al. (2004) and bunching onion TSAs (GA/CT 8.7%, GT/CA 6.5%) Sun et al. (2016), whereas it differs in Chinese chive TSAs (GT/CA 13.6% and GA/CT 11.5%) Zhou et al. (2015).

In plants, the most frequent motif in the trinucleotide SSRs is AAG/CTT, whereas CCG/GCC is the most common in cereals (Li et al. 2002). Tsukazaki et al. (2015) reported that the most abundant motif in the trinucleotide SSRs was AAG/CTT in bunching onion (11.0%, Table 8.3). Since similar results were obtained for bulb onion (12.5%, Kuhl et al. 2004; 5.9%, Han et al. 2016), bunching onion (7.8%, Liu et al. 2014; 14.7%, Sun et al. 2016), and Chinese chive (13.3%, Zhou et al. 2015), AAG/CTT motif seems to be the most frequent in *Alliums*.

The GC content of bunching onion nuclear DNA was previously estimated at 37.9% (Kirk et al. 1970) and 39.8% (Ricroch et al. 2005), making it one of the lowest known values for angiosperms (Kirk et al. 1970; Stack and Coming 1979). The GC content of bulb onion ESTs was estimated at 41.9% and 30.4% in total and intron regions, respectively (Kuhl et al. 2004). They concluded that *Asparagales* (bulb onion) are more similar to eudicots (*Arabidopsis*) than to *Poales* (rice) for the genomic characteristics such as exon usage and mean GC content, GC distribution, and relative GC content at each codon position (Kuhl et al. 2004). It is considered that these reasons are in accordance with the difference of the most frequent trinucleotide motif between *Allium* and cereals.

8.4.2 Polymorphisms and Transferabilities

The sequences flanking the SSR motifs in the genome are conserved within species and often across the species within a genus and even across related genera (Gupta and Varshney 2000). Both genomic and genic-SSR markers can be transferred across species. However, genic-SSR markers are expected to have a high transferability due to the conservation of transcribed regions among related species (Kalia et al. 2011). On the other hand, the polymorphism of genic-SSRs seems to lower than that of gSSRs for distinguishing the closely related genotypes in general. Thus, gSSRs are superior over genic SSRs for fingerprinting or cultivar classification study especially in intraspecies.

Studies on the polymorphisms of SSR markers in *Allium* are listed in Table 8.4. gSSRs are often recalcitrant to application across species due to their low transferabilities. Nunome et al. (2009) reported that the more than 90% of eggplant (*Solanum melongena*) gSSR markers were well amplified in related species (*S. linnaeanum* and *S. incanum*), while the transferabilities decreased in tomato (55%) and pepper (31%). In *Allium*, bulb onion EST-SSR markers gave high (>75%) transferability in bunching onion and garlic, whereas less (40%) in Chinese chive (Tsukazaki et al., unpublished data). In contrast, transferability of bunching onion gSSRs decreased in these species (50% in bulb onion and garlic, and 16% in Chinese chive) using 360 primer sets (Tsukazaki et al., unpublished data). The transferability of bulb onion gSSRs in bunching onion inbred lines was low (43.3%), whereas that of bulb onion EST was 75.1% (Tsukazaki et al. 2008). Baldwin et al. (2012a) reported that only 25% of the bulb onion gSSR markers amplified in bunching onion and *A. roylei*. Khar et al. (2011) used 30 bulb onion gSSRs developed by Fischer and Bachmann (2000) to evaluate the genetic diversity of Indian onion, and only three were able to amplify in bulb onion lines as compared to bulb onion EST-SSR where 16 of the 30 primers were amplified.

Tsukazaki et al. (2008) reported that the efficiency of polymorphism between bunching onion inbred lines using bunching onion gSSRs was 28.3%. The frequency of polymorphic markers between bunching onion inbred lines in

Table 8.4 Polymorphisms of SSR markers in *Alliums*

Species	Origin	Number of markers tested	Number of accessions / lines	No. of alleles	PIC	*He*	References	Remarks
A. cepa	gSSRs	15	83 from 13 species	18.3	-	-	Fischer and Bachmann (2000)	
	gSSRs	20	24 bulb onion populations (12 individuals / cultivar)	7.9	0.45	0.49	Baldwin et al. (2012a)	Selected polymorphic markers
	EST-SSRs	24	16	2.7	-	-	Mahajan et al. (2009)	
	gSSRs and EST-SSRs	3 gSSRs, 16 EST-SSRs	34 bulb onion cultivars and 12 wild species	4.3	0.81	-	Khar et al. (2011)	
A. fistulosum	gSSRs	33	9	3.3	0.59	-	Song et al. (2004)	
	gSSRs	14	8 bunching onion cultivars (33 individuals / cultivar)	4.3	0.59	-	Tsukazaki et al. (2006)	
	gSSRs	96	5	3.2	0.55		Tsukazaki et al. (2007)	Including 7 monomorphic markers
	gSSRs	33	32 cultivars (24 individuals / cultivar)	10.4	-	0.71	Tsukazaki et al. (2010)	Including 1 bulb onion gSSR and 3 bulb onion EST-SSRs
	gSSRs	153	11	3.7	0.50	-	Tsukazaki et al. (2011)	
	TSA-SSRs	395	8	2.4	-	0.35	Tsukazaki et al. (2015)	
A. sativum	gSSRs	8	90	8.0	0.62	0.66	Ma et al. (2009)	Selected polymorphic markers
	gSSRs	10	75	4.4	0.52	0.58	Cunha et al. (2012)	Selected polymorphic markers
	gSSRs	8	613	14.1	0.63		Zhao et al. (2011)	
	gSSRs	24	130	5.3	-	0.56	Cunha et al. (2014)	
	EST-SSRs	31	26	5.0	-	0.60	Ipek et al. (2015)	EST sequences was obtained from garlic EST database (Kim et al. (2009))
A. mongolicum	gSSRs	13	60	4.3	-	0.71	Zhang et al. (2014)	

bulb onion EST-SSRs was low (11.0%) and it was comparable to that in bulb onion gSSRs (15.4%). On the other hand, Ma et al. (2009) reported that only eight of 128 primer pairs produced reproducible polymorphic bands in the garlic germplasm. It is considered that the low efficiency of polymorphic markers is due to its clonal reproduction habit. Nevertheless, these eight polymorphic SSR loci detected on an average 8.0 alleles per locus with an average PIC value 0.62.

The degree of polymorphisms in gSSRs tends to be higher than in genic-SSRs, although it is dependent upon the number of markers and population size. For example, in the case of bunching onion inbred lines, the average number of alleles and PIC value or degree of heterogeneity (*He*) were 2.4 and 0.35 using TSA-SSRs in contrast to 3.2–3.7 and 0.50–0.59 using gSSRs (Song et al. 2004; Tsukazaki et al. 2007, 2011, 2015), respectively. These values increased when population size became larger with wide genetic diversity in *Allium* (Tsukazaki et al. 2006, 2010; Zhao et al. 2011, Baldwin et al. 2012a) (Table 8.4).

8.4.3 Application Using SSR Markers

As mentioned above, SSRs are powerful molecular markers because of their codominant mode of inheritance, higher reliability, and abundance in genomes. Thus, SSRs are used for constructing linkage maps, marker-assisted selection, phylogenetic studies, genomic synteny analysis, and cultivar identification (Jones et al. 1997; Bredemeijer et al. 1998).

8.4.4 Linkage Map and QTL Analysis

The studies of linkage map constructions and/or QTL analysis using SSR markers in *Allium* are listed in Table 8.5. These studies are limited to bulb onion and bunching onion. The first public linkage map in *Allium* was developed by King et al. (1998). Although this bulb onion linkage map was based on RAPD and RFLP markers, Martin et al. (2005) subsequently added 35 EST-SSRs, 43 single nucleotide polymorphisms (SNPs), and 4 insertion–deletion (InDel) markers on this map. Martin et al. (2005) also assigned each linkage group to bulb onion chromosome by using a complete set of *Allium* monosomic addition lines (AMALs; $2n = 16 + 1$, FF + 1A to FF + 8A) developed by Shigyo et al. (1996).

McCallum et al. (2006) constructed a partial onion linkage map based on bulb onion EST-SSRs and compared to a linkage map of Martin et al (2005). This map was subsequently used to detect QTLs for fructan content and bulb pungency (McCallum et al. 2006, 2007).

In bunching onion, Ohara et al. (2005) constructed linkage maps based primarily on AFLP markers with 13 gSSRs using reciprocally backcrossed progenies. Twenty-four SSR markers were subsequently added on these maps and conducted QTL analysis of seedling growth (Ohara et al. 2009).

Tsukazaki et al. (2008) constructed a gSSR-based linkage map consisting of 17 groups (Fig. 8.3). This is the first report of a linkage map mainly based on SSR markers in the genus *Allium*. With the 104 anchor markers, including 82 bunching onion gSSRs and 11 bulb onion EST-SSRs, whose chromosome assignments were identified in *A. cepa* and/or *A. fistulosum*, via the use of a complete set of AMALs (Shigyo et al. 1996) and six single-alien deletion lines (Hang et al. 2004; Yaguchi et al. 2009), all linkage groups were connected to the 8 basic *Allium* chromosomes. This linkage map was then used as a reference for constructing linkage maps of the other bunching onion populations (Tsukazaki et al. 2011; Wako et al. 2016), and QTL analysis of pseudostem pungency (Tsukazaki et al. 2012) and bolting time (Wako et al. 2016). In addition, Tsukazaki et al. (2015) reported a new bunching onion linkage map mainly based on genic and genomic SSRs which consisted of 17 linkage groups (Fig. 8.4). This map was then applied for QTL analysis of morphological traits and pseudostem pigmentation (Tsukazaki et al. 2017). More than 200 genic-markers derived from both bunching onion TSAs and bulb onion ESTs were located on this map, however, a trial

Table 8.5 Linkage map and/or QTL analysis using SSR markers in *Alliums*

	Mapping Population	Number of mapped SSR markers	Application	References
A. cepa	F2	35 bulb onion EST-SSRs with RAPDs, RFLPs and EST-derived SNPs and InDels	Linkage map	Martin et al. (2005)
	F2	Not described in detail	QTL analysis for bulb fructan content	McCallum et al. (2006)
	F2	Not described in detail	QTL analysis for bulb pungency	McCallum et al. (2007)
	F2	17 bulb onion EST-SSRs with EST-derived CAPSs, HRM and InDels	Linkage map	Baldwin et al. (2012b)
A. fistulosum	BC1	13 bulb onion gSSRs, 1 bulb onion gSSR with AFLPs	Linkage map	Ohara et al. (2005)
	F2	212 bunching onion gSSRs, 1 bulb onion gSSR, 16 bulb onion EST-SSRs	Linkage map	Tsukazaki et al. (2008)
	BC1	28 bulb onion gSSRs, 1 bulb onion gSSR, 10 bulb onion EST-SSRs with AFLPs	QTL analysis for seedling growth	Ohara et al. (2009)
	F2	199 bunching onion gSSRs, 8 bulb onion EST-SSRs	Linkage map and chromosome assignment	Tsukazaki et al. (2011)
	F2	207 bunching onion gSSRs, 8 bulb onion EST-SSRs with RAPDs	QTL analysis for pseudostem pungency	Tsukazaki et al. (2012)
	F2	81 bunching onion TSA-SSRs, 96 bunching onion gSSRs, 33 bulb onion EST-SSRs with TSA-derived CAPSs amd InDels	Linkage map	Tsukazaki et al. (2015)
			QTL analysis for morphological traits and pseudostem pigmentation	Tsukazaki et al. (2017)
	F2	249 bunching onion gSSRs, 2 bulb onion EST-SSRs	QTL analysis for bolting time	Wako et al. (2016)

to compare linkage map between bunching onion and bulb onion (Duangjit et al. 2013; Baldwin et al. 2014) via genic-markers was very difficult because of few orthologous markers.

8.4.5 Phylogeny Analysis, Genomic Diversity, and Cultivar Identification

SSR markers are more polymorphic than the other markers because they are multi-allelic, which are different from bi-allelic markers (SNPs, etc.). In the case of bunching onion, the degree of polymorphism within inbred lines could be ranked as follows: gSSRs > TSA-SSRs > TSA-derived InDels (Tsukazaki et al. 2007, 2015). Hence, the polymorphic data using SSR markers is useful for characterization of accessions in plant germplasm collections and taxonomic studies. Although gSSRs are often recalcitrant to application across species due to their lower transferabilities, some phylogenic studies have been conducted not only in intraspecies but also among *Allium spp.*

The studies of phylogeny analysis and/or cultivar classification using SSR markers in *Allium* are listed in Table 8.6. Although Fischer and Bachmann (2000) reported the first phylogeny analysis using SSR markers, these studies have been mainly conducted in garlic and related

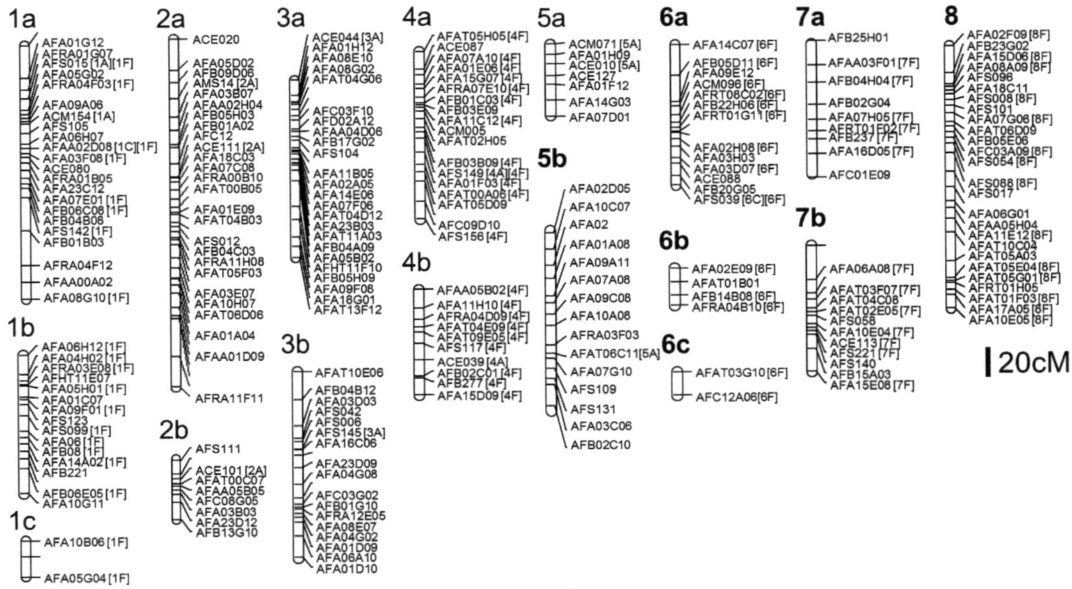

Fig. 8.3 Bunching onion chromosome map in the DJ population mainly based on bunching onion genomic simple sequence repeat (gSSR) markers (Revised from Tsukazaki et al. 2008). It consists of 17 linkage groups with 230 SSRs (212 bunching onion SSRs and 17 bulb onion SSRs) and 26 bulb onion-derived non-SSR markers (4 InDels, 17 CAPSs, and 4 dCAPSs) covering 2116 cM. The genetic markers used to assign linkage groups to their relevant chromosomes are marked with [1A] to [8A] (assigned to *A. cepa* chromosomes) and [1F] to [8F] (to *A. fistulosum* chromosomes)

species because of its clonal reproduction habit. Ma et al. (2009) applied eight gSSRs to 90 garlic accessions and clustered into two distinct groups which were moderate related to geographical location. Ipek et al. (2015) analyzed 31 garlic accessions including *A. longicuspis* using 26 EST-SSRs. Twenty-one of these accessions had been previously analyzed by Ipek et al. (2003) who applied AFLP, RAPD, and isozyme markers. The clusters based on genetic relationships between garlic accessions by EST-SSRs were reflected in those by AFLP markers.

On the other hand, Fischer and Bachmann (2000) applied 30 gSSRs to 83 onion accessions and 24 from 16 species including three sections *Cepa*, *Schoenoprasm* and *Annuloprasum* within the subgenus *Rhizirideum*. A dendrogram based on 83 onion accessions at 15 SSR loci partly reflected the geographical origins of landraces. McCallum et al. (2008) used 56 EST-SSRs and four gSSRs for 89 inbred and open-pollinated populations of wide geographical adaptation, and principal coordinate analysis was performed. In addition, Baldwin et al. (2012a) scored the 20 bulb onion gSSR markers in 12 individuals from 24 bulb onion populations covering latitude 44° N to 37°S and used for discriminant analysis of principal components analysis. They revealed high degree of within-population heterozygosity (H_e= 0.19–0.47) and within-population variation (F_{st} = 0.26).

In bunching onion, Li et al. (2008) applied both gSSR markers and sequence-related amplified polymorphism (SRAP) markers for analyzing genetic diversity. Tsukazaki et al. (2010) conducted a population diversity study using SSR markers. They applied 30 bunching onion gSSR to every 24 individuals from 30 bunching onion open pollination cultivars originating mostly from Japan. The analysis of molecular variance (AMOVA) revealed that 77% of molecular variance was within cultivars (variation among cultivars was 23%). From a cluster analysis based on genetic distance from allele frequencies of each locus, bunching onion cultivars were classified into three clusters. These

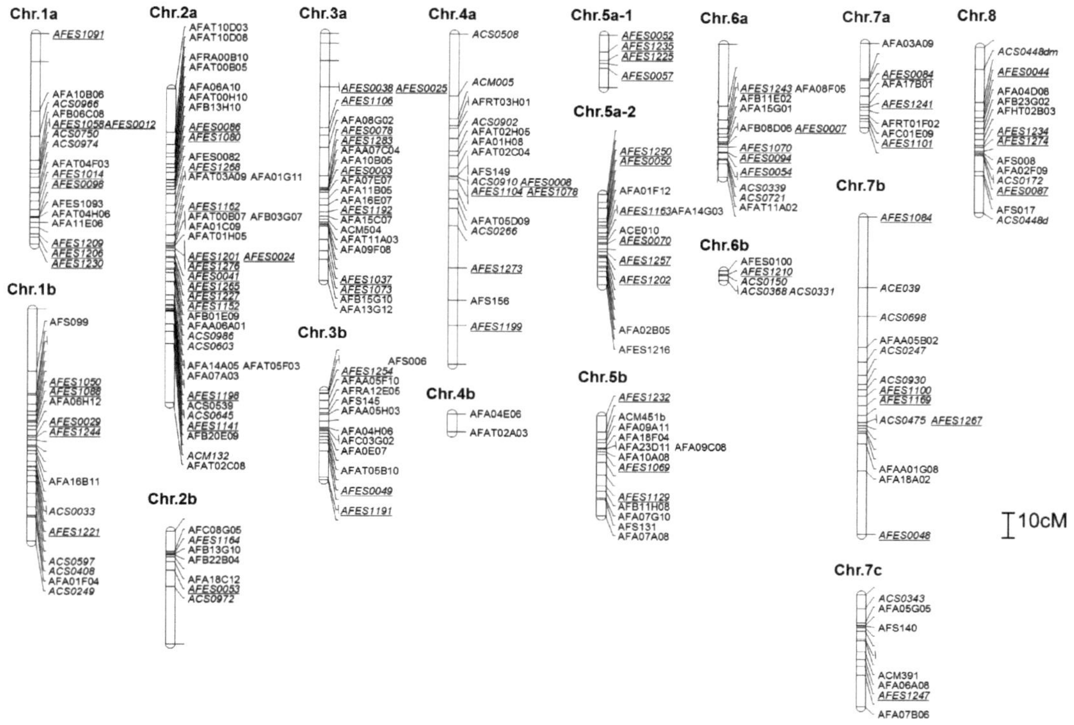

Fig. 8.4 Bunching onion linkage map in the FA population mainly based on bunching onion Transcriptome shotgun assembly (TSA)-derived markers (revised from Tsukazaki et al. 2015). It consists of 17 linkage groups with 367 markers (190 bunching onion TSAs, 96 bunching onion genomic SSRs, 39 bulb onion ESTs, and 4 other markers) covering 1150 cM. Bunching onion TSA-derived SSRs are indicated in *italic* with underlines and bulb onion EST-SSRs are indicated in *italic*

Table 8.6 Phyrogenic analysis and cultivar classification using SSR markers in *Alliums*

	Population size	Number of SSR markers used	Application	References
A. cepa	83 bulb onion accessions	15 bulb onion gSSRs	Phylogeny analysis	Fischer and Bachmann (2000)
	50< seedlings from 35 bulb onion populations	58 bulb onion EST-SSRs	Phylogeny analysis	Jakše et al. (2005)
	82 lines	56 bulb onion EST-SSRs, 4 bulb onion gSSRs	Phylogeny analysis	McCallum et al. (2008)
	16 lines	24 bulb onion EST-SSRs	Phylogeny analysis	Mahajan et al. (2009)
	12 individuals from 24 bulb onion populations	20 bulb onion gSSRs	Phylogeny analysis	Baldwin et al. (2012a)
A. fistulosum	20 lines	13 bulb onion gSSRs, 4 bulb onion gSSR	Phylogeny analysis	Li et al. (2008)
	24 individuals from 32 bunching onion cultivars	33 bunching onion gSSRs	Cultivar classification and cultivar identification	Tsukazaki et al. (2010)

(continued)

Table 8.6 (continued)

	Population size	Number of SSR markers used	Application	References
A. sativum	90 accessions	8 garlic gSSRs	Phylogeny analysis	Ma et al. (2009)
	95 core set including 3 related species	8 garlic gSSRs	Phylogeny analysis	Zhao et al. (2011)
	130 accessions	24 garlic gSSRs	Phylogeny analysis	Cunha et al. (2014)
	26 accessions	31 garlic EST-SSRs	Phylogeny analysis	Ipek et al. (2015)

clusters corresponded to the expected three cultivar groups, that is, "Kaga", "Senju", and "Kujo".

Tsukazaki et al. (2010) also tried to identify cultivar by assignment test using these genotype data. In the assignment test at the individual level, 89.1% of the individuals were assigned to their original cultivar. When the assignment was conducted based on groups each consisting of four individuals, the percentage of correct assignments was considerably improved (99.3%). These results suggested that the assignment test approach will be useful for cultivar identification in allogamous bunching onions, which have large within-cultivar genetic diversity.

8.5 Conclusions

As described above, SSRs are powerful molecular markers because of their codominant mode of inheritance, higher reliability, and abundance in genomes, and therefore, SSRs are widely applied for constructing linkage maps, QTL analysis, phylogenetic studies, and cultivar identification in *Allium* studies.

NGS technologies have enabled large amounts of sequence data to be acquired in *Alliums*. These technologies will be a mainstream for developing new SSR markers. In addition, the public NGS data should also be fully exploited for developing SSR markers. However, *Alliums* have a large and highly heterozygous genome making the sharing of germplasm and analysis of sequencing data complicated. Recently, Khosa et al. (2016) reported a reference transcriptome catalogue from multi organs of doubled haploid bulb onion line, "CUDH2107". However, the author did not describe SSRs in this study, several thousands of SSR markers may be obtained by screening of sequence data.

To date, detailed comparison of linkage maps from bunching onion and bulb onion has been hindered thus far by the limited numbers of common markers. It is considered that the macrosynteny is highly conserved between bunching onion and bulb onion by assigning the common markers (Tsukazaki et al. 2008). However, it was reported that a minimum of 2 heteromorphic bivalents were observed in meiotic analysis of interspecific hybrids between bunching onion and bulb onion (Peffley 1986), suggesting that chromosomal mutations such as translocation and inversion might be involved in the speciation between bunching onion and bulb onion. For detailed studies of such chromosomal mutations, microsynteny will be facilitated at the genome-wide level by plotting many more common (orthologous) markers onto the maps of both species, and these markers will also help to further our understanding of *Allium* comparative genomics.

References

Abdelkrim J, Robertson BC, Stanton JAL, Gemmell NJ (2009) Fast, cost-effective development of species-specific microsatellite markers by genomic sequencing. Biotechniques 46:185–192

Baldwin S, Pither-Joyce M, Wright K, Chen L, McCallum J (2012a) Development of robust genomic simple sequence repeat markers for estimation of genetic diversity within and among bulb onion (*Allium cepa* L.) populations. Mol Breed 30:1401–1411

Baldwin S, Revanna R, Thomson S, Pither-Joyce M, Wright K, Crowhurst R, Fiers M, Chen L, Macknight R, McCallum JA (2012b) Toolkit for bulk PCR-based marker design from next-generation sequence data: application for development of a framework linkage map in bulb onion (*Allium cepa* L.). BMC Genom 13:637

Baldwin S, Revanna R, Pither-Joyce M, Shaw M, Wright K, Thomson S, Moya L, Lee R, Macknight R, McCallum JA (2014) Genetic analyses of bolting in bulb onion (*Allium cepa* L.). Theor Appl Genet 127:535–547

Bradeen JM, Havey MJ (1995) Randomly amplified polymorphic DNA in bulb onion and its use to assess inbred integrity. J Am Soc Hortic Sci 120:752–758

Bredemeijer GMM, Arens P, Wouters D, Visser D, Vosman B (1998) The use of semi-automated fluorescent microsatellite analysis for tomato cultivar identification. Theor Appl Genet 97:584–590

Castoe TA, Poole AW, Gu W, Konin APJ, Daza JM, Smith EN, Pollock DD (2010) Rapid identification of thousands of copperhead snake (*Agkistrodon contortrix*) microsatellite loci from modest amounts of 454 shotgun genome sequence. Mol Ecol Res 10:341–347

Crockett PA, Bhalla PL, Lee CK, Singh MB (2000) RAPD analysis of seed purity in a commercial hybrid cabbage (*Brassica oleracea* var. *capitata*) cultivar. Genome 43:317–321

Cunha CP, Hoogerheide ES, Zucchi MI, Monteiro M, Pinheiro JB (2014) SSR-based genetic diversity and structure of garlic accessions from Brazil. Genetica 142:419–431

Cunha CP, Resende FV, Zucchi MI, Pinheiro JB (2012) New microsatellite markers for garlic, *Allium sativum* (*Alliaceae*). Am J Bot 99:e17–e19

Duangjit J, Bohanec B, Chan AP, Town CD, Havey MJ (2013) Transcriptome sequencing to produce SNP-based genetic maps of onion. Theor Appl Genet 126:2093–2101

Edwards KJ, Barker JHA, Daly A, Jones C, Karp A (1996) Microsatellite libraries enriched for several microsatellite sequences in plants. Biotechniques 20:758–760

Fischer D, Bachmann K (1998) Microsatellite enrichment in organisms with large genomes (*Allium cepa* L.). Biotechniques 24:796–802

Fischer D, Bachmann K (2000) Onion microsatellites for germplasm analysis and their use in assessing intra- and interspecific relatedness within the subgenus *Rhizirideum*. Theor Appl Genet 101:153–164

Frary A, Xu YM, Liu JP, Mitchell S, Tedeschi E, Tanksley S (2005) Development of a set of PCR-based anchor markers encompassing the tomato genome and evaluation of their usefulness for genetics and breeding experiments. Theor Appl Genet 111:291–312

Gupta PK, Varshney RK (2000) The development and use of microsatellite markers for genetic analysis and plant breeding with emphasis on bread wheat. Euphytica 113:163–165

Han J, Thamilarasan SK, Natarajin S, Park JI, Chung MY, Nou IS (2016) de novo assembly and transcriptome analysis of bulb onion (*Allium cepa* L.) during cold acclimation using constasting genotypes. PLoS ONE 11:e0161987

Hang TTM, Shigyo M, Yamauchi N, Tashiro Y (2004) Production and characterization of alien chromosome additions in shallot (*Allium cepa* L. Aggregatum group) carrying extra chromosome(s) of Japanese bunching onion (*A. fistulosum* L.). Genes Genet Syst 79:263–269

Ipek M, Ipek A, Simon PW (2003) Comparison of AFLPs, RAPD markers, and isozymes for diversity assessment of garlic and detection of putative duplicates in germplasm collections. J Am Soc Hortic Sci 128:246–252

Ipek M, Ipek A, Almquist SG, Simon PW (2005) Demonstration of linkage and development of the first low-density genetic map of garlic, based on AFLP markers. Theor Appl Genet 110:228–236

Ipek M, Sahin N, Ipek A, Cansev A, Simon PW (2015) Development and validation of new SSR markers from expressed regions in the garlic genome. Sci Agric 72:41–46

Jakše J, Martin W, McCallum J, Havey MJ (2005) Single nucleotide polymorphisms, indels, and simple sequence repeats for onion cultivar identification. J Am Soc Hortic Sci 130:912–917

Jones CJ, Edwards KJ, Castaglione S, Winfield MO, Sala F, van de Wiel C, Bredemeijer G, Vosman B, Matthes M, Daly A, Brettschneider R, Bettini R, Buiatti M, Maestri E, Malcevschi A, Marmiroli N, Aert R, Volckaert G, Rueda J, Linacero R, Vazquez A, Karp A (1997) Reproducibility testing of RAPD, AFLP and SSR markers in plants by a network of European laboratories. Mol Breed 3:381–390

Kalia RK, Rai MK, Kalia S, Singh R, Dhawan AK (2011) Microsatellite markers: an overview of the recent progress in plants. Euphytica 177:309–334

Khar A, Lawande KE, Negi KS (2011) Microsatellite marker based analysis of genetic diversity in short day tropical Indian onion and cross amplification in related *Allium* spp. Genet Resour Crop Evol 58:741–752

Kim DW, Jung TS, Nam SH, Kwon HR, Kim A, Chae SH, Choi SH, Kim DQ, Kim RN, Park HS (2009) GarlicESTdb: an online database and mining tool for garlic EST sequences. BMC Plant Biol 9:61

King JJ, Bradeen JM, Bark O, McCallum JA, Havey MJ (1998) A low-density map of onion reveals a role for tandem duplication in the evolution of an extremely large diploid genome. Theor Appl Genet 96:52–62

Kirk JTO, Rees H, Evans G (1970) Base composition of nuclear DNA within the genus *Allium*. Heredity 25:507–512

Khosa JS, Lee R, Bräuning S, Lord J, Pither-Joyce M, McCallum J, Macknight RC (2016) Doubled haploid 'CUDH2107' as a reference for bulb onion (*Allium cepa* L.) research: development of a transcriptome catalogue and identification of transcripts associated with male fertility. PLoS ONE 11:e0166568

Kuhl JC, Cheung F, Yuan Q, Martin W, Zewdie Y, McCallum J, Catanach A, Rutherford P, Sink KC, Jenderk M, Prince JP, Town CD, Havey MJ (2004) A unique set of 11,008 onion expressed sequence tags reveals expressed sequence and genomic differences between the monocot orders *Asparagales* and *Poales*. Plant Cell 16:114–125

Lagercrantz U, Ellegren H, Andersson L (1993) The abundance of various polymorphic microsatellite motifs differs between plants and vertebrates. Nucleic Acid Res 21:1111–1115

Li HZ, Yin YP, Zhang CQ, Zhang M, Li JM (2008) Comparison of characteristics of SRAP and SSR markers in genetic diversity analysis of cultivars in *Allium fistulosum* L. Seed Sci Tech 36:423–434

Li YC, Korol AB, Fahima T, Beiles A, Nevo E (2002) Microsatellites: genomic distribution, putative functions and mutational mechanisms: a review. Mol Ecol 11:2453–2465

Li YC, Korol AB, Fahima T, Beiles A, Nevo E (2004) Microsatellites within genes: structure, function, and evolution. Mol Biol Evol 21:991–1007

Liu Q, Wen C, Zhao H, Zhang L, Wang J, Wang Y (2014) RNA-seq reveals leaf cuticular wax-related genes in Welsh onion. PLoS ONE 9:e113290

Ma KH, Kwang JG, Zhao W, Dixit A, Lee GA, Kim HH, Chung IM, Kim NS, Lee JS, Ji JJ, Kim TS, Park YJ (2009) Isolation and characteristics of eight novel polymorphic microsatellite loci from the genome of garlic (*Allium sativum* L.). Sci Hortic 122:355–361

Mahajan V, Jakse J, Havey MJ, Lawande KE (2009) Genetic fingerprinting of onion cultivars using SSR markers. Indian J Hortic 66:62–68

Martin WJ, McCallum J, Shigyo M, Jakše J, Kuhl JC, Yamane N, Pither-Joyce M, Gökçe AF, Sink KC, Town CD, Havey MJ (2005) Genetic mapping of expressed sequences in onion and in silico comparisons with rice show scant colinearity. Mol Gen Genomics 274:197–204

McCallum J, Clarke A, Pither-Joyce M, Shaw M, Butler R, Brash D, Scheffer J, Sims I, van Heusden S, Shigyo M, Havey M (2006) Genetic mapping of a major gene affecting onion bulb fructan content. Theor Appl Genet 112:958–967

McCallum J, Pither-Joyce M, Shaw M, Kanel F, Davis S, Butler R, Schffer J, Jakše J, Havey MJ (2007) Genetic mapping of sulfur assimilation genes a QTL for bulb onion pungency. Theor Appl Genet 114:815–822

McCallum J, Thompson S, Pither-Joyce M, Kenel F (2008) Genetic diversity analysis and single-nucleotide polymorphism marker development in cultivated bulb onion based on expressed sequence tag-simple sequence repeat markers. J Am Soc Hortic Sci 133:810–818

Nunome T, Negoro S, Miyatake K, Yamaguchi H, Fukuoka H (2006) A protocol for the construction of microsatellite enriched genomic library. Plant Mol Biol Rep 24:305–312

Nunome T, Negoro S, Kono I, Kanamori H, Miyatake K, Yamaguchi H, Ohyama A, Fukuoka H (2009) Development of SSR markers derived from SSR-enriched genomic library of eggplant (*Solanum melongena* L.). Theor Appl Genet 119:1143–1153

Ohara T, Song YS, Tsukazaki H, Wako T, Nunome T, Kojima A (2005) Construction of a genetic linkage map of Japanese bunching onion (*Allium fistulosum*) based on AFLP and SSR markers. Euphytica 144:255–263

Ohara T, Tsukazaki H, Song YS, Wako T, Yamashita K, Kojima A (2009) Mapping of quantitative trait loci controlling seedling growth in bunching onion *(Allium fistulosum* L.). J Jpn Soc Hortic Sci 78:436–442

Ohyama A, Asamizu E, Negoro S, Miyatake K, Yamaguchi H, Tabata S, Fukuoka H (2009) Characterization of tomato SSR markers developed using BAC-end and cDNA sequences from genome databases. Mol Breed 23:685–691

Peffley EB (1986) Evidence for chromosomal differentiation of *Allium fistulosum* and *A. cepa*. J Am Soc Hortic Sci 111:126–129

Powell W, Machray GC, Provan J (1996) Polymorphism revealed by simple sequence repeats. Trends Plant Sci 1:215–222

Ricroch A, Yockteng R, Brown SC, Nadot S (2005) Evolution of genome size across some cultivated *Allium* species. Genome 48:511–520

Shigyo M, Tashiro Y, Issiki S, Miyazaki S (1996) Establishments of a series of alien monosomic addition lines of Japanese bunching onion (*Allium fistulosum* L.) with extra chromosomes from shallot (*A. cepa* L. Aggregatum group). Genes Genet Syst 71:363–371

Song YS, Suwabe K, Wako T, Ohara T, Nunome T, Kojima A (2004) Development of microsatellite markers in bunching onion (*Allium fistulosum* L.). Breed Sci 54:361–365

Stack SM, Coming DE (1979) The chromosomes an DNA of *Allium cepa*. Chromosoma 70:161–181

Sun X, Zhou S, Meng F, Liu S (2012) De novo assembly and characterization of the garlic (*Allium sativum*) bud transcriptome by Illumina sequencing. Plant Cell Rep 31:1823–1828

Sun XD, Yu XH, Zhou SM, Liu SQ (2016) De novo assembly and characterization of the Welsh onion (*Allium fistulosum* L.) bud transcriptome using Illumina technology. Mol Genet Genomics 291:647–659

Tsukazaki H, Fukuoka H, Song YS, Yamashita K, Wako T, Kojima A (2006) Considerable heterogeneity in commercial F_1 varieties of bunching onion (*Allium fistulosum*) and the proposal of a breeding scheme for conferring genetic traceability using SSR markers. Breed Sci 56:321–326

Tsukazaki H, Nunome T, Fukuoka H, Kanamori H, Kono I, Yamashita K, Wako T, Kojima A (2007) Isolation of 1,796 SSR clones from SSR-enriched DNA libraries of bunching onion (*Allium fistulosum*). Euphytica 157:83–94

Tsukazaki H, Yamashita K, Yaguchi S, Masuzaki S, Fukuoka H, Yonemaru J, Kanamori H, Kono I, Hang TTH, Shigho M, Kojima A, Wako T (2008) Construction of SSR-based chromosome map in bunching onion (*Allium fistulosum*). Theor Appl Genet 117:1213–1223

Tsukazaki H, Honjo M, Yamashita K, Ohara T, Kojima A, Ohsawa R, Wako T (2010) Classification and identification of bunching onion (*Allium fistulosum*) varieties based on SSR markers. Breed Sci 60:139–152

Tsukazaki H, Yamashita K, Yaguchi S, Yamashita K, Hagihara T, Shigyo M, Kojima A, Wako T (2011) Direct determination of the chromosomal location of bunching onion and bulb onion markers using bunching onion-shallot monosomic additions and allotriploid-bunching onion single alien deletions. Theor Appl Genet 122:501–510

Tsukazaki H, Yaguchi S, Yamashita K, Shigyo M, Kojima A, Wako T (2012) QTL analysis for pseudostem pungency in bunching onion (*Allium fistulosum*). Mol Breed 30:1689–1698

Tsukazaki H, Yaguchi S, Sato S, Hirakawa H, Katayose Y, Kanomori H, Kurita K, Itoh T, Kumagai M, Mizuno S, Hamada M, Fukuoka H, Yamashita K, McCallum JA, Shigyo M, Wako T (2015) Development of transcriptome shotgun assembly-derived markers in bunching onion (*Allium fistulosum*). Mol Breed 35:55

Tsukazaki H, Yaguchi S, Yamashita K, Wako T (2017) QTL analysis of morphological traits and pseudostem pigmentation in bunching onion (*Allium fistulosum*). Euphytica 213:152

van Heusden AW, van Ooijen JW, Vrielink-van Ginkel R, Verbeek WHK, Wietsma WA, Kik C (2000) A genetic map of an interspecific cross *Allium* based on amplified fragment length polymorphism (AFLP) markers. Theor Appl Genet 100:118–126

Wako T, Tsukazaki H, Yaguchi S, Yamashita K, Ito S, Shigyo M (2016) Mapping of quantitative trait loci for bolting time in bunching onion (*Allium fistulosum* L.). Euphytica 209:537–546

Weising K, Winter P, Huttel B, Kahl G (1998) Microsatellite markers for molecular breeding. In: Basra AS (ed) Crop science: recent advances. The Haworth Press, pp 113–143

Wilkie SE, Isaac PG, Slater RJ (1993) Random amplified polymorphic DNA (RAPD) markers for genetic analysis in *Allium*. Theor Appl Genet 86:497–504

Yaguchi S, Hang TTM, Tsukazaki H, Hoa VQ, Masuzaki S, Wako T, Masamura N, Onodera S, Shimoi N, Yamauchi N, Shigyo M (2009) Molecular and biochemical identification of alien chromosome additions in shallot (*Allium cepa* L. Aggregatum group) carrying extra chromosome(s) of bunching onion (*A. fistulosum* L.). Genes Genet Syst 84:43–55

Zane I, Bergelloni L, Patarnello T (2002) Strategies for microsatellite isolation: a review. Mol Ecol 11:1–16

Zhang TY, Chen HK, Zhang CR (2014) Polymorphic microsatellite markers for *Allium mongolicum* Regel (*Amaryllidaceae*). Genes Genet Syst 89:133–136

Zhao WG, Chung JW, Lee GA, Ma KH, Kim HH, Kim KT, Chung IM, Lee JK, Kim NS, Kim SM, Park YJ (2011) Molecular genetic diversity and population structure of a selected core set in garlic and its relatives using novel SSR markers. Plant Breed 130:46–54

Zhou SM, Chen LM, Liu SQ, Wang XF, Sun XD (2015) De novo assembly and annotation of the Chinese chive (*Allium tuberosum* Rottler ex Spr.) transcriptome using the Illumina platform. PLoS ONE 10: e0133312

Gene Annotation

9

Hideki Hirakawa

Abstract

Due to the large genome size in *Allium* species, annotated genome sequence information is limited to the organelle genomes of *Allium cepa*. On the other hand, there are several annotated transcriptome sequences analyzed by using NGS (RNA-Seq) in *A. cepa* (onion), *A. fistulosum* (Japanese bunching onion), *A. tuberosum* (Chinese chive), *A. cepa* Aggregatum group (shallot), and doubled haploid *A. cepa*. These genomic and genic data have been released from the databases of *Allium* species, such as AlliumMap, PlantNATsDB, AlliumTDB, some databases in NCBI, PGDBj as integrated database, and so on. As the genomic and genic information are accumulated, the studies of understanding gene functions in *Allium* species would be accelerated.

9.1 Introduction

With the progress of NGS (Next Generation Sequencing) technologies, the genome sequences of different varieties have been determined in diverse plant species. Currently, short reads are mainly available by MiniSeq/MiSeq/NextSeq/HiSeq/NovaSeq platforms using the single molecule array technology developed by Illumina (https://www.illumina.com). As for long reads, PacBio RSII and Sequel using Single Molecule, Real-Time (SMRT) sequencing technology have been released from Pacific Biosciences (http://www.pacb.com), and MinION/GridION/PromethION using nanopore technology have been released from Oxford Nanopore Technologies (https://nanoporetech.com). To reconstruct the genome sequence, the reads sequenced by these NGS technologies are assembled, and then contigs and scaffolds are constructed. After the genome assembly, gene annotations, such as prediction of gene structures, transcripts, genes, proteins, ribosomal RNAs (rRNAs), and transfer RNAs (tRNAs), are generally carried out. Subsequently, the functions of the genes are inferred by similarity searches against databases (DBs).

As for the genome sequences of *Allium* species, the chloroplast genome sequence of normal (N) male-fertile and male-sterile (S) *Allium cepa* (onion) (von Kohn et al. 2013) and the mitochondrial genome sequence of *A. cepa* (Kim et al. 2016) have been determined. Gene annotations such as gene prediction and prediction of gene functions in the chloroplast and mitochondrial genomes were conducted. Although the organelle genome sequences of *A. cepa* have been determined, the nuclear genome sequence has not been determined yet because of its

H. Hirakawa (✉)
Kazusa DNA Research Institute, Kazusa-kamatari, Kisarazu, Chiba 292-0818, Japan
e-mail: hh@kazusa.or.jp

M. Shigyo et al. (eds.), *The Allium Genomes*, Compendium of Plant Genomes,
https://doi.org/10.1007/978-3-319-95825-5_9

enormous size. According to the C-value database (http://data.kew.org/cvalues/CvalServlet?querytype=1), the estimated genome sizes of *Allium* species range from 7.4 Gb in *A. sibiricum* to 72.9 Gb in *A. validum*. As well-known species, the genome size of *A. fistulosum* (Japanese bunching onion) was estimated to be 12.2 Gb, and that of *A. cepa* was estimated to be 16.4 Gb.

In the PAG (Plant & Animal Genome XXIII in 2015) conference, the genome sequence of the doubled haploid line DH066619 of *A. cepa* was presented by Richard Finkers at Wageningen UR Plant Breeding in the Netherlands (https://pag.confex.com/pag/xxiii/webprogram/Paper17794.html) under the SEQUON project (http://www.oniongenome.net). In the study, the genome size of bulb onion (*A. cepa* Common onion group) was estimated to be 16 Gb. De novo assembly was carried out using paired-end reads of Illumina HiSeq by the MaSuRCA program (Zimin et al. 2013), and the total genome size was 10.8 Gb and consisted of 6.2 M contigs (>500 bp) with an N50 length of 2776 bp. In addition, the genome sequence of shallot (*A. cepa* Aggregatum group) is currently being determined by Yamaguchi University in Japan.

Due to the large size of the genome in *Allium* species, many transcriptome analysis studies have been subject to research targets. Currently, de novo assembly of RNA-Seq analysis is carried out to clarify gene contents at the genome-wide level, and then the unigenes are constructed. After the construction of unigenes, gene annotations, such as prediction of gene structures, transcripts, genes, proteins, ribosomal RNAs (rRNAs), and transfer RNAs (tRNAs), are generally carried out.

In *Allium* species, there are no DBs related to genome sequences. On the other hand, there are some DBs related to transcripts, such as AlliumMap (McCallum et al. 2012) (http://alliumgenetics.org) for providing genetic maps and marker data from multiple *Allium* species and populations, PlantNATsDB (Chen et al. 2012) (http://bis.zju.edu.cn/pnatdb/) for providing natural antisense transcripts (NATs), the Onion Transcriptome Database (http://onion.riceblast.snu.ac.kr) for providing transcriptome assembly and annotation data of *A. cepa* H6 and SP3B, and so on. In this chapter, details of AlliumTDB (http://alliumtdb.kazusa.or.jp) released from Yamaguchi University, and some DBs at the NCBI (National Center for Biotechnology Information) (https://www.ncbi.nlm.nih.gov), and PGDBj (Plant Genome DataBase Japan) (http://pgdbj.jp) are described.

9.2 Gene Annotation of the *A. cepa* Organelle Genome

9.2.1 Chloroplast Genome

The chloroplast genome sequence of normal (N) male-fertile and male-sterile (S) of *A. cepa* has been determined by assembling the reads sequenced by the 454 FLX platform (Roche Diagnostics, Germany) using Newbler (Roche Diagnostics, Germany), and the total lengths of N and S were 153,538 bp (accession: KF728080) and 153,355 bp (accession: KF728079), respectively. Gene annotation was carried out by using Dogma (Wyman et al. 2004), and the numbers of genes, rRNAs, and tRNAs were the same between the N and S cytoplasms, i.e., 135 genes, 8 rRNAs, and 38 tRNAs (von Kohn et al. 2013). By comparing the chloroplast genome sequences, polymorphisms that were useful for classifying onion cytoplasms between N and S were identified (von Kohn et al. 2013).

9.2.2 Mitochondrial Genome

The mitochondrial genome sequence of *A. cepa* has been determined by assembling the paired-end reads sequenced by NextSeq 500 platform (Illumina, Hayward, CA) using a CLC genome assembler v4.06 beta (CLC Inc., Aarhus, Denmark) and GapCloser program v1.12 (BGI, Shenzhen, P.R. China), and the total length was 316,363 bp (accession: KU3186712). Gene annotation was performed by using the MITOFY program (Alverson et al. 2010), and 39 genes (25,497 bp length), 3 ribosomal RNAs (5104 bp length), and 16 transfer RNAs (893 bp length) were extracted. The functions of the genes were predicted by sequence alignments against the orthologous genes cloned from other plant species (Kim et al. 2016).

The completed *A. cepa* chloroplast and mitochondrial genome sequences with annotation have been used as fundamental information for elucidating the evolution of onion Cytoplasmic Male Sterility (CMS).

9.3 Gene Annotation of Unigenes

There are several reports related to the sequencing of transcripts using short reads by RNA-Seq methods in *Allium* species, i.e., 54,165 unigenes in *A. cepa* (Kim et al. 2015), 42,881 unigenes in *A. fistulosum* (Liu et al. 2014), 293,475 unigenes in *A. cepa* (Rajkumar et al. 2015), 60,031 unigenes in *A. tuberosum* (Zhou et al. 2015), 53,837 unigenes in *A. fistulosum* (Sun et al. 2016), 117,189 unigenes in *A. cepa* (Zheng et al. 2016), 54,047 unigenes in *A. cepa* (Han et al. 2016), 271,665 unigenes in the doubled haploid of *A. cepa* (Khosa et al. 2016), and 56,161 unigenes in the doubled haploid of *A. cepa* Aggregatum group (Abdelrahman et al. 2017). The gene annotations conducted in these studies are summarized in Table 9.1. The number of unigenes assembled by Rajkumar et al. was 293,475 in *A. cepa*, which was highest among the species investigated, while that assembled by Sum et al. was 53,837 in *A. fistulosum*, which was lowest among the studied species. In most of the studies, Trinity was used for the assembly. The databases used for functional annotation were non-redundant proteins (NR) (ftp://ncbi.nlm.nih.gov/blast/db/FASTA/nr.gz), NT (https://www.ncbi.nlm.nih.gov/nucleotide/), RefSeq (https://www.ncbi.nlm.nih.gov/refseq/), and KOG (COG) (Tatusov et al. 2003) in NCBI, Swiss-Prot (https://www.ebi.ac.uk/uniprot/) and UniProt (http://www.uniprot.org) in EMBL, KEGG (http://www.genome.jp/kegg/), Gene Ontology (GO) (http://geneontology.org), InterPro (https://www.ebi.ac.uk/interpro/), Pfam (http://pfam.xfam.org/), The Arabidopsis Information Resource (TAIR) (https://www.arabidopsis.org), and the International Rice Genome Sequencing Project (RAP-DB) (http://rapdb.dna.affrc.go.jp).

As an example of an annotation procedure in transcriptome analysis, the detailed strategy of gene annotation in transcriptome analysis of the *A. cepa* Aggregatum group conducted by Yamaguchi University in Japan is presented below.

9.4 Unigene of the *Allium cepa* Aggregatum Group

To understand physiological traits such as the biosynthetic pathway of saponins in *Allium*, expression analyses of monosomic addition lines (MALs) of the Japanese bunching onion (*A. fistulosum*, FF) with an extra chromosome from shallot (*A. cepa* Aggregatum group, AA) were investigated (Abdelrahman et al. 2017). The RNA-Seq reads sampled from bulbs of AA were sequenced by a HiSeq 2500 platform (Illumina) (accession: DRA005096) for use as reference sequences in the expression analyses. The reads including the adaptor sequence and low-quality nucleotides were excluded, and the remaining reads were assembled in each sample by Trinity r20121005 (Grabherr et al. 2011). The contigs obtained in each sample were clustered and representative contigs in each cluster were further assembled; the resultant contigs were defined as unigenes. As a result, 56,161 unigenes were constructed and submitted to similarity searches against the databases of TAIR (TAIR10) and RAP-DB (IRGSP-1.0; http://rgp.dna.affrc.go.jp/IRGSP/), and NCBI NR using the BLASTX (Altschul et al. 1997) program with an *E*-value cutoff of 1E-10. The results showed that 29,746 (52.9%), 29,307 (52.1%), and 33,365 (59.4%) genes were similar to the genes of TAIR, RAP-DB, and NR, respectively. The gene annotations were carried out to the unigenes of *A. cepa* Aggregatum group, the genes of *Oryza sativa* and *Arabidopsis thaliana* based on the Gene Ontology (GO) (http://geneontology.org). The 56,161 unigenes of *A. cepa* Aggregatum group, 35,626 genes of *O. sativa* (IRGSP-1.0), and 27,655 genes of *A. thaliana* (Araport11; https://www.araport.org) were conducted to

Table 9.1 Gene annotation of unigene in *Allium* species

	A. cepa (onion)				Doubled haploid *A. cepa*	*A. fistulosum* (Welsh onion)		*A. tuberosum* (Chinese chive)	*A. cepa* aggregatum group (shallot)
Cultivar (line)	Doubled haploid (H6), inbred line (SP3B)	–	Chalinghuangpi	36,122 (cold tolerant), 36101 (cold susceptible)	CUDH2107	BianGan (BG), glossy BianGan (GLBG)	–	–	Monosomic line
Samples	Whole seedlings	Bulbs	Leaves	Leaves	leaves, floral buds, unopened florets, open florets, older flowers, roots	Leaves	Leaves, false stem, basal plate, root	Leaves, shoots, roots	Roots, bulbs, leaves
Platform	HiSeq 2000	HiSeq 2000	HiSeq 2500	HiSeq 2000	HiSeq 2000	HiSeq 2000	HiSeq 2000	HiSeq 2000	HiSeq 2500
Raw reads	15.0 Gb (H6), 11.6 Gb (SP3B)	9,90,74,309	–	45,21,94,370	–	22,410,309 (BG), 24,894,222 (GLBG)	5,79,52,072	5,70,19,902	2,62,13,534
Clean reads	–	8,30,46,820	10,28,74,740	45,08,21,674	396 million	47.30 million	5,33,78,674	5,19,68,882	2,47,51,242
Total clean nucleotides	10.1 Gb (H6), 7.9 Gb (SP3B)	~20 Gb	12.86 Gb	–	–	4.52 Gb (BG), 5.03 Gb (GLBG)	4,80,40,80,660	4,67,71,99,380	2,22,76,11,780
Number of unigenes	54,165 (combined)	2,93,475	1,17,189	54,047	2,71,665	42,881 (combined)	53,837	60,031	56,161
Length of unigenes	59.6 Mb (combined)	28,08,82,036	74.03 Mb	–	17,74,24,188	33,760,304 (combined)	3,33,16,474	3,79,00,663	3,16,63,532
Average length (bp)	1100.4 (combined)	957	632	1331	653	787.3 (combined)	619	631	564
N50 length (bp)	–	1594	–	1625	1055	1304 (combined)	832	900	829
Annotated genes	27,421	1,15,251	39,472	30,926	56,805	22,289	36,155	36,523	–

(continued)

Table 9.1 (continued)

	A. cepa (onion)				Doubled haploid *A. cepa*	*A. fistulosum* (Welsh onion)		*A. tuberosum* (Chinese chive)	*A. cepa* aggregatum group (shallot)
Cultivar (line)	Doubled haploid (H6), inbred line (SP3B)	–	Chalinghuangpi	36,122 (cold tolerant), 36101 (cold susceptible)	CUDH2107	BianGan (BG), glossy BianGan (GLBG)	–	–	Monosomic line
Annotated genes (%)	50.6	39.2	33.7	61.5	20.9	52.0	67.2	60.8	–
Assembler	Velvet v1.2.08, Oases v0.2.06	Velvet v1.2.10, Oases v0.2.8, CD-HIT	Trinity	Trinity	Trinity	Trinity	Trinity r2013-02-25	Trinity r2013-02-05	Trinity r20121005, TGI Clustering Tool (TGICL) v2.1, Phrap 23.0
Functional annotation (DBs)	NR, RefSeq, Uniprot, InterPro	NR, GO	NR, NT, Swiss-Prot, KOG, KEGG, GO, PFAM	NR, NT, COG, Swiss-Prot, KEGG, GO	KOG, GO	NR, NT, COG, Swiss-Prot, KEGG, GO	NR, NT, COG, Swiss-Prot, KEGG, GO	NR, NT, COG, Swiss-Prot, KEGG, GO	GO, TAIR, RAP-DB, NR
Number of SSRs (software)	–	–	–	5180 (Msatcommnader)	–	1558 (MISA)	2014 (MISA)	2279 (MISA)	–
SRA accession	SRP041918	SRX547958	SRA347079	SRP064878	5898004-9	SRR1609126, SRR1609976	SRR1023632	SRR1020564	DRA005096
TSA accession	GBRQ01000000 (combined), GBRO01000000 (H6), GBRN01000000 (SP3B)	–	GEOY00000000,	GETF00000000	–	–	–	–	–
Database	OTD (Onion Transcriptome Database; http://onion.snu.ac.kr)	–	–	–	–	–	–	–	AlliumTDB (http://alliumtdb.kazusa.or.jp)

(continued)

Table 9.1 (continued)

	A. cepa (onion)				Doubled haploid *A. cepa*	*A. fistulosum* (Welsh onion)		*A. tuberosum* (Chinese chive)	*A. cepa* aggregatum group (shallot)
Cultivar (line)	Doubled haploid (H6), inbred line (SP3B)	–	Chalinghuangpi	36,122 (cold tolerant), 36101 (cold susceptible)	CUDH2107	BianGan (BG), glossy BianGan (GLBG)	–	–	Monosomic line
Publication	Kim et al. (2015) *DNA Research*, 22, 19–27	Rajkumar et al. (2015) *PLOS ONE*, 10:e0135387	Zheng et al. (2016) *PLOS ONE*, 11: e0157871	Han et al. (2016) *PLOS ONE*, 11: e0161987	Khosa et al. (2016) *PLOS ONE*, 11: e0166568	Liu et al. (2014), *PLOS ONE*, 9: e113290	Sun et al. (2016) *Mol. Genet. Genomics*, 291:647–659	Zhou et al. (2015) *PLOS ONE*, 10: e0133312	Abdelrahman et al. (2017) *PLOS ONE*, 12:e0181784

InterProScan (Hunter et al. 2009) searches against InterPro database. As results, 16,943 (30.1%) of 56,161 unigenes in *A. cepa* Aggregatum group, 20,926 (58.7%) of 35,626 genes in *O. sativa*, and 17,093 (61.8%) of 27,655 genes in *A. thaliana* were assigned to GO functional categories. The unigenes of *A. cepa* Aggregatum group were constructed by de novo assembly of RNA-Seq reads and contains partially predicted genes, while the genes of *O. sativa* and *A. thaliana* are basically full-length as they are predicted from the genome sequences. Therefore, the percentage of the unigenes assigned to GO categories in *A. cepa* Aggregatum group was lower than those in *O. sativa* and *A. thaliana*. The distribution of the genes assigned to biological processes (BP), cellular components (CC), and molecular functions (MF) in *A. cepa* Aggregatum group, *O. sativa*, and *A. thaliana* are shown in Fig. 9.1. The distributions of the GO categories were similar among them, which indicated that the functional characterization of the unigenes of *A. cepa* Aggregatum group is considered to be valid. The most frequently observed gene functions in *A. cepa* Aggregatum group were "metabolic processes (7352 genes)" and "cellular process (6410)" in BP, "cell (4471)", "intracellular (2696)", and "membrane (2118)" in CC, and "binding (9547)" and "catalytic activity (7069)" in MF. The number of genes categorized into the GO is shown in parentheses. Gene annotation was also carried out for unigenes by quantifying the expression levels in each organ. RNA-Seq data from 30 cDNA libraries of roots, bulbs, and leaves of AA, MALs, and FF were mapped onto the unigenes by Bowtie 2 v2.2.0 (Langmead and Salzberg 2012) in end-to-end mode. Based on the number of reads mapped onto the unigenes, RPKM (Reads Per Kilobase per Million mapped reads) values were calculated for each gene. The genes with different expression levels in the roots, bulbs, and leaves of AA and MALs were compared to those of FF, then the RPKM fold change and false discovery rate were calculated to find differently expressed genes. Gene annotation was performed using the AlliumTDB (http://alliumtdb.kazusa.or.jp). By mapping the unigenes onto the KEGG metabolic pathways, 50 unigenes in the roots, bulbs, and leaves of AA and MALs were found to be related to saponin biosynthesis. The RPKM values were also calculated for the MALs and their parental lines (FF1A, FF2A, FF3A, FF4A, FF5A, FF6A, FF7A, FF8A, AA, FF) sampled from bulbs, roots, and leaves against unigenes of DHA_Bulb. Here, FF1A means that the chromosome additional line, in which chromosome 1 of AA was added to FF. The differential gene expression analysis revealed many up-regulated genes in the biosynthetic pathway of steroidal saponins in the extra chromosome 2 from the shallot (FF2A) line (Abdelrahman et al. 2017).

9.5 Genome Sequencing of the Doubled Haploid of Shallot

Gene annotation in *Allium* species has been carried out in several studies for the following plant species: *A. cepa* (Kim et al. 2015; Rajkumar et al. 2015; Zheng et al. 2016; Han et al. 2016; Khosa et al. 2016), *A. fistulosum* (Liu et al. 2014; Sun et al. 2016), *A. tuberosum* (Zhou et al. 2015), and *A. fustulosum* with an extra chromosome from the *A. cepa* Aggregatum group (Abdelrahman et al. 2017). In these studies, de novo assemblies of RNA-Seq data were conducted to understand the genes, but it is considered that the construction of unigenes is not sufficient to understand the gene structures, since it is difficult to reconstruct the complete gene structure merely by assembling short reads. Recently, full-length cDNA sequences were made available by the Iso-Seq method (http://www.pacb.com/applications/rna-sequencing/) using PacBio reads, but only the genes expressed in the samples were sequenced. To more completely clarify the overall gene structures and expression patterns, information on the whole genome sequence is considered to be necessary.

Currently, de novo assembly of the doubled haploid of shallot is being carried out using Illumina short reads by Yamaguchi University in Japan. The total lengths of contigs and scaffolds

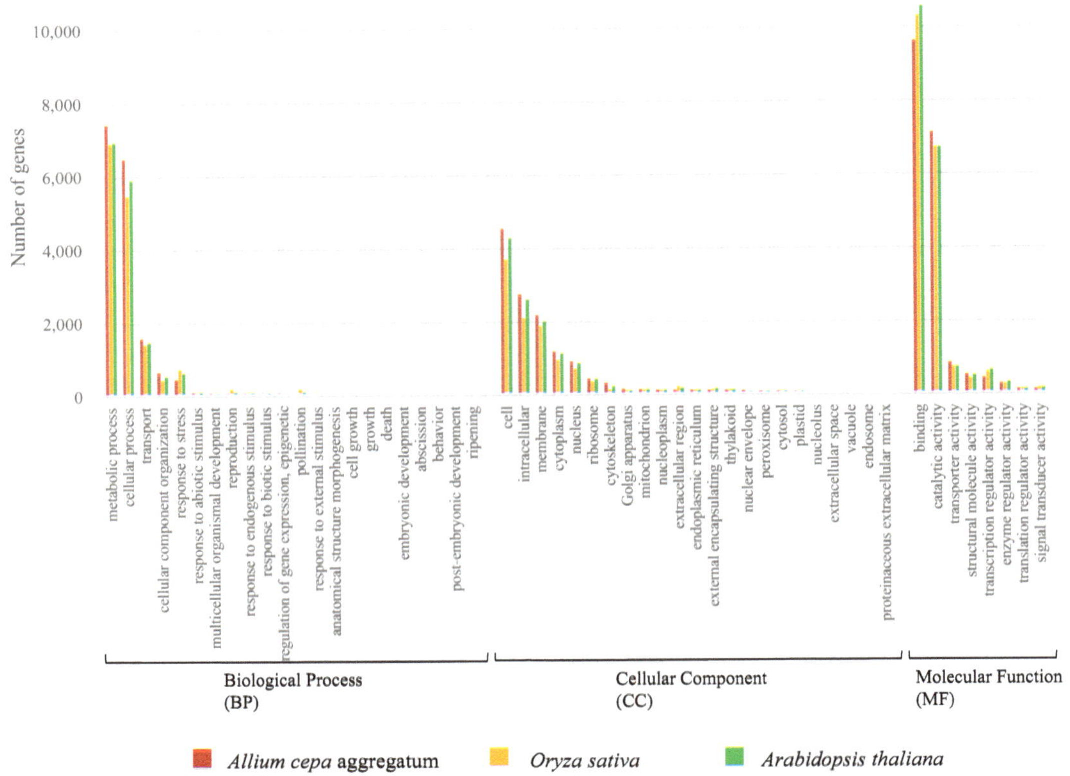

Fig. 9.1 Distribution of GO categories for the unigenes in DHA_Bulb. The number of unigenes of *Allium cepa* Aggregatum group assigned to GO categories are shown in red bar. The numbers of the genes of *Oryza sativa* and *Arabidopsis thaliana* are shown in orange and green, respectively

obtained are close to the estimated genome size (16 Gb), but the N50 length of the scaffolds is still short because many repetitive sequences are contained in the genome. Recently, as long read sequencers, Sequel (http://www.pacb.com/products-and-services/pacbio-systems/sequel/) has been released from Pacific Biosciences (http://www.pacb.com), and MinION (https://nanoporetech.com/products/minion), GridION (https://nanoporetech.com/products/gridion), and PromethION (https://nanoporetech.com/products/promethion) have been released from Oxford Nanopore Technologies (https://www.nanoporetech.com). All of these tools are useful to extend the assembled genome sequence. In addition, as a new technology, BioNano Genomics Irys (https://bionanogenomics.com/products/irys/) and Saphyr (https://bionanogenomics.com/products/saphyr/) Systems (http://bionanogenomics.com) using optical mapping technology have been developed, and have been applied to scaffolding or the detection of structural variations. Chromium has been released from 10x Genomics (http://www.10xgenomics.com), which uses GemCode technology for phasing, scaffolding, and detecting structural variations. Dovetail Genomics (http://dovetailgenomics.com) has developed Chicago method to scaffold with a wide range of information within chromatin structures, which are especially useful for anchoring the scaffolds. In particular, NRGene (http://www.nrgene.com) has developed the DenovoMAGIC assembler (http://www.nrgene.com/technology/denovomagic/), which successfully assembles many plant and animal genomes, such as *Aegilops tauschii* (wheat ancestor), bovine, maize, strawberry (heterozygote octoploid), canola, and ryegrass (heterozygous) using Illumina short reads. This software can be applied to any genomes with high heterozygosity or high polyploidy. By using the technologies described above, it will be possible to improve the scaffolds of the shallot genome sequence currently in progress.

9.6 Databases

9.6.1 AlliumTDB

To provide the transcript data in *Allium* species accumulated in the project led by Yamaguchi University, AlliumTDB (Allium Transcript DataBase; http://alliumtdb.kazusa.or.jp) has been constructed. The top page of AlliumTDB is shown in Fig. 9.2. In AlliumTDB, unigenes constructed in each of 12 plant materials, i.e., the doubled haploid of onion (DHC) sampled from leaves (dataset name: DHC_Leaf), the doubled haploid of shallot (DHA) sampled from leaves, bulbs, and roots (DHA_Leaf, DHA_Bulb, DHA_Root), an F_1 hybrid of DHA × DHC sampled from leaves, bulbs, and stems (F_1_Leaf, F_1_Bulb, F_1_Root), *A. roylei* sampled from leaves, bulbs, and stems (A.roylei_Leaf, A.roylei_Bulb, A.roylei_Root), *A. fistulosum* sampled from stems (FFStem), and *A. fistulosum* with shallot 5A chromosome sampled from stems (FF_5AStem), have been released. Users can perform a BLAST search at the "BLAST" page (http://alliumtdb.kazusa.or.jp/blast.html) by selecting the databases, i.e., unigenes constructed from each of the 12 plant materials described above, unigenes of *A. cepa* constructed from EST sequences assembled by the Kazusa DNA Research Institute (GI_kazusaEST), and unigenes of *A. fistulosum* (A.fistulosum_Unigene) assembled with the reads of DHA_Bulb, FFStem, and FF_5AStem. In the "KEYWORD" page (http://alliumtdb.kazusa.or.jp/keyword.html), users can search the genes by keywords in the results of BLAST searches against the TAIR, RAP-DB, and NCBI NR databases. Here, for example, we searched for the sequence information of CL1023.Contig2_DHA_Bulb. Users can browse the sequence of the unigene (Fig. 9.3a) and the expression profiles of the unigene in the 12 samples (Fig. 9.3b). In the profile, RPKM values of the MALs and parents (FF1A, FF2A, FF3A, FF4A, FF5A, FF6A, FF7A, FF8A, AA, FF) sampled from bulbs, roots, and leaves are listed. The results of the BLAST searches against the TAIR, RAP-DB, and NCBI NR databases are summarized in Fig. 9.3c, and the corresponding unigenes in the other samples are shown in Fig. 9.3d. In the table, the unigene set name (DB), unigene sequence name (AC), %identity (Iden), %positive (Posi), BLAST score (Score), *E*-value (E-value), and unigene length (Annotation) are listed. At the top of the "LIST" page (http://alliumtdb.kazusa.or.jp/list.html), the table of the RPKM values calculated for the doubled haploid of shallot (DHA_Bulb) of the 12 samples described above is available. These tables can be downloaded at the "LIST" page.

9.6.2 NCBI

The sequence information of *Allium* species has been registered in GenBank (https://www.ncbi.nlm.nih.gov/genbank/), dbEST (https://www.ncbi.nlm.nih.gov/dbEST/), Gene (https://www.ncbi.nlm.nih.gov/gene/), Protein (https://www.ncbi.nlm.nih.gov/protein/), and so on. Genic data such as complete, partial CDSs or RNAs of *A. cepa* (garden onion, 578,293 sequences), *A. sativum* (cultivated garlic, 79,605), *A. fistulosum* (welsh onion, 55,680), *A. munzii* (Munz's onion, 242), *A. przewalskianum* (228), *A. ampeloprasum* (broadleaf wild leek, 222), *A. wallichii* (194), *A. cyathophorum* (188), *A. tuberosum* (Chinese chives, 183), *A. canadense* (meadow garlic, 135), *A. prattii* (109), and *A. spicatum* (108) has been registered. In dbEST, EST (Expression Sequence Tag) sequences of *A. sativum* (cultivated garlic, 22,771) and *A. cepa* (garden onion, 20,204) have been registered. NGS (Next Generation Sequencing) data derived from the genome sequences of *A. cepa* (garden onion, 5), *A. sativum* (cultivated garlic, 2 entries), and *A. fistulosum* (welsh onion, 1), and that derived from

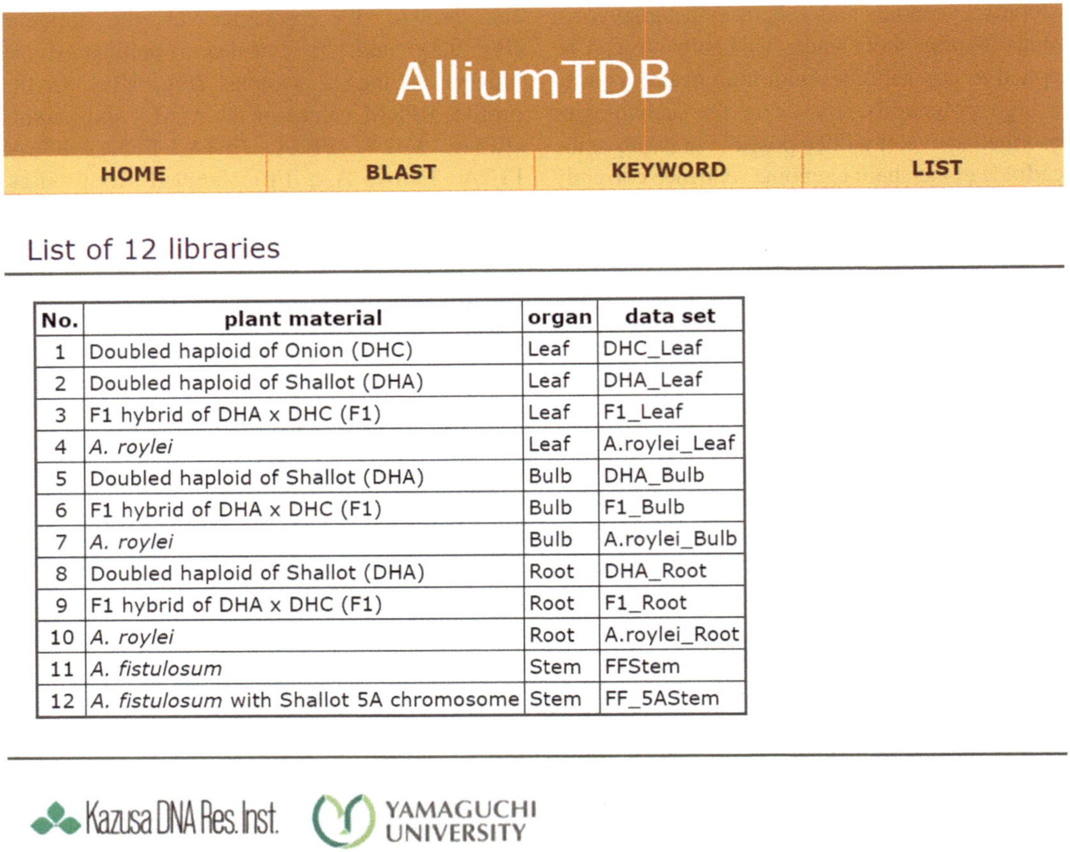

No.	plant material	organ	data set
1	Doubled haploid of Onion (DHC)	Leaf	DHC_Leaf
2	Doubled haploid of Shallot (DHA)	Leaf	DHA_Leaf
3	F1 hybrid of DHA x DHC (F1)	Leaf	F1_Leaf
4	*A. roylei*	Leaf	A.roylei_Leaf
5	Doubled haploid of Shallot (DHA)	Bulb	DHA_Bulb
6	F1 hybrid of DHA x DHC (F1)	Bulb	F1_Bulb
7	*A. roylei*	Bulb	A.roylei_Bulb
8	Doubled haploid of Shallot (DHA)	Root	DHA_Root
9	F1 hybrid of DHA x DHC (F1)	Root	F1_Root
10	*A. roylei*	Root	A.roylei_Root
11	*A. fistulosum*	Stem	FFStem
12	*A. fistulosum* with Shallot 5A chromosome	Stem	FF_5AStem

Fig. 9.2 Top page of the AlliumTDB

transcript sequences of *A. cepa* (garden onion, 11), *A. sativum* (cultivated garlic, 9), *A. fistulosum* (welsh onion, 2), *A. cepa* var. Aggregatum (shallot, 2), *A. fistulosum* with *A. cepa* monosomic additional line (1), *A. ampeloprasum* (broadleaf wild leek, 1), *A. amphibolum* (1), *A. ascalonicum* (wild onion, 1), *A. bidentatum* (1), *A. chinense* (rakkyo, 1), *A. commutatum* (1), *A. macrostemon* (1), and *A. tuberosum* (Chinese chives, 1) have been registered at the NCBI SRA database (https://www.ncbi.nlm.nih.gov/sra/). The genome sequence of *A. cepa* has not been published, but the genome sequence of *A. cepa* could be constructed by using the NGS data registered in the SRA database. Once the genome sequence of *A. cepa* has been determined, it would be useful for finding genome-wide polymorphisms such as SNPs and InDels or for expression analysis. By using these data, gene annotations in *Allium* species could be performed for gene finding or functional analyses. In addition, the common names written here were taken from the PLANTS Database released by the Natural Resources Conservation Service at the USDA (United States Department of Agriculture; https://plants.usda.gov).

9.6.3 PGDBj (Plant Genome DataBase Japan)

Along with the advances in the studies of plant sciences, the genomic and genic sequences and DNA markers of a variety of types of plants have been clarified, and numerous studies on these topics have been published. After the genome

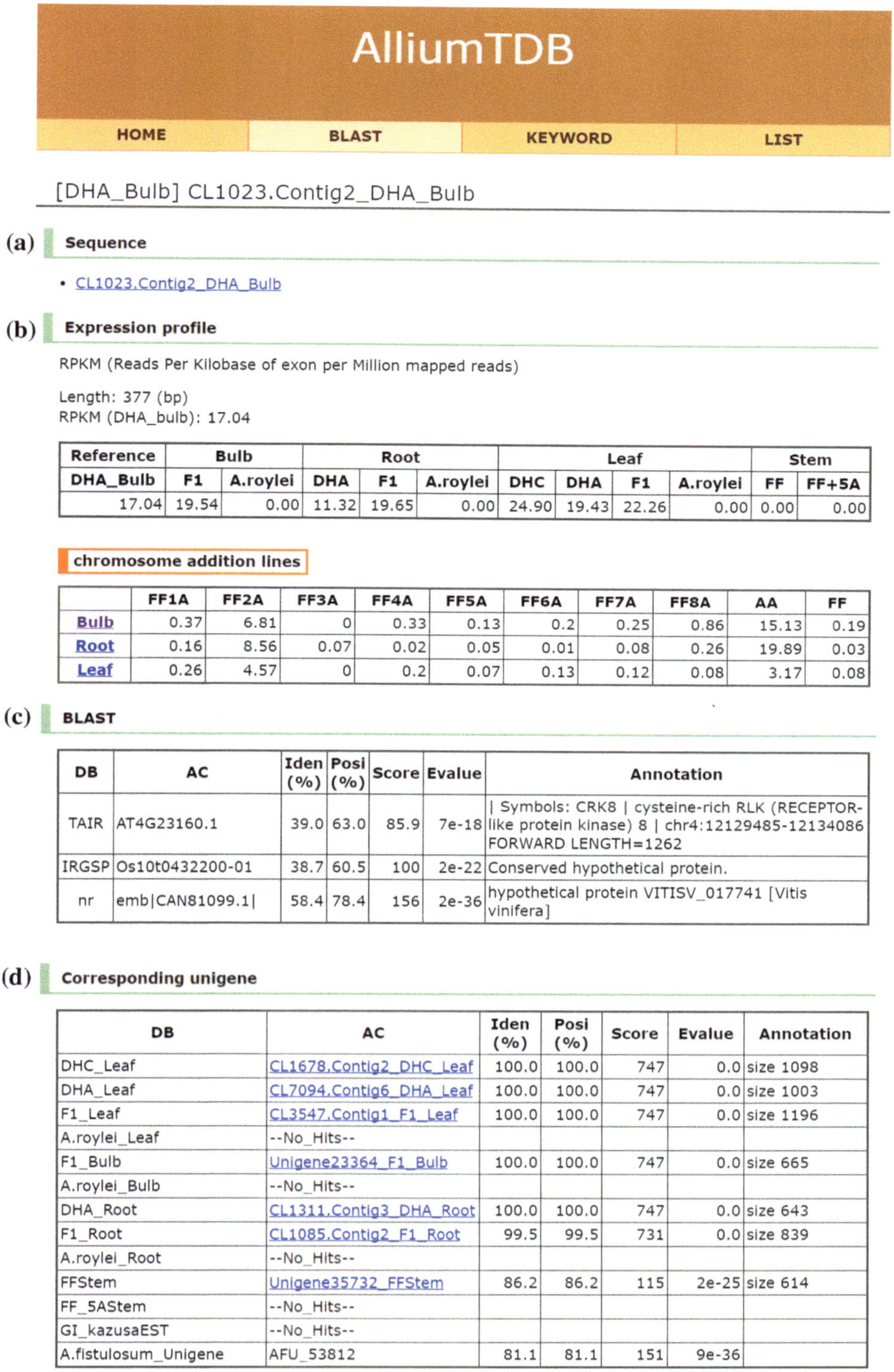
AlliumTDB

HOME | BLAST | KEYWORD | LIST

[DHA_Bulb] CL1023.Contig2_DHA_Bulb

(a) **Sequence**

- CL1023.Contig2_DHA_Bulb

(b) **Expression profile**

RPKM (Reads Per Kilobase of exon per Million mapped reads)

Length: 377 (bp)
RPKM (DHA_bulb): 17.04

Reference	Bulb		Root			Leaf				Stem	
DHA_Bulb	F1	A.roylei	DHA	F1	A.roylei	DHC	DHA	F1	A.roylei	FF	FF+5A
17.04	19.54	0.00	11.32	19.65	0.00	24.90	19.43	22.26	0.00	0.00	0.00

chromosome addition lines

	FF1A	FF2A	FF3A	FF4A	FF5A	FF6A	FF7A	FF8A	AA	FF
Bulb	0.37	6.81	0	0.33	0.13	0.2	0.25	0.86	15.13	0.19
Root	0.16	8.56	0.07	0.02	0.05	0.01	0.08	0.26	19.89	0.03
Leaf	0.26	4.57	0	0.2	0.07	0.13	0.12	0.08	3.17	0.08

(c) **BLAST**

DB	AC	Iden (%)	Posi (%)	Score	Evalue	Annotation
TAIR	AT4G23160.1	39.0	63.0	85.9	7e-18	| Symbols: CRK8 | cysteine-rich RLK (RECEPTOR-like protein kinase) 8 | chr4:12129485-12134086 FORWARD LENGTH=1262
IRGSP	Os10t0432200-01	38.7	60.5	100	2e-22	Conserved hypothetical protein.
nr	emb|CAN81099.1|	58.4	78.4	156	2e-36	hypothetical protein VITISV_017741 [Vitis vinifera]

(d) **Corresponding unigene**

DB	AC	Iden (%)	Posi (%)	Score	Evalue	Annotation
DHC_Leaf	CL1678.Contig2_DHC_Leaf	100.0	100.0	747	0.0	size 1098
DHA_Leaf	CL7094.Contig6_DHA_Leaf	100.0	100.0	747	0.0	size 1003
F1_Leaf	CL3547.Contig1_F1_Leaf	100.0	100.0	747	0.0	size 1196
A.roylei_Leaf	--No_Hits--					
F1_Bulb	Unigene23364_F1_Bulb	100.0	100.0	747	0.0	size 665
A.roylei_Bulb	--No_Hits--					
DHA_Root	CL1311.Contig3_DHA_Root	100.0	100.0	747	0.0	size 643
F1_Root	CL1085.Contig2_F1_Root	99.5	99.5	731	0.0	size 839
A.roylei_Root	--No_Hits--					
FFStem	Unigene35732_FFStem	86.2	86.2	115	2e-25	size 614
FF_5AStem	--No_Hits--					
GI_kazusaEST	--No_Hits--					
A.fistulosum_Unigene	AFU_53812	81.1	81.1	151	9e-36	

Fig. 9.3 Gene information in AlliumTDB. The gene information of CL1023.Contig2_DHA_Bulb in DHA_-Bulb is shown. **a** Unigene name. **b** Expression profiles of the unigene. **c** RPKM values of the chromosome addition lines. **d** Corresponding unigenes in the other samples

Fig. 9.4 Top page of the PGDBj (Plant Genome DataBase Japan)

sequencings, many databases related to the determined sequences were released. To integrate the plant genome information, the Plant Genome DataBase Japan (PGDBj: http://pgdb.jp) was developed (Asamizu et al. 2014; Nakaya et al. 2017). As shown in Fig. 9.4, the top page (http://pgdbj.jp/en/) has a cross-search system against the three internal databases (DBs) of PGDBj (Ortholog DB, Plant Resource DB, and DNA marker DB) and external DBs related to metabolome [KNapSAck (http://kanaya.naist.jp/KNApSAcK/) and MassBase (http://webs2.kazusa.or.jp/massbase/)]. In Ortholog DB, 1.14 million genes of 40 plant species in *Viridiplantae* and 0.8 million genes of 213 Cyanobacteria have been registered. Orthologous genes were defined based on the top hits of all-vs-all BLAST searches. Currently, *Allium* species are not included in the 40 plant species, but their orthologous genes can be searched based on the genes of Poaceae (*Brachypodium distachyon*, *Oryza brachyantha*, *Oryza sativa*, *Setaria italica*). In the Plant Resource DB, 0.85 million genes of six plant species and citrus resources for approximately 900 individuals have been registered. In the DNA marker DB, 260,000 DNA markers such as SSLPs (Simple Sequence Length Polymorphisms) and SNPs (Single Nucleotide Polymorphisms) of 65 plant species (http://pgdbj.jp/plantdb/plantdb.html?ln=en) and 14,000 QTLs of 45 plant species are available. These markers have been manually curated from the literature. In PGDBj, *A. cepa* and *A. fistulosum* have been registered. The summary of the contents for *A. cepa* and *A. fistulosum* have been released at http://pgdbj.jp/plantdb/plantinfo.html?ln=en&cmd=entry&ppid=t4679 and http://pgdbj.jp/plantdb/plantinfo.html?ln=ja&cmd=entry&ppid=t35875, respectively. The contents of *A. cepa* are shown in Fig. 9.5. Currently, 220 markers, including 68 CAPS, 17 InDel, 2 SCAR, 46 SNP, 113 SSR, and 24 others, and 59 QTLs have been registered in *A. cepa*, and 33 SSR markers have been registered in *A. fistulosum*. By using the cross-search system, PGDBj would be helpful for researchers to obtain genomic information including gene annotation efficiently.

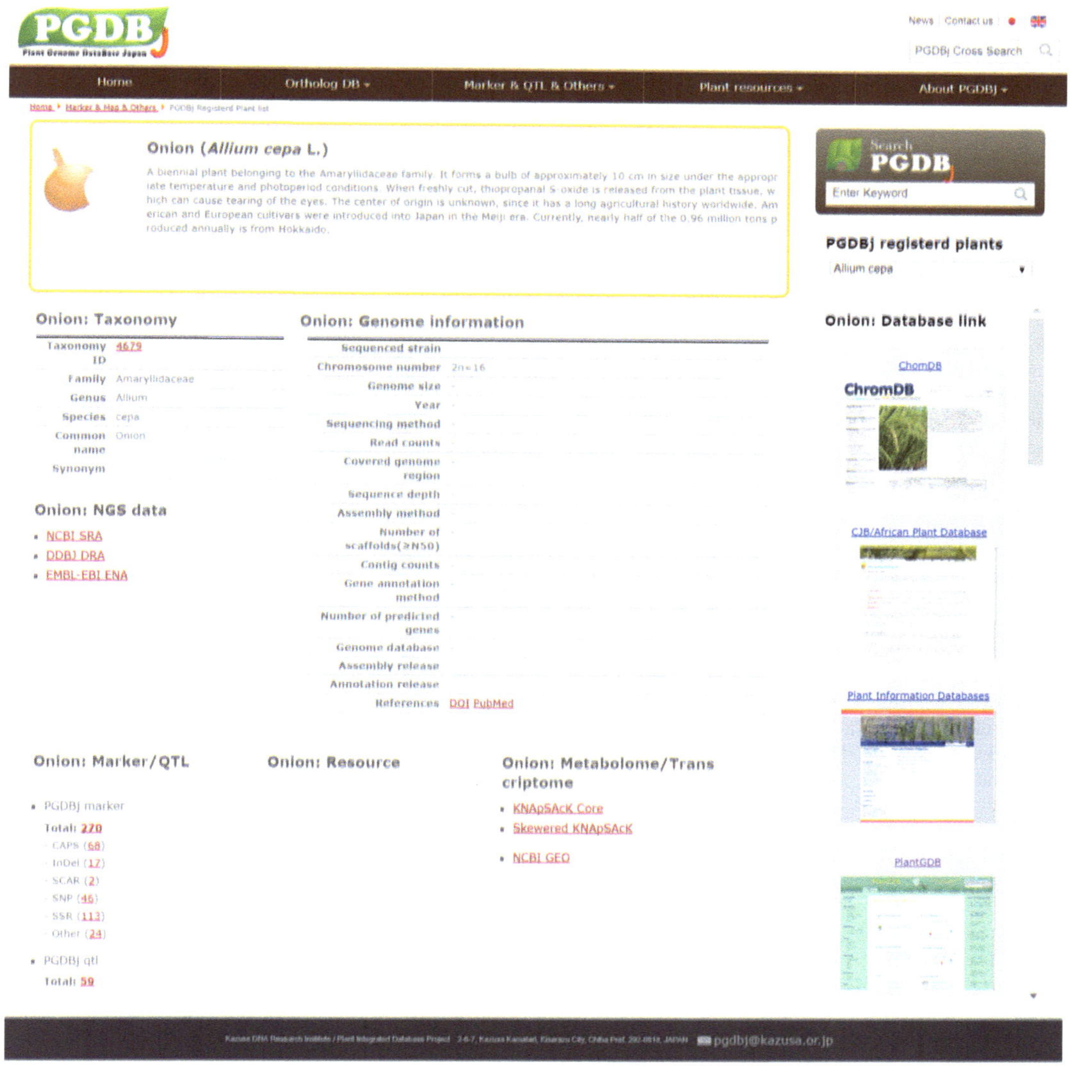

Fig. 9.5 Summary of the genome information registered for *Allium cepa* in PGDBj

9.7 Conclusion

Currently, there is no publication on the genome sequences of *Allium* species. On the other hand, transcript sequencing has been carried out using NGS for mainly *A. cepa*, *A. fistulosum*, and *A. tuberosum*, and genome information such as the sequences of unigenes and gene annotations are available in the literature or databases. Although the transcript sequences of the other *Allium* species are available, the number of sequences is not sufficient, because they were sequenced for EST or cDNA by the Sanger method. In the future, transcript sequences of *Allium* species will be accumulated in a public database, and more functional information such as orthologous or paralogous genes will be clarified. In addition, the whole genome sequence of *Allium* species will be determined by NGS or new technologies, and then novel genes could be identified and their functions investigated in more detail. By referring to the genomic and genic information, the studies of gene functions in *Allium* species would be accelerated.

References

Abdelrahman M, El-Sayed M, Sato S, Hirakawa H, Ito S-I, Tanaka K, Mine Y, Sugiyama N, Suzuki M, Yamauchi N, Shigyo M (2017) RNA-sequencing-based transcriptome and biochemical analyses of steroidal saponin pathway in a complete set of *Allium fistulosum-A. cepa* monosomic addition lines. PLoS ONE 12:e0181784

Altschul SF, Madden TL, Schäffer AA (1997) Gapped BLAST and PSI-BLAST: a new generation of protein database search programs. Nucl Acids Res 25:3389–3402

Alverson AJ, Wei X, Rice DW, Stern DB, Barry K, Palmer JD (2010) Insights into the evolution of mitochondrial genome size from complete sequences of *Citrullus lanatus* and *Cucurbita pepo* (*Cucurbitaceae*). Mol Biol Evol 27:1436–1448

Asamizu E, Ichihara H, Nakaya A, Nakamura Y, Hirakawa H, Ishii T, Tamura T, Fukami-Kobayashi K, Nakajima Y, Tabata S (2014) Plant Genome DataBase Japan (PGDBj): a portal website for the integration of plant genome-related databases. Plant Cell Physiol 55:e8

Chen D, Yuan C, Zhang J, Zhang Z, Bai L, Meng Y, Chen LL, Chen M (2012) PlantNATsDB: a comprehensive database of plant natural antisense transcripts. Nucleic Acids Res 40(Database issue):D1187–D1193

Grabherr MG, Haas BJ, Yassour M, Levin JZ, Thompson DA, Amit I, Adiconis X, Fan L, Raychowdhury R, Zeng Q, Chen Z, Mauceli E, Hacohen N, Gnirke A, Rhind N, di Palma F, Birren BW, Nusbaum C, Lindblad-Toh K, Friedman N, Regev A (2011) Full-length transcriptome assembly from RNA-Seq data without a reference genome. Nat Biotechnol 2011:29:644–652.

Han J, Thamilarasan SK, Natarajan S, Park JI, Chung MY, Nou IS (2016) De novo assembly and transcriptome analysis of bulb onion (*Allium cepa* L.) during cold acclimation using contrasting genotypes. PLoS ONE 11:e0161987

Hunter S, Apweiler R, Attwood TK, Bairoch A, Bateman A, Binns D, Bork P, Das U, Daugherty L, Duquenne L, Finn RD, Gough J, Haft D, Hulo N, Kahn D, Kelly E, Laugraud A, Letunic I, Lonsdale D, Lopez R, Madera M, Maslen J, McAnulla C, McDowall J, Mistry J, Mitchell A, Mulder N, Natale D, Orengo C, Quinn AF, Selengut JD, Sigrist CJ, Thimma M, Thomas PD, Valentin F, Wilson D, Wu CH, Yeats C (2009) InterPro: the integrative protein signature database. Nucleic Acids Res 37:D211–D215

Khosa JS, Lee R, Bräuning S, Lord J, Pither-Joyce M, McCallum J, Macknight RC (2016) Doubled haploid 'CUDH2107' as a reference for bulb onion (*Allium cepa* L.) research: development of a transcriptome catalogue and identification of transcripts associated with male fertility. PLoS ONE 11:e0166568

Kim S, Kim MS, Kim YM, Yeom SI, Cheong K, Kim KT, Jeon J, Kim S, Kim DS, Sohn SH, Lee YH, Choi D (2015) Integrative structural annotation of de novo RNA-Seq provides an accurate reference gene set of the enormous genome of the onion (*Allium cepa* L.). DNA Res 22:19–27

Kim B, Kim K, Yang TJ, Kim S (2016) Completion of the mitochondrial genome sequence of onion (*Allium cepa* L.) containing the CMS-S male-sterile cytoplasm and identification of an independent event of the ccmF N gene split. Curr Genet 62:873–885

Langmead B, Salzberg SL (2012) Fast gapped-read alignment with Bowtie 2. Nat Methods 9:357–359

Liu Q, Wen C, Zhao H, Zhang L, Wang J, Wang Y (2014) RNA-Seq reveals leaf cuticular wax-related genes in Welsh onion. PLoS ONE 9:e113290

McCallum J, Baldwin S, Shigyo M, Deng Y, van Heusden S, Pither-Joyce M, Kenel F (2012) AlliumMap-A comparative genomics resource for cultivated *Allium* vegetables. BMC Genom 13:168

Nakaya A, Ichihara H, Asamizu E, Shirasawa S, Nakamura Y, Tabata S, Hirakawa H (2017) Plant Genome DataBase Japan (PGDBj). Methods Mol Biol 1533:45–77

Rajkumar H, Ramagoni RK, Anchoju VC, Vankudavath RN, Syed AU (2015) De novo transcriptome analysis of *Allium cepa* L. (Onion) bulb to identify allergens and epitopes. PLoS ONE 10:e0135387

Sun XD, Yu XH, Zhou SM, Liu SQ (2016) De novo assembly and characterization of the Welsh onion (*Allium fistulosum* L.) transcriptome using Illumina technology. Mol Genet Genomics 291:647–659

Tatusov RL, Fedorova ND, Jackson JD, Jacobs AR, Kiryutin B, Koonin EV, Krylov DM, Mazumder R, Mekhedov SL, Nikolskaya AN, Rao BS, Smirnov S, Sverdlov AV, Vasudevan S, Wolf YI, Yin JJ, Natale DA (2003) The COG database: an updated version includes eukaryotes. BMC Bioinform 4:41

von Kohn C, Kiełkowska A, Havey MJ (2013) Sequencing and annotation of the chloroplast DNAs and identification of polymorphisms distinguishing normal male-fertile and male-sterile cytoplasms of onion. Genome 56:737–742

Wyman SK, Jansen R, Boore JL (2004) Automatic annotation of organellar genomes with DOGMA. Bioinformatics 20:3252–3255

Zheng X, Tang S, Zhu S, Dai Q, Liu T (2016) Identification of an NAC transcription factor family by deep transcriptome sequencing in onion (*Allium cepa* L.). PLoS ONE 11:e0157871

Zhou SM, Chen LM, Liu SQ, Wang XF, Sun XD (2015) De novo assembly and annotation of the Chinese chive (*Allium tuberosum* Rottler ex Spr.) transcriptome using the Illumina platform. PLoS ONE 10: e0133312

Zimin AV, Marçais G, Puiu D, Roberts M, Salzberg SL, Yorke JA (2013) The MaSuRCA genome assembler. Bioinformatics 29:2669–2677

Gene Family Evolution in *Allium* Species

10

Jiffinvir Khosa, Robyn Lee, John McCallum and Richard Macknight

Abstract

During evolution, plant genomes have undergone duplications, deletions and rearrangements resulting in a wide variation in genome size and number of gene family members between different species. The variation in gene families is an important mechanism for adaptation to different environmental conditions. *Allium* species, such as bulb onion (*Allium cepa*), have a large unsequenced genome. However, high throughput transcriptome sequencing datasets are now available which provide an efficient way to identify the genes present in different *Allium* species. With this knowledge, strategies to accelerate physiological and genetic analysis for enhanced breeding can be developed. In this chapter, we will describe how RNA sequencing is providing a better understanding of *Allium* genetics and survey the diversity of gene families involved in bulbing, flowering, male fertility, flavonoid biosynthesis and sulphur assimilation in bulb onion. In general, we found that onion has a similar number of gene family members to other monocots, such as rice, which have much smaller genomes. This is consistent with the large genome size of *Allium* being due to a massive expansion of repetitive DNA.

J. Khosa · R. Lee · R. Macknight (✉)
Department of Biochemistry, University of Otago, Dunedin, New Zealand
e-mail: richard.macknight@otago.ac.nz

J. McCallum · R. Macknight
New Zealand Institute for Plant and Food Research Ltd., University of Otago, Dunedin, New Zealand

10.1 Introduction

Over the last 17 years, the sequencing of plant genomes has provided tremendous resources for identifying genes underlying agriculturally important traits and equipping breeders with new tools to enhance the breeding of new cultivars (Varshney et al. 2014). The sequencing of the bulb onion genome and other *Allium* species has been hampered by their very large genome size (16 Gbp) and complex structure (McCallum 2007; Khosa et al. 2016b). The bulb onion genome is about 40 times larger than the rice genome (0.4 Gbp), which is well annotated and provides an excellent reference for comparing to other grasses (McCallum 2007; Jakse et al. 2008). Analysis of the genomes of other plant species has revealed that increases in genome size are largely due to whole-genome duplications and an increase in repetitive DNA. For example, it is thought that the ancestor of grasses had five chromosomes, and genome duplication resulted in first 10 chromosomes, then the current 12 chromosomes in rice. The large increase in the genome size of other grasses (for example,

M. Shigyo et al. (eds.), *The Allium Genomes*, Compendium of Plant Genomes,
https://doi.org/10.1007/978-3-319-95825-5_10

2.4 Gbp for maize and 17 Gbp for wheat) has been the result of duplication and increase in repetitive sequences (Wendel et al. 2016).

The *Allium* genus is very large, comprising 972 accepted species (see theplantlist.org) that are widely distributed over the Holarctic region from the dry subtropics to the boreal zone (Fritsch and Friesen 2002). While most species grow in open, sunny, rather dry sites in arid and moderately humid climates, *Alliums* have also evolved to grow in diverse habitats. Ornamental and vegetable *Alliums* exhibit correspondingly wide variability in stature, branching, storage organs, and floral structures (Hanelt 1990). Bulb onion (*Allium cepa*) has been cultivated for over 4000 years, and carvings of onions exist on the pyramid walls highlighting their importance in the Egyptian diet. Crop domestication resulted in strong genetic selection for key genes (and specific variants) that control the plant structures and physiological attributes which distinguish crop plants from their wild relatives (Doebley et al. 2006; Baldwin et al. 2013). Analysing the diversity of key gene families involved in plant architecture and adaptation within and among domesticated *Allium* and their wild allies may provide insights into the key genes under selection during domestication and dispersal (Ross-Ibarra et al. 2007).

In this chapter, we compare gene families involved in the regulation of key traits of bulb onion, such as bulbing, flowering, male fertility, flavonoid biosynthesis, and sulphur assimilation with the corresponding gene families present in other plants. The comparison suggests that the large genome in genus *Allium* is not due to the widespread expansion of individual gene families. This is consistent with a recent prediction that bulb onion has about 35,000 protein-coding genes, similar to other plants that have small genomes such as *Arabidopsis* (27,416) and rice (37,544) (Wendel et al. 2016; Sohn et al. 2016). While in cases, individual gene families have undergone expansion, other gene families contain fewer members than other plant species. Thus, rather than large-scale gene or genome duplications, the large genome in genus *Allium* is likely to be the result of a massive expansion of the amount of repetitive DNA (King et al. 1998; Jakse et al. 2008).

10.2 RNA Sequencing for Gene Identification in Genus *Allium*

RNA sequencing (RNA-Seq) provides a powerful way to identify the genes in species with large genomes (Mutz et al. 2013; Martin et al. 2013). RNA-Seq has been used in *Allium* species, such as bulb onion, garlic, bunching onion and chives to identify candidate genes regulating various traits (Khosa et al. 2016b). A wide range of genotypes and tissues have been sequenced using different platforms (Table 10.1). The focus of these RNA-Seq studies has been on developing molecular markers and discovering candidate genes, involved in bulbing, flowering, flower development, fertility restoration, sulphur biosynthesis and allergenicity (Table 10.1). In addition, transcriptome data has been generated in bulb onion treated with cold temperatures in an effort to better understand freezing tolerance mechanisms (Han et al. 2016). Recently, a comparative transcriptome analysis of nine different *Allium* species has been carried out to better understand morphogenesis and evolution of fistular leaves in genus *Allium* (Zhu et al. 2017).

The number of protein-encoding genes in bulb onion has been estimated to be about 35,000. However, the number of predicted transcripts from these studies is often very high (up to 293,000 unique transcripts), due to multiple transcripts being derived from a single gene and the presence of transcripts that do not encode functional proteins (Table 10.1) (Sohn et al. 2016). These large transcriptome datasets provide an opportunity to compare the number of genes in particular gene families with those of other plant species. The number of distinct genes in each transcription factor gene family varies between plant species. However, overall bulb onion has a similar number of transcription factors to rice (Table 10.2). The variation of gene

Table 10.1 Summary of published RNA sequencing projects in different *Allium* species

Species	Sequencing platform	Genotype	Gene identification	Number of transcripts	Tissue	References
Bulb onion	Illumina HiSeq 2000 platform	CUDH2107	Genes involved in flower development	271,665	Leaves, floral buds from unexpanded umbels, unopened florets from expanded umbels, open florets with pollen, older flowers and roots	Khosa et al. (2016a)
	Illumina HiSeq 2000	36,122 and 36,101	Cold tolerance	93,637	Leaves	Han et al. (2016)
	Illumina HiSeq™ 2500	Chalinghuangpi	NAC transcription factor	117,189	Leaves	Zheng et al. (2016a)
	Illumina HiSeq 2000	Pusa Madhavi	Identification of allergens and epitopes	293,475	Bulb	Rajkumar et al. (2015)
	HisSeq™ 2500,	Utah Yellow Sweet Y1351	Carbohydrate metabolism	79,376	Bulb	Zhang et al. (2016)
	Illumina HiSeq 2000	H6, SP3B	–	165,179	Six weeks whole seedlings	Kim et al. (2014)
	Illumina HiSeq 2000	Bravo, Jumbo, Babosa, California Red, Pukekohe Longkeepe, Rio Tinto, Rumba, Sapporo Yellow Globe and South Port White Globe	–	46,596; 36,897; 99,010; 81,574; 99,761; 81,975; 103,178; 69,206 76,187	Leaves of 4–8-week-old plants	Scholten et al. (2016)
	454 sequencing	CUDH2150	Identification of *FT* gene family	24,106	Leaves and shoot meristem at the 4–5 leaf stage	Baldwin et al. (2012)
	PacBio RSII system	Eumjinara and Sinsunhwang	–	99,247	Flower, leaf, bulb and root	Sohn et al. (2016)
	ABI PRISM 3730XL analyzer	506L and H6	Male fertility restorer genes	32,674	Unopened flowers	Kim et al. (2015)
	Roche 454 FLX	OH1 and 5225	–	27,065; 33,254	Vernalized bulbs, tissue from leaves, unopened umbels, bulbs and roots	Duangjit et al. (2013)
	Illumina HiSeq 2500	–	–	117,189	Leaves	Zhu et al. (2017)
	Illumina Hiseq 2000	Orlando	PCD related genes	45,891	Bulb scales	Galsurker et al. (2017)
Shallot	Illumina HiSeq 2500	–	–	83,186	Leaves	Zhu et al. (2017)
Garlic	Illumina HiSeqTM 2000	Cangshan 15	Genes involved in organic sulphur biosynthesis	127,933	Vegetative buds	Sun et al. (2012)

(continued)

Table 10.1 (continued)

Species	Sequencing platform	Genotype	Gene identification	Number of transcripts	Tissue	References
	–	Cangshan 15	Shoot apex sprouting	45,363	Dormant and sprouting garlic shoot apex	Sun et al. (2013)
		Fertile garlic clone #87	Photoperiodic flowering; flower development and organosulfur metabolism	240,000	Root, Basal plate, Leaf, Clove, Inflorescence and Flowers	Kamenetsky et al. (2015)
		Fertile #87 (F87) and male-sterile #96 (MS96)	Genes involved in male sterility and fertility	–	Flower (Early, Mid and late)	Shemesh-Mayer et al. (2015)
	HiSeqTM 2500	ChalingZiPiSuan		135,360	Bulbs, and whole plant	Liu et al. (2015)
	Illumina HiSeq 2500	–	–	132,225	Leaves	Zhu et al. (2017)
Bunching onion	GS-FLX; Illumina HiSeq 2000	Ki," "F" and "A."	–	42,511;121,354	2-week-old seedlings, leaf, roots, basal meristem, immature flower bract, mature bract (about 1 week before anthesis), opened flowers, immature fruits and sliced pseudostem	Tsukazaki et al. (2015)
	Illumina HiSeq 2000	Bian Gan and glossy BianGan	Waxy cuticle biosynthesis	73,128	Leaves	Liu et al. (2014)
	Illumina HiSeq 2000	Zhangqiu	Genes involved in sulphur and selenium metabolism	103,286	Leaves, false stem, basal plate and root were collected from 14-day old seedlings	Sun et al. (2016)
	Illumina HiSeq 2500	–	–	128,372	Leaves	Zhu et al. (2017)
Chinese Chive	Illumina HiSeq 2000	–	–	150,154	Leaves, shoots and roots	Zhou et al. (2015)
A. ascalonicum	Illumina HiSeq 2500		–		Leaves	Zhu et al. (2017)
A. chinense	Illumina HiSeq 2500	–	–	121,008	Leaves	Zhu et al. (2017)
A. macrostemon	Illumina HiSeq 2500		–	161,681	Leaves	Zhu et al. (2017)
A. tuberosum	Illumina HiSeq 2500		–	148,715	Leaves	Zhu et al. (2017)
A. porrum	Illumina HiSeq 2500		–	189,713	Leaves	Zhu et al. (2017)
F_1 (*A. roylei* and *A. fistulosum*)	Illumina HiSeq 2000	PRI 91021–8	–	10,361	Leaves of 4–8-week-old plants	Scholten et al. (2016)

Table 10.2 Number of members in different transcription factor families in monocots

Gene family	Bulb onion[a]	Rice (*indica*)[b]	Wheat[b]	Maize[b]	Sorghum[b]	*Phalaenopsis* (orchid)[b]
bHLH	162	169	324	308	297	96
NAC	147	158	263	189	180	85
ERF	132	138	181	204	172	91
MYB	121	121	263	203	145	108
WRKY	109	109	171	161	134	67
C2H2	105	113	224	179	140	93

[a]Khosa et al. (2016a)
[b]http://planttfdb.cbi.pku.edu.cn/ on 15-3-2017

family sizes in transcription factors is most likely due to gene duplication and deletions that alter gene family sizes (Guo 2013).

10.3 Evolution of the *FLOWERING LOCUS T* (*FT*) Gene Family in Bulb Onion

The *FT* gene family is found in all taxa of plants and encodes phosphatidylethanolamine-binding domain proteins (PEBP) (Kardailsky et al. 1999; Turck et al. 2008). The number of *PEBP*-like genes vary greatly between different plant species; in model plant *Arabidopsis*, six *FT*-like genes, *FT*, *TSF* (*Twin sister of FT*), *MFT* (*Mother of FT*), *BFT* (*Brother of FT*), *TFL1* (*Terminal flower like 1*) and *ATC* (*Arabidopsis thaliana relatives* of *centroradialis*) have been found, whereas in rice, 13 *FT*-like genes have been identified (Table 10.3) (Turck et al. 2008; Zheng et al. 2016b). Phylogenetic analyses suggest that the PEBP gene family can be divided into three subfamilies: *MFT*, *TFL1*, and *FT* (Karlgren et al. 2011). The *FT*- and *TFL1*-like genes are highly conserved in sequence, but exhibit antagonistic functions: *FT* acts as an activator of flowering, whereas *TFL1* acts as a repressor (Turck et al. 2008; Wickland and Hanzawa 2015). *FT* is produced in the leaves under inductive environmental conditions, and the protein is transported to the shoot apical meristem for flowering initiation (Kardailsky et al. 1999; Turck et al. 2008). The average number of *FT*-like genes in monocots is approximately six times higher than in eudicots (Table 10.3). The expansion of *FT*-like gene families in recent lineages might be due to tandem and segmental duplication in their genomes (Zheng et al. 2016b). In bulb onion, we identified six *FT*-like genes, and their phylogenetic relationship with other monocot FTs revealed that these *FT*-like genes belong to the *FT*-like group (Lee et al. 2013).

FTs can act as universal flowering signals in plants, but some members are involved in a diverse range of functions (Pin and Nilsson 2012). *FT* genes have been found to play an important role in the regulation of poplar growth, heterosis in tomato, stomata opening, potato tuberization and bulbing in onions (Hsu et al. 2011; Krieger et al. 2010; Kinoshita et al. 2011; Navarro et al.

Table 10.3 Number of *FT*-like genes and their functions in different plant species

Species	*FT*-like genes	Functions	References
Bulb onion	6	Bulbing and flowering	Lee et al. (2013)
Rice	13	Flowering	Zheng et al. (2016b)
Maize	15	Flowering	Zheng et al. (2016b)
Sorghum	11	Flowering	Zheng et al. (2016b)
Potato	3	Tuberization and flowering	Navarro et al. (2011)
Poplar	2	Bud cessation and growth	Böhlenius et al. (2006)

2011; Lee et al. 2013). Three *FT*-like genes of bulb onion (*AcFT1*, *AcFT2* and *AcFT4*) influence flowering time in the *Arabidopsis ft-1* mutant (Lee et al. 2013). *AcFT1* acts as a mobile signal to vigorously promote flowering, whereas *AcFT4* represses flowering, and *AcFT2* only marginally alters *Arabidopsis* flowering time. Expression analysis of *FT*-like genes in bulb onion indicates that, during the juvenile stage of the bulb onion life cycle, *AcFT4* is upregulated and acts antagonistically to repress *AcFT1* under both non-inductive, short day (SD) and inductive long day (LD) length conditions. When the onion plant reaches a certain developmental stage, under inductive day length conditions, *AcFT4* is downregulated, and *AcFT1* is upregulated to induce bulbing. Further, *AcFT4*-overexpressing onion plants never form bulbs and have a similar appearance to leeks; *AcFT1*-overexpressing plantlets form bulb-like structures whilst still in tissue culture. Expression and functional studies indicate that *AcFT1* acts as a promoter and that *AcFT4* act as an inhibitor of bulbing. *AcFT2* might act as a flowering promoter as it is only expressed in vernalized and flowering onion plants (Lee et al. 2013). Among other alliums, expression of *FTs* in garlic (*Allium sativum* L.) has been studied and show that *AsFT1* is expressed at higher levels in bulbing plants, and *AsFT4* acts antagonistically to it, however, it is not known whether they are involved in bulbing regulation (Shalom et al. 2015). B-box sequences (regions which determine *FT* function) of *FT1*, *FT2* and *FT4* among different *Allium* species show a high degree of conservation (Khosa et al. unpublished). This indicates that the role of *FTs* in the regulation of storage organ formation and flowering might be conserved in different alliums.

10.4 Circadian Clock and Bulbing

There are a number of parallels between the photoperiodic control of onion bulb formation and the photoperiodic induction of flowering in other plants (Brewster 2008). The circadian clock enables plants to respond to changes in seasons and to regulate different development processes (Sanchez and Kay 2016). In the photoperiodic flowering pathway of *Arabidopsis*, expression of clock genes *GIGANTEA* (*GI*) and *FLAVIN-BINDING, KELCH REPEAT, F-BOX1* (*FKF1*) coincides in long days (LD), enabling the GI-FKF1 proteins to form a complex that binds to CYCLING DOF FACTOR 1 and 2 (CDF1 and CDF2), targeting them for ubiquitin-dependent degradation, thereby releasing *CONSTANS* (*CO*) from repression. CO then accumulates in LD and initiates transcription of *FT*, which in turn initiates flowering after the FT protein moves from the leaves to the shoot apical meristem (Sawa et al. 2007; Johansson and Staiger 2015). In the majority of plant species, single copies of *GI* and *FKF1* are found, but in plants with a highly duplicated genome, such as maize and orchid, they occur in multiple copies (Table 10.4). In bulb onion, there are two copies of *GI* and one copy of *FKF1*. This indicates that certain genomic regions in bulb onion genome have undergone duplication, leading to multiple copies of some genes, but not all (Taylor et al. 2010). Similar to bulb onion, potato responds to day length for the formation of underground storage organs (tubers), and it has been shown that *StGI* and *StFKF1* regulate circadian clock genes to activate the potato *FT* orthologue, *StSP6A*, expression for tuber initiation (Kloosterman et al. 2013). This indicates that photoperiodic pathway genes have evolved to regulate diverse developmental stages, such as underground storage organ formation. Similar to *Arabidopsis*, bulb onion *AcGI* and *AcFKF1* show maximum expression at dusk; however, it is not yet known whether they are involved in bulbing (Taylor et al. 2010; Lee et al. 2013). Furthermore, *AcFKF1* is differentially expressed in SD and LD onions, and this might be responsible for day length sensitivity and timing-of-bulbing differences between SD and LD onions, but it needs further investigation (Taylor et al. 2010).

Table 10.4 The number of *GIGANTEA* and *FKF1* homologues identified in various species

Plants	*GIGANTEA*	*FKF-1*	References
Arabidopsis	1	1	Higgins et al. (2010)
Rice	1	1	Murakami et al. (2007), Higgins et al. (2010)
Brachypodium	1	1	Higgins et al. (2010)
Barley	1	1	Higgins et al. (2010)
Maize	2	2	Dong et al. (2012)
Soybean	3	2	Li et al. (2013)
Medicago	1	–	Kim et al. (2013)
Bulb onion	2	1	Taylor et al. (2010)
Oncidium orchid	2	2	Chang et al. (2011)
Cymbidium sinense	2	–	Zhang et al. (2013)

10.5 Flower Development

The genes involved in the specification of floral organ identity and flower development have been identified and characterized in a wide range of plant species. These studies indicate that common mechanisms of flower development occur across plant species (Thomson et al. 2017). The majority of floral identity genes belong to the MADS box gene family, with a few exceptions, such as *APETALA2* (*AP2*) (Ng and Yanofsky 2001; Becker and Theißen 2003; Wellmer et al. 2014; Thomson et al. 2017). These floral organ identity genes are grouped into five classes (A, B, C, D and E) based on their role in flower development. Genes in class A + E specify sepals, A + B + E petals, B + C + E stamens, C + E carpels, and D ovules, have been identified in different plant species (Wellmer et al. 2014; Thomson et al. 2017). The diversification (number and expression pattern) of MADS box genes involved in flower development contributes towards the floral diversity in the plant kingdom (Irish and Litt 2005; Heijmans et al. 2012). Alliums as a genus are notable and highly valued for their floral diversity (Hanelt 1990).

The bulb onion flower consists of fives whorls of floral organs viz: outer perianth, inner perianth, outer stamens, inner stamens, and carpels (Jones and Emsweller 1936). Recently, we identified various genes involved in flower development including MADS box genes (Khosa et al. 2016a). The phylogenetic relationship of bulb onion transcripts encoding ABCDE-like genes with those of other monocots indicates these transcripts belong to the ABCDE model (Fig. 10.1). Transcripts encoding B class genes *AP3* and *PI* show a close relationship with each other, while C and D class transcripts cluster together in the same clade. Bulb onion E class genes, *AGAMOUS LIKE6* (*AGL6*) and *SEPALLATA3*-like (*SEP3*-like), grouped in separate clades. *AP3* and *PISTILLATA* (*PI*) are members of a paralogous gene lineage, whereas C and D class genes are thought to have arisen from gene duplications during the evolution of angiosperms. E class genes diverged from other MADS box genes most recently (Becker and Theißen 2003; Theißen et al. 2016; Chanderbali et al. 2016). The number of genes encoding class B proteins is higher in bulb onion and other Asparagales (orchids) than in grasses and *Arabidopsis* (Table 10.5). Different studies indicate that ABCDE genes underwent duplication in Asparagales to result in diverse floral structures (Mondragón-Palomino 2013; Tsai et al. 2014; Cai et al. 2015; Otani et al. 2016; Dodsworth 2017). The bulb onion MADS-box genes are highly expressed in floral organs (Khosa et al. 2016a). In the future, it would be desirable to study the expression of ABCDE genes to have a better understanding of their role in flower development and whether any of the paralogous members play a divergent role.

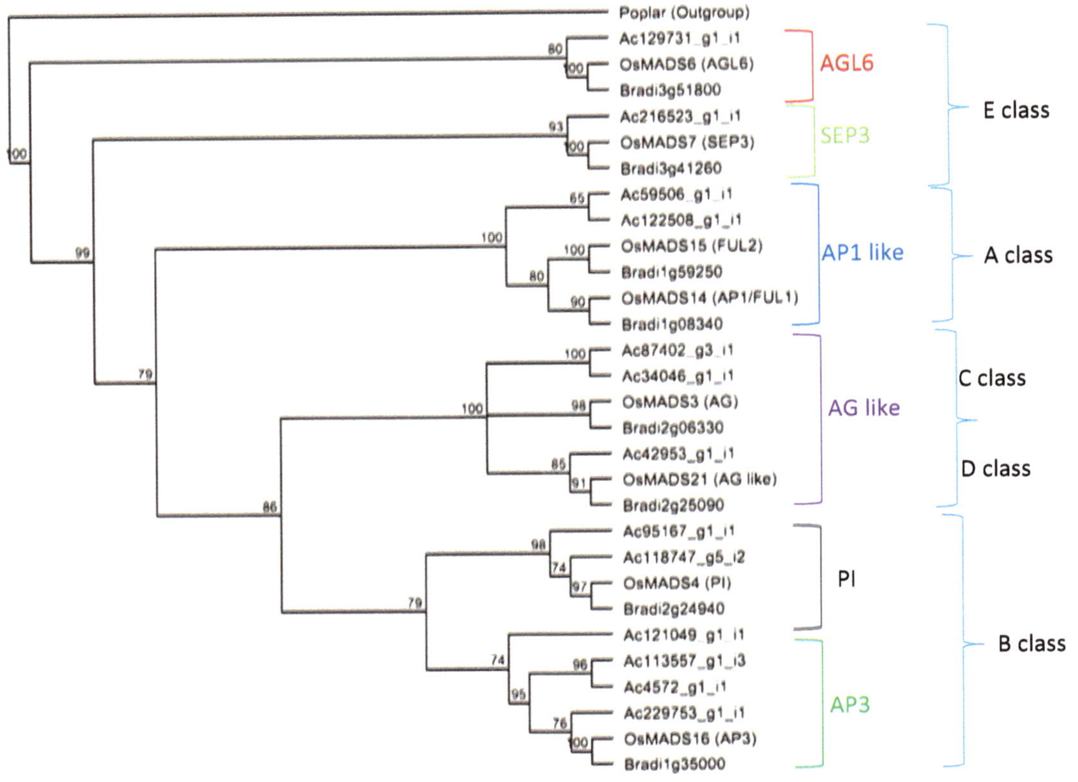

Fig. 10.1 Phylogenetic relationship of bulb onion transcripts encoding ABCDE-like genes with those of other monocots

Table 10.5 The distribution of gene homologues involved in ABCDE model of flower development in different plant species

Class	Gene name	Bulb onion	*Erycina pusilla* (orchid)	*Cymbidium sinense* (orchid)	*Arabidopsis*	Rice
A	*AP1/FUL*	2	3	1	4	4
B	*AP3*	4	3	6	1	1
B	*PI*	2	1	1	1	2
C	*AG*	3	4	3	4	5
E	*AGL6*	1	3	1	2	2
E	*SEP3*	1	2	1	1	2

Adapted from Khosa et al. (2016a), Lin et al. (2016), Xu and Kong (2007)

LEAFY (LFY) is a plant-specific transcription factor involved in floral meristem identity and in the initiation of floral development in plants (Moyroud et al. 2010). In genus *Allium*, the *LFY* orthologue has been identified in bulb onion, shallot, bunching onion and garlic (Rotem et al. 2007; Yang et al. 2016). The *LFY* homologs of different *Allium* species exhibit a high level of sequence conservation and a close phylogenetic relationship. In garlic, two alternatively spliced transcripts of *LFY* have been identified, and only the unspliced variant is associated with the transition from vegetative to floral organ differentiation (Rotem et al. 2007). Similar to other species, *LFY* in garlic and bulb onion is highly expressed during floral transition and floral organ development (Neta et al. 2011; Yang et al. 2016). *Arabidopsis* plants overexpressing *AcLFY* exhibit early flowering and a series of morphological malformations (Yang et al. 2016). In the future, it would be interesting to determine the interaction of *LFY* with other floral integrators to activate genes involved in floral development.

10.6 Male Sterility and Fertility Restoration

Hybrid development of bulb onion varieties became economically feasible with the discovery of cytoplasmic male-sterility (CMS) systems. Various physiological and genetic studies have given us better insights into the origin, distribution and factors regulating CMS in different plant species (Chen and Liu 2014). In many species, most restorer genes (restorers of fertility, Rf or *Rf*) cloned so far belong to pentatricopeptide repeat (PPR) protein family, which contain a characteristic PPR motif of 35 amino acids (Schnable and Wise 1998; Chase 2007; Chen and Liu 2014; Islam et al. 2014). In bulb onion, a wide range of molecular markers have been identified for the identification of male-sterile and maintainer (male-fertile sister) lines (Khosa et al. 2016a). The floral parts of male-fertile and -sterile plants were used for transcriptome sequencing (RNA-Seq), and bulked segregant analysis (BSA) to identify candidate genes involved in fertility restoration (Kim et al. 2015). Candidate genes involved in male fertility restoration gene (*Rf*) in bulb onion have been identified.

A total of 483 transcripts containing the PPR motif have been identified, and among them, 41 transcripts of *Rf*-like PPR genes were found. The number of PPR and *Rf*-like genes identified in bulb onion is consistent with that in other plant species (Fujii et al. 2011; Sykes et al. 2017). However, the bulb onion *Rf*-like genes identified so far are not in linkage disequilibrium (LD) with the male sterility locus (*Ms*). Instead of a PPR-like gene, *AcPMS1* (a regulator of DNA mismatch repair, MMR) seems to be the most plausible candidate gene responsible for the restoration of male fertility in onion CMS. In *Arabidopsis* and tomato mismatch repair genes are responsible for reduced pollen development and induction of CMS (Li et al. 2009; Dion et al. 2007; Sandhu et al. 2007). Also, in other crop plants, various non-PRR genes control fertility restoration. (Chen and Liu 2014). It is possible that the real causal gene for fertility restoration has been missed during screening for homozygous SNPs and differentially expressed genes. Deep RNA, or whole genome, sequencing of mitochondrial sequences from male-sterile and fertile plants is required for better understanding of the evolution and molecular basis of CMS in onion (Kim et al. 2016).

The identification of fertile garlic flowering clones enabled researchers to identify the genetic basis of male sterility. RNA seq of male-fertile and male-sterile developing flowers led to the identification of >16,000 differentially expressed genes. The genes involved in the development of reproductive organs such as *AP3*, *MMD1*, *MS2* and *glycerol-3-phosphate acyltransferase 2* (GPAT2) show high expression in male-fertile plants, whereas genes involved in energy flow, respiration and mitochondrial functions show high expression in male-sterile plants (Shemesh-Mayer et al. 2015). Similar results have been reported during RNA-Seq of Japanese bunching onion male-sterile and fertile plants (Liu et al. 2016). These studies suggest that respiratory restrictions and/or non-regulated programmed cell death of the tapetum can lead to energy deficiency and consequent pollen abortion (Shemesh-Mayer et al. 2015; Liu et al. 2016). In future, functional characterization of the putative role of these genes in male sterility and fertility restoration will give us new clues about CMS system in genus *Allium*.

10.7 Flavonoid Biosynthesis

RNA-Seq data has been used to identify key genes in flavonoid production (Khosa et al. 2016b; Schwinn et al. 2016). Onions have a wide variation in bulb colour due to variation in the production of different flavonoid compounds, such as red anthocyanins, pale yellow flavanols, and bright yellow chalcones. Anthocyanins, flavanols, and chalcones are all products of the flavonoid biosynthetic pathway. The key genes involved in anthocyanin biosynthesis pathway such as *anthocyanidin synthase* (*ANS*) gene, *dihydroflavonol 4-reductase* (*DFR*), chalcone synthase (*CHS*) and *chalcone isomerase* (*CHI*) have been identified, and mutation in these genes leads to different bulb colour (Masuzaki et al. 2006; Khar et al. 2008). In the bulb onion genome, *DFR* and *ANS* occur in single copies, however, in some other plants they are found in multiple copies (Table 10.6). The multiple copies of these genes also show differential expression in different organs, and different catalytic activities, which indicates that gene duplication might play a role in diversifying the functions of DFR enzymes (Huang et al. 2012).

Anthocyanin synthesis is regulated by a transcriptional activation complex consisting of *R2R3-MYB*, bHLH and a WD-repeat (WDR) proteins. This complex acts directly upon the promoters of flavonoid biosynthetic genes (Xu et al. 2015). Recently, four *R2R3-MYB* factors that putatively regulate flavonoid production have been identified in bulb onion transcriptomic datasets (Schwinn et al. 2016). The *MYB* genes from bulb onion and lilies cluster with the SG1/PAP1 clade and these sequences are absent in Poaceae and Orchidaceae indicating divergence of these transcription factors within Asparagales. The *MYB1* and anthocyanin biosynthetic genes, *DFR* and *CHS*, exhibit high expression in red onions. Transient overexpression, and knockdown experiments of the *MYB1* gene leads to induction and inhibition of anthocyanin production, respectively (Schwinn et al. 2016). Bulb onion *MYB1*, overexpressed in transgenic garlic, results in strong red pigmentation in the callus, leaves and leaf bases, but not in control plants. Onion *MYB1* is closely related to that of dicots and complements the anthocyanin *MYB* mutant of snapdragon (Schwinn et al. 2016).

10.8 Sulphur Assimilation and Metabolism

The genes involved in sulphate assimilation pathway are present in photosynthetic organisms, fungi, and many bacteria for the synthesis of sulphur-containing amino acids, and a range of other metabolites such as glucosinolates and alliins (Takahashi et al. 2011). Different *Allium* species have a high content of organosulphur compounds, and in recent years the genomic basis of sulphur assimilation has been studied, especially in bulb onion. Various key genes involved in organosulphur biosynthesis pathways have been reported in bulb onion, garlic and

Table 10.6 The distribution of key gene homologues involved in the anthocyanin biosynthesis pathway

Species	*Dihydroflavonol-4-reductase*	*Anthocyanidin synthase*	References
Arabidopsis	1	1	https://www.arabidopsis.org/
Tomato	1	1	Bongue-Bartelsman et al. (1994)
Potato	1	1	De Jong et al. (2003)
Medicago	2	1	Xie et al. (2004), Pang et al. (2007)
Lotus japonicus	6	–	Shimada et al. (2005)
Rice	1	2	Shih et al. (2008)
Populus trichocarpa	2	2	Huang et al. (2012)
Bulb onion	1	1	Kim et al. (2004a, b)

bunching onion (Brewster 2008; McManus et al. 2012; Kamenetsky et al. 2015; Sun et al. 2016).

In the sulphur assimilation pathway, sulphite reductase (SiR) is a key enzyme involved in the reduction of sulphite for the synthesis of sulphur compounds in bulb onion and other plants (Takahashi et al. 2011; McManus et al. 2012). Bulb onion have single functional *SiR* copy, whereas the rice and poplar genomes have two copies (Kopriva 2006; McManus et al. 2012). Further, phylogenetically, the *Allium* SiR groups with those of Eudicots; the SiR genes of Poales form an independent clade. This is consistent with large-scale phylogenetic relationships, based on EST and genome sequences (Kuhl et al. 2004). Recently, a large number of isoforms encoding different enzymes involved in sulphur metabolism have been identified in a garlic transcriptome dataset (Kamenetsky et al. 2015). The majority of these genes were expressed in a wide range of organs except reproductive organs. This indicates that cysteine sulphoxides are synthesized in leaves and roots, and then translocated to the underground storage organ (Kamenetsky et al. 2015).

10.9 Future Prospects

Over the last decade or so, large amount of genomic data has been generated for different plant species providing us with insights into the genome, gene family and evolution. Advances in technology have led to significant reductions in the cost of generating this data, and offer new avenues for gene discovery. While there is still limited genomic data for *Allium* species such as bulb onion. Transcriptomic datasets have now been generated for bulb onion, and other *Allium* species, leading to gene discovery and marker development. These resources have been utilized to identify the different gene families involved in regulation of bulbing, flowering, bulb colour, fertility restoration and sulphur metabolism. Insights into the evolution and neofunctionalization of genes have been garnered, and there is a lot more to discover yet. In coming years, RNA-Seq along, with whole genome sequencing, will lead to a comprehensive understanding of the different gene families in bulb onion and the *Allium* genus as a whole.

References

Baldwin S, Revanna R, Thomson S, Pither-Joyce M, Wright K, Crowhurst R, Fiers M, Chen L, Macknight R, McCallum J (2012) A toolkit for bulk PCR-based marker design from next-generation sequence data: application for development of a framework linkage map in bulb onion (*Allium cepa* L.). BMC Genom 13:637

Baldwin S, Revanna R, Pither-Joyce M, Shaw M, Wright K, Thomson S, Moya L, Lee R, Macknight R, McCallum J (2013) Genetic analyses of bolting in bulb onion (*Allium cepa* L). Theor App Genet 127:535–547

Becker A, Theißen G (2003) The major clades of MADS-box genes and their role in the development and evolution of flowering plants. Mol Phylogenet Evol 29:464–489

Böhlenius H, Huang T, Charbonnel-Campaa L, Brunner AM, Jansson S, Strauss SH, Nilsson O (2006) CO/FT regulatory module controls timing of flowering and seasonal growth cessation in trees. Science 312:1040–1043

Bongue-Bartelsman M, O'Neill SD, Tong Y, Yoder JI (1994) Characterization of the gene encoding dihydroflavonol 4-reductase in tomato. Gene 138:153–157

Brewster JL (2008) Onions and other vegetable *Alliums*. CABI, Oxfordshire, UK

Cai J, Liu X, Vanneste K, Proost S, Tsai W-C, Liu K-W, Chen L-J, He Y, Xu Q, Bian C, Zheng Z, Sun F, Liu W, Hsiao Y-Y, Pan Z-J, Hsu C-C, Yang Y-P, Hsu Y-C, Chuang Y-C, Dievart A, Dufayard J-F, Xu X, Wang J-Y, Wang J, Xiao X-J, Zhao X-M, Du R, Zhang G-Q, Wang M, Su Y-Y, Xie G-C, Liu G-H, Li L-Q, Huang L-Q, Luo Y-B, Chen H-H, Peer YV, Liu Z-J (2015) The genome sequence of the orchid *Phalaenopsis equestris*. Nat Genet 47:65–72. https://doi.org/10.1038/ng.3149

Chanderbali AS, Berger BA, Howarth DG, Soltis PS, Soltis DE (2016) Evolving ideas on the origin and evolution of flowers: new perspectives in the genomic era. Genetics 202:1255–1265. https://doi.org/10.1534/genetics.115.182964

Chang YY, Chu YW, Chen CW, Leu WM, Hsu HF, Yang CH (2011) Characterization of oncidium 'Gower Ramsey' transcriptomes using 454 GS-FLX pyrosequencing and their application to the identification of genes associated with flowering time. Plant Cell Physiol 52:1532–1545. https://doi.org/10.1093/pcp/pcr101 Epub 2011 Jul 23

Chase CD (2007) Cytoplasmic male sterility: a window to the world of plant mitochondrial–nuclear interactions. Trends Genet 23:81–90

Chen L, Liu YG (2014) Male sterility and fertility restoration in crops. Annu Rev Plant Biol 65:579–606

De Jong WS, De Jong DM, De Jong H, Kalazich J, Bodis M (2003) An allele of dihydroflavonol 4-reductase associated with the ability to produce red anthocyanin pigments in potato (*Solanum tuberosum* L.). Theor Appl Genet 107:1375–1383

Dion E, Li L, Jean M, Belzile F (2007) An *Arabidopsis MLH1* mutant exhibits reproductive defects and reveals a dual role for this gene in mitotic recombination. Plant J 51:431–440

Dodsworth S (2017) Petal, Sepal, or Tepal? B-genes and monocot flowers. Trends Plant Sci 22:8–10. https://doi.org/10.1016/j.tplants

Doebley JF, Gaut BS, Smith BD (2006) The molecular genetics of crop domestication. Cell 127:1309–1321

Dong Z, Danilevskaya O, Abadie T, Messina C, Coles N, Cooper M (2012) A gene regulatory network model for floral transition of the shoot apex in maize and its dynamic modeling. PLoS ONE 7(8):e43450. https://doi.org/10.1371/journal.pone.0043450

Duangjit J, Bohanec B, Chan AP, Town CD, Havey MJ (2013) Transcriptome sequencing to produce SNP-based genetic maps of onion. Theor Appl Genet 126:2093–2101

Fritsch RM, Friesen N (2002) Evolution, domestication and taxonomy. In: Rabinowitch HD, Currah L (eds) Allium crop science: recent advances. CABI, Wallingford, UK, pp 5–30

Fujii S, Bond CS, Small ID (2011) Selection patterns on restorer-like genes reveal a conflict between nuclear and mitochondrial genomes throughout angiosperm evolution. Proc Natl Acad Sci USA 108:1723–1728

Galsurker O, Doron-Faigenboim A, Teper-Bamnolker P, Daus A, Fridman Y, Lers A, Eshel D (2017) Cellular and molecular changes associated with onion skin formation involvement of programmed cell death. Front Plant Sci 7:2031. https://doi.org/10.3389/fpls.2016.02031

Guo YL (2013) Gene family evolution in green plants with emphasis on the origination and evolution of *Arabidopsis thaliana* genes. Plant J 73:941–951. https://doi.org/10.1111/tpj.12089

Han J, Thamilarasan SK, Natarajan S, Park J-I, Chung M-Y, Nou I-S (2016) De novo assembly and transcriptome analysis of bulb onion (*Allium cepa* L.) during cold acclimation using contrasting genotypes. PLoS ONE 11:e0161987. https://doi.org/10.1371/journal.pone.0161987

Hanelt P (1990) 'Taxonomy, evolution and history' in onions and allied crops. In: Rabinowitch HD, Brewster JL (eds) Botany, physiology, and genetics, vol 1. CRC Press, Boca Raton, pp 1–26

Heijmans K, Morel P, Vandenbussche M (2012) MADS-box genes and floral development: the dark side. J Exp Bot 63:5397–5404. https://doi.org/10.1093/jxb/ers233

Higgins JA, Bailey PC, Laurie DA (2010) Comparative genomics of flowering time pathways using *Brachypodium distachyon* as a model for the temperate grasses. PLoS ONE 5(4):e10065. https://doi.org/10.1371/journal.pone.0010065

Hsu CY, Adams JP, Kim H, No K, Ma C, Strauss SH, Drnevich J, Vandervelde L, Ellis JD, Rice BM, Wickett N, Gunter LE, Tuskan GA, Brunner AM, Page GP, Barakat A, Carlson JE, dePamphilis CW, Luthe DS, Yuceer C (2011) FLOWERING LOCUS T duplication coordinates reproductive and vegetative growth in perennial poplar. Proc Natl Acad Sci USA 108:10756–10761

Huang Y, Gou J, Jia Z, Yang L, Sun Y, Xiao X, Song F, Luo K (2012) Molecular cloning and characterization of two genes encoding dihydroflavonol-4-reductase from populus trichocarpa. PLoS ONE 7(2):e30364. https://doi.org/10.1371/journal.pone.0030364

Irish VF, Litt A (2005) Flower development and evolution: gene duplication, diversification and redeployment. Curr Opin Genet Dev 15:454–460

Islam MS, Studer S, Møller IM, Asp T (2014) Genetics and biology of cytoplasmic male sterility and its applications in forage and turf grass breeding. Plant Breeding 133:299–312

Jakse J, Meyer JDF, Suzuki G, McCallum J, Cheung F, Town CD, Havey MJ (2008) Pilot sequencing of onion genomic DNA reveals fragmented transposable elements, low gene densities, and significant gene enrichment after methy filtration. Mol Genet Genomics 280:287–292

Johansson M, Staiger D (2015) Time to flower: interplay between photoperiod and the circadian clock. J Exp Bot 66:719–730. https://doi.org/10.1093/jxb/eru441

Jones HA, Emsweller SL (1936) Development of the flower and macro-gametophyte of *Allium cepa*. J Agric Sci 10:415–428

Kamenetsky R, Faigenboim A, Mayer ES, Michael TB, Gershberg C, Kimhi S, Esquira I, Shalom SR, Eshel D, Rabinowitch HD, Sherman A (2015) Integrated transcriptome catalogue and organ-specific profiling of gene expression in fertile garlic (*Allium sativum* L.). BMC Genom 16:12

Kardailsky I, Shukla VK, Ahn JH, Dagenais N, Christensen SK, Nguyen JT, Chory J, Harrison MJ, Weigel D (1999) Activation tagging of the floral inducer FT. Science 286:1962–1965

Karlgren A, Gyllenstrand N, Källman T, Sundström JF, Moore D, Lascoux M, Lagercrantz U (2011) Evolution of the PEBP gene family in plants: functional diversification in seed plant evolution. Plant Physiol 156:1967–1977

Khar A, Jakse J, Havey MJ (2008) Segregations for onion bulb colors reveal that red is controlled by at least three loci. J Am Soc Hortic Sci 133:42–47

Khosa JS, Lee R, Bräuning S, Lord J, Pither-Joyce M, McCallum J, Macknight RC (2016a) Doubled Haploid 'CUDH2107' as a reference for bulb onion (*Allium cepa* L.) research: development of a transcriptome catalogue and identification of transcripts associated with male fertility. PLoS ONE 11(11):e0166568. https://doi.org/10.1371/journal.pone.0166568

Khosa JS, McCallum J, Dhatt AS, Macknight R (2016b) Enhancing onion breeding using molecular tools. Plant Breed 135:9–20

Kim B, Kim K, Yang TJ, Kim S (2016) Completion of the mitochondrial genome sequence of onion (*Allium cepa* L.) containing the CMS-S male-sterile cytoplasm and identification of an independent event of the ccmF N gene split. Curr Genet 62:873–885

Kim MY, Kang YJ, Lee T, Lee S (2013) Divergence of flowering-related genes in three legume species. Plant Genome. https://doi.org/10.3835/plantgenome2013.03.0008

Kim S, Kim CW, Park M, Choi D (2015) Identification of candidate genes associated with fertility restoration of cytoplasmic male-sterility in onion (*Allium cepa* L.) using a combination of bulked segregant analysis and RNA-seq. Theor Appl Genet 128:2289–2299

Kim S, Binzel ML, Park SH, Yoo KS, Pike LM (2004a) Inactivation of DFR (dihydroflavonol 4-reductase) gene transcription results in blockage of anthocyanin production in yellow onions (*Allium cepa*). Mol Breed 14:253–263. https://doi.org/10.1023/B:MOLB.0000047770.92977.04

Kim S, Binzel ML, Yoo KS, Park S, Pike LM (2004b) Pink (P), a new locus responsible for a pink trait in onions (*Allium cepa*) resulting from natural mutations of anthocyanidin synthase. Mol Genet Genomics 272:18–27. https://doi.org/10.1007/s00438-004-1041-5

Kim S, Kim MS, Kim YM, Yeom SI, Cheong K, Kim KT, Jeon J, Kim S, Kim DS, Sohn SH, Lee YH, Choi D (2014) Integrative structural annotation of de novo RNA-Seq provides an accurate reference gene set of the enormous genome of the onion (*Allium cepa* L.). DNA Res 22:19–27

King JJ, Bradeen JM, Bark O, McCallum J, Havey MJ (1998) A low density genetic map of onion reveals a role for tandem duplication in the evolution of extremely large diploid genome. Theor Appl Genet 96:52–62

Kinoshita T, Ono N, Hayashi Y, Morimoto S, Nakamura S, Soda M, Kato Y, Ohnishi M, Nakano T, Inoue S, Shimazaki K (2011) FLOWERING LOCUST regulates stomatal opening. Curr Biol 21:1232–1238

Kloosterman B, Abelenda JA, Carretero-Gomez M, Oortwijn M, De Boer JM, Kowitwanich K, Horvath BM, Eck HJ, Smaczniak C, Prat S, Visser RGF, Bachem CWB (2013) Naturally occurring allele diversity allows potato cultivation in northern latitudes. Nature 495:246–250

Kopriva S (2006) Regulation of sulfate assimilation in *Arabidopsis* and beyond. Ann Bot 97:479–495

Krieger U, Lippman ZB, Zamir D (2010) The flowering gene SINGLE FLOWER TRUSS drives heterosis for yield in tomato. Nat Genet 42:459–463

Kuhl J, Cheung F, Yuan Q, Martin W, Zewdie Y, McCallum J, Catanach A, Rutherford P, Sink KC, Jenderek M, Prince JP, Town CD, Havey MJ (2004) A unique set of 11,008 expressed sequence tags (EST) reveals expressed sequence and genomic differences between monocot order *asparagales* and *poales*. Plant Cell 16:114–125

Lee R, Baldwin S, Kenel F, McCallum J, Macknight R (2013) FLOWERING LOCUS T genes control onion bulb formation and flowering. Nat Commun 4:2884

Li F, Zhang X, Hu R, Wu F, Ma J, Meng Y, Fu YF (2013) Identification and molecular characterization of FKF1 and GI homologous genes in soybean. PLoS ONE 8 (11):e79036. https://doi.org/10.1371/journal.pone.0079036

Li L, Dion E, Richard G, Domingue O, Jean M, Belzile F (2009) The Arabidopsis DNA mismatch repair gene PMS_1 restricts somatic recombination between homeologous sequences. Plant Mol Biol 69:675–684

Lin CS, Hsu CT, Liao DC, Chang WJ, Chou ML, Huang YT, Chen JJ, Ko SS, Chan MT, Shih MC (2016) Transcriptome-wide analysis of the MADS-box gene family in the orchid *Erycina pusilla*. Plant Biotechnol J 14:284–298. https://doi.org/10.1111/pbi.12383

Liu Q, Lan Y, Wen C, Zhao H, Wang J, Wang Y (2016) Transcriptome sequencing analyses between the cytoplasmic male sterile line and its maintainer line in welsh onion (*Allium fistulosum* L). IntJ Mol Sci 17:1058. https://doi.org/10.3390/ijms17071058

Liu Q, Wen C, Zhao H, Zhang L, Wang J, Wang Y (2014) RNA-Seq reveals leaf cuticular wax-related genes in Welsh onion. PLoS ONE 11:e113290. https://doi.org/10.1371/journal.pone.0113290

Liu T, Zeng L, Zhu S, Chen X, Tang Q, Mei S, Tang S (2015) Large-scale development of expressed sequence tag-derived simple sequence repeat markers by deep transcriptome sequencing in garlic (*Allium sativum* L.). Mol Breed 35:204

Martin LB, Fei Z, Giovannoni JJ, Rose JK (2013) Catalyzing plant science research with RNA-seq. Front Plant Sci 4:66. https://doi.org/10.3389/fpls.2013.00066/full

Masuzaki S, Shigyo M, Yamauchi N (2006) Complete assignment of structural genes involved in flavonoid biosynthesis influencing bulb color to individual chromosomes of the shallot (*Allium cepa* L.). Genes Genet Syst 81:255–263

McCallum J (2007) Onion. In: Kole C (ed) Genome mapping and molecular breeding in plants, vol 5. Springer, Heidelberg, Berlin, New York, pp 331–342

McManus MT, Joshi S, Searle B, Pither-Joyce M, Shaw M, Leung S, Albert N, Shigyo M, Jakse J, Havey MJ, McCallum J (2012) Genotypic variation in sulfur assimilation and metabolism of onion (*Allium cepa* L.) III. Characterization of sulfite reductase. Phytochem 83:34–42

Mondragón-Palomino M (2013) Perspectives on MADS-box expression during orchid flower evolution and development. Front Plant Sci 4:377. https://doi.org/10.3389/fpls.2013.00377

Moyroud E, Kusters E, Monniaux M, Koes R, Parcy F (2010) LEAFY blossoms. Trends Plant Sci 15:346–352

Murakami M, Tago Y, Yamashino T, Mizuno T (2007) Comparative overviews of clock-associated genes of *Arabidopsis thaliana* and *Oryza sativa*. Plant Cell Physiol 48:110–121

Mutz KO, Heilkenbrinker A, Lönne M, Walter JG, Stahl F (2013) Transcriptome analysis using next-generation sequencing. Curr Opin Biotechnol 24:22–30

Navarro C, Abelenda JA, Cruz-Oro E, Cuellar CA, Tamaki S, Silva J, Shimamoto K, Prat S (2011) Control of flowering and storage organ formation in potato by FLOWERING LOCUS T. Nature 478:119–122

Neta R, David-Schwartz R, Peretz Y, Sela I, Rabinowitch HD, Flaishman M, Kamenetsky R (2011) Flower development in garlic: the ups and downs of gaLFY expression. Planta 233:1063–1072. https://doi.org/10.1007/s00425-011-1361-8

Ng M, Yanofsky MF (2001) Function and evolution of the plant MADS-box gene family. Nat Rev Genet 2:186–195

Otani M, Sharifi A, Kubota S, Oizumi K, Uetake F, Hirai M, Hoshino Y, Kanno A, Nakano M (2016) Suppression of B function strongly supports the modified ABCE model in *Tricyrtis* sp. (Liliaceae). Sci Rep 6:24549. https://doi.org/10.1038/srep24549

Pang Y, Peel GJ, Wright E, Wang Z, Dixon RA (2007) Early steps in proanthocyanidin biosynthesis in the model legume *Medicago truncatula*. Plant Physiol 145:601–615

Pin P, Nilsson O (2012) The multifaceted roles of FLOWERING LOCUS T in plant development. Plant Cell Environ 35:1742–1755

Rajkumar H, Ramagoni RK, Anchoju VC, Vankudavath RN, Syed AUZ (2015) De novo transcriptome analysis of *Allium cepa* L. (onion) bulb to identify allergens and epitopes. PLoS ONE 10(8):e0135387. https://doi.org/10.1371/journal.pone.0135387

Ross-Ibarra J, Morrell PL, Gaut BS (2007) Plant domestication, a unique opportunity to identify the genetic basis of adaptation. Proc Natl Acad Sci 104(suppl 1):8641–8648

Rotem N, Shemesh E, Peretz Y, Akad F, Edelbaum O, Rabinowitch HD, Sela I, Kamenetsky R (2007) Reproductive development and phenotypic differences in garlic are associated with expression and splicing of LEAFY homologue gaLFY. J Exp Bot 58:1133–1141

Sanchez SE, Kay SA (2016) The plant circadian clock: from a simple timekeeper to a complex developmental manager. Cold Spring Harb Perspect Biol 8:12. https://doi.org/10.1101/cshperspect.a027748

Sandhu AP, Abdelnoor RV, Mackenzie SA (2007) Transgenic induction of mitochondrial rearrangements for cytoplasmic male sterility in crop plants. Proc Natl Acad Sci USA 104:1766–1770

Sawa M, Nusinow DA, Kay SA, Imaizumi T (2007) FKF1 and GIGANTEA complex formation is required for day-length measurement in *Arabidopsis*. Science 318:261–265

Schnable PS, Wise RP (1998) The molecular basis of cytoplasmic male sterility and fertility restoration. Trends Plant Sci 3:175–180

Scholten OE, van Kaauwen MPW, Shahin A, Hendrickx PM, Paul Keizer LC, Burger K, van Heusden AW, van der Linden CG, Vosman B (2016) SNP-markers in *Allium* species to facilitate introgression breeding in onion. BMC Plant Biol 16:187

Schwinn KE, Ngo H, Kenel F, Brummell DA, Albert NW, McCallum JA, Pither-Joyce M, Crowhurst NS, Eady C, Davies KM (2016) The Onion (*Allium cepa* L.) *R2R3-MYB* Gene *MYB1* regulates anthocyanin biosynthesis. Front Plant Sci 7:1865. https://doi.org/10.3389/fpls.2016.01865

Shalom SR, Gillett D, Zemach H, Kimhi S, Forer I, Zutahy Y, Tam Y, Teper-Bamnolker P, Kamenetsky R, Eshel D (2015) Storage temperature controls the timing of garlic bulb formation via shoot apical meristem termination. Planta 242:951–962

Shemesh-Mayer E, Ben-Michael T, Rotem N, Rabinowitch HD, Doron-Faigenboim A, Kosmala A, Perlikowski D, Sherman A, Kamenetsky R (2015) Garlic (*Allium sativum* L.) fertility: transcriptome and proteome analyses provide insight into flower and pollen development. Front Plant Sci 6:271. https://doi.org/10.3389/fpls.2015.00271

Shih CH, Chu H, Tang LK, Sakamoto W, Maekawa M, Chu IK, Wang M, Lo C (2008) Functional characterization of key structural genes in rice flavonoid biosynthesis. Planta 228:1043–1054

Shimada N, Sasaki R, Sato S, Kaneko T, Tabata S, Aoki T, Ayabe SA (2005) Comprehensive analysis of six dihydroflavonol 4-reductases encoded by a gene cluster of the *Lotus japonicus* genome. J Exp Bot 56:2385–2573

Sohn SH, Ahn YK, Lee TH, Lee JE, Jeong MH, Seo CH, Chandra R, Kwon YS, Kim CW, Kim DS, Won SY, Kim JS, Choi D (2016) Construction of a draft reference transcripts of onion (*Allium cepa*) using long-read sequencing. Plant Biotechnol Rep 10:383–390

Sun X, Zhou S, Meng F, Liu S (2012) De novo assembly and characterization of the garlic (*Allium sativum*) bud transcriptome by Illumina sequencing. Plant Cell Rep 31:1823–1828

Sun XD, Ma GQ, Cheng B, Li H, Qi S (2013) Identification of differentially expressed genes in shoot apex of garlic (*Allium sativum* L.) using Illumina sequencing. J Plant Stud 2:136–148

Sun XD, Yu XH, Zhou SM, Liu SQ (2016) De novo assembly and characterization of the Welsh onion (*Allium fitulosum* L.) transcriptome using Illumina technology. Mol Genet Genomics 291:647–659

Sykes T, Yates S, Nagy I, Asp T, Small I, Studer B (2017) In silico identification of candidate genes for fertility restoration in cytoplasmic male sterile perennial ryegrass (*Lolium perenne* L.). Genome Biol Evol 9:351–362

Takahashi H, Kopriva S, Giordano M, Saito K, Hell R (2011) Sulfur assimilation in photosynthetic organisms: molecular functions and regulations of transporters and assimilatory enzymes. Ann Rev of Plant Biol 62:157–184

Taylor A, Massiah AJ, Thimas B (2010) Conservation of *Arabidopsis thaliana* photoperiodic flowering genes in onion (*Allium Cepa* L.). Plant Cell Physiol 51:1638–1647

Theißen G, Melzer R, Rümpler F (2016) MADS-domain transcription factors and the floral quartet model of flower development: linking plant development and evolution. Development 143:3259–3271. https://doi.org/10.1242/dev.134080

Thomson B, Zheng B, Wellmer F (2017) Floral organogenesis: when knowing your ABCs is not enough. Plant Physiol 173:56–64

Tsai WC, Pan ZJ, Hsiao YY, Chen LJ, Liu ZJ (2014) Evolution and function of MADS-box genes involved in orchid floral development. J Syst Evol 52:397–410

Tsukazaki H, Yaguchi S, Sato S, Hirakawa H, Katayose Y, Kanamori H, Kurita K, Itoh T, Kumagai M, Mizuno S, Hamada M, Fukuoka H, Yamashita K, McCallum JA, Shigyo M, Wako T (2015) Development of transcriptome shotgun assembly-derived markers in bunching onion (*Allium fistulosum*). Mol Breed 35:55

Turck F, Fornara F, Coupland G (2008) Regulation and identity of florigen: FLOWERING LOCUS T moves center stage. Annu Rev Plant Biol 59:573–594

Varshney RK, Terauchi R, McCouch SR (2014) Harvesting the promising fruits of genomics: applying genome sequencing technologies to crop breeding. PLoS Biol 12:e1001883

Wellmer F, Graciet E, Riechmann JL (2014) Specification of floral organs in *Arabidopsis*. J Exp Bot 65:1–9

Wendel JF, Jackson SA, Meyers BC, Wing RA (2016) Evolution of plant genome architecture. Genome Biol 17:37. https://doi.org/10.1186/s13059-016-0908-1

Wickland DP, Hanzawa Y (2015) The *FLOWERING LOCUS T/TERMINAL FLOWER 1* gene family: Functional evolution and molecular mechanisms. Mol Plant 8:983–997

Xie DY, Jackson LA, Cooper JD, Ferreira D, Paiva NL (2004) Molecular and biochemical analysis of two cDNA clones encoding dihydroflavonol-4-reductase from *Medicago truncatula*. Plant Physiol 134:979–994

Xu G, Kong H (2007) Duplication and divergence of floral MADS-box genes in grasses: evidence for the generation and modification of novel regulators. J Integr Plant Biol 49:927–939

Xu W, Dubos C, Lepiniec L (2015) Transcriptional control of flavonoid biosynthesis by MYB-bHLH-WDR complexes. Trends Plant Sci 20:176–185

Yang C, Ye Y, Song C, Chen D, Jiang B, Wang Y (2016) Cloning and functional identification of the *AcLFY* gene in *Allium cepa*. Biochem Biophys Res Commun 473:1100–1105

Zhang C, Zhang H, Zhan Z, Liu B, Chen Z, Liang Y (2016) Transcriptome analysis of sucrose metabolism during bulb swelling and development in onion (*Allium cepa* L.). Front Plant Sci 7:1425. https://doi.org/10.3389/fpls.2016.01425

Zhang J, Wu K, Zeng S, Teixeira da Silva JA, Zhao X, Tian C, Xia H, Duan J (2013) Transcriptome analysis of *Cymbidium sinense* and its application to the identification of genes associated with floral development. BMC Genom 14:279

Zheng X, Tang S, Zhu S, Dai Q, Liu T (2016a) Identification of an NAC transcription factor family by deep transcriptome sequencing in onion (*Allium cepa* L.). PLoS ONE 11(6):e0157871. https://doi.org/10.1371/journal.pone.0157871

Zheng XM, Wu FQ, Zhang X, Lin QB, Wang J, Guo XP, Lei CL, Cheng ZJ, Zou C, Wan JM (2016b) Evolution of the PEBP gene family and selective signature on *FT*-like clade. J Syst Evol 54:502–510

Zhou SM, Chen LM, Liu SQ, Wang XF, Sun XD (2015) De novo assembly and annotation of the Chinese chive (*Allium tuberosum* Rottler ex Spr.) transcriptome using the Illumina platform. PLoS ONE 10(7):e0133312. https://doi.org/10.1371/journal.pone.0133312

Zhu S, Tang S, Tan Z, Yu Y, Dai Q, Liu T (2017) Comparative transcriptomics provide insight into the morphogenesis and evolution of fistular leaves in *Allium*. BMC Genom 18:60. https://doi.org/10.1186/s12864-016-3474-8

Structural and Functional Genomic Resources Developed

11

Go Suzuki, Turgut Yigit Akyol and Shusei Sato

Abstract

Structural genomic resources of *Allium* plants have been less developed because *Allium* species have a relatively large genome size. In this background, there have been few reports of construction of BAC libraries of *Allium* species. Nevertheless, the BAC clones of *Allium cepa* (bulb onion) developed as a partial library can be used for not only molecular cloning of novel onion genes but also used as FISH probes in molecular cytogenetic studies. The BAC sequencing and BAC FISH analyses in *A. cepa* revealed that the highly repetitive elements, such as retrotransposon, are major components of the large chromosomes. With the development of next-generation sequencing technologies, transcriptome level sequencing and assembly has been applied to several *Allium* species, such as bulb onion, shallot, garlic, bunching onion, and Chinese chive. To activate the accumulating transcriptome information as practical genomic resources of *Allium* research, a number of attempts have been made for anchoring unigenes onto the genetic map of *Allium* species. Several databases are available on the Internet that provide information about *Allium* transcriptome, genetic map, and DNA markers.

11.1 Introduction

Allium species have a relatively large genome size (Labani and Elkington 1987; Ricroch et al. 2005; Zonneveld et al. 2005), resulting in technical difficulty of genomic analysis in molecular genetic studies. Because of large genome size, the structural genomic resources of *Allium* plants have been less developed. A bacterial artificial chromosome (BAC) library was one of the most important genomic tools in the genome era, when genome projects and comparative genomics studies were done manually, using the molecular cloning method (Thomas et al. 2000; Zhang and Wu 2001). In the last two decades, there have been few reports of the construction of BAC libraries of *Allium* species: to our knowledge, BAC libraries of garlic (*Allium sativum*) and bulb onion (*Allium cepa*) to date have been reported by Suzuki et al. (2001) and Lee et al. (2003).

G. Suzuki
Laboratory of Plant Molecular Genetics, Division of Natural Science, Osaka Kyoiku University, 4-698-1 Asahigaoka, Kashiwara, Osaka 582-8582, Japan
e-mail: gsuzuki@cc.osaka-kyoiku.ac.jp

T. Y. Akyol · S. Sato (✉)
Graduate School of Life Sciences, Tohoku University, 2-1-1 Katahira, Aoba-ku, Sendai 980-8577, Japan
e-mail: shuseis@ige.tohoku.ac.jp

T. Y. Akyol
e-mail: tyakyol@dc.tohoku.ac.jp

M. Shigyo et al. (eds.), *The Allium Genomes*, Compendium of Plant Genomes,
https://doi.org/10.1007/978-3-319-95825-5_11

These valuable *Allium* BAC clones can be efficiently used as fluorescence in situ hybridization (FISH) probes in molecular cytogenetic studies. In addition, several gene-specific BAC clones were screened in *A. cepa*, revealing genomic organization of bulb onion genes (Do et al. 2004; Jakse et al. 2008; Masamura et al. 2012; Suzuki et al. 2013). In the first half of this chapter, construction of the BAC library of *A. cepa* and its usage will be summarized in short.

Large genome size of *Allium* species also prevented the development of resources for functional genomic analysis. Analysis of expressed sequence information was the practical approach for creation of functional genomic resources in *Allium* species, and limited information was accumulated in the era of conventional Sanger sequencing. The development of next-generation sequencing (NGS) technology increased the amount of sequence data and reduced the cost of data production. NGS made expressed sequence analysis into the so-called transcriptome analysis, and the transcriptomic studies expanded functional genomic resources for *Allium* species. In the second half of this chapter, numbers of transcriptome analyses applied for *Allium* genomic research are summarized. In order to use the transcriptome information as an actual functional genomic resource, it is important to link the gene level information with functional annotation to genetic map information. These approaches are also summarized in this chapter. To utilize the accumulated information as functional genomic resource, the system to support data access is also essential, and thus the databases providing the information related to functional genomic studies in *Allium* species are listed.

11.2 Structural Genomic Resources Developed

11.2.1 Construction and Stock of the BAC Library of *Allium Cepa*

The haploid genome size of *A. cepa* is about 15,000 Mb, which is almost 120 times larger than *Arabidopsis thaliana* (Arumuganathan and Earle 1991; Lysak et al. 1998). A partial BAC library of *A. cepa* consisting of 48,000 clones with average 100-kb insert size DNA, which represent approximately 0.32 genome equivalent, was constructed using Korean cultivar Cheonjudaego (Suzuki et al. 2001). To construct this BAC library, high molecular weight (HMW) DNA embedded in agarose plugs was prepared from protoplast cells isolated from leaves. The purified plugs were sticky, and the HMW DNA was difficult to be digested (data not shown). However, relatively longer incubation (1 h in 37 °C) with *Hin*dIII lead to successful partial digestion. The size-fractionated DNA ranging from ca. 100 to 150 kb (Fig. 11.1a) was then inserted into the pBeloBAC11 vector (Kim et al. 1996), and was electroporated into *Escherichia coli* strain DH10B. The average insert size of 100 kb was determined by *Not*I digestion of randomly selected 270 clones (Fig. 11.1b).

These 48,000 BAC clones were pooled into 480 glycerol stocks, corresponding to only five 96-well plates and kept in −80 °C (Suzuki et al. 2002). Although this BAC library did not cover the whole genome, some targeted BAC clones were obtained by using PCR screening from three-dimensional (3D) superpools of BAC DNAs (Fig. 11.2a). At first, with gene-specific primer sets, PCR screening was performed using 5 plate superpools, 12 column superpools, and 8 row superpools as template. When a candidate pool was identified, positive clone(s) could be screened by a colony PCR method using colonies obtained from a glycerol stock of the positive pool (100 clones are pooled in each glycerol stock) (Fig. 11.2b). Under this strategy, BACs containing a root alliinase gene (*ALL1*), lachrymatory factor synthase (LFS) genes, and a RAD21 gene were successfully identified and analyzed (Table 11.1) (Do et al. 2004; Masamura et al. 2012; Suzuki et al. 2013).

11.2.2 Molecular Cloning of Genomic Regions Containing Bulb Onion Genes

Do et al. (2004) reported that a novel root alliinase gene, *ALL1* (alliinase-like 1), was identified from the BAC screening in *A. cepa*. The first PCR

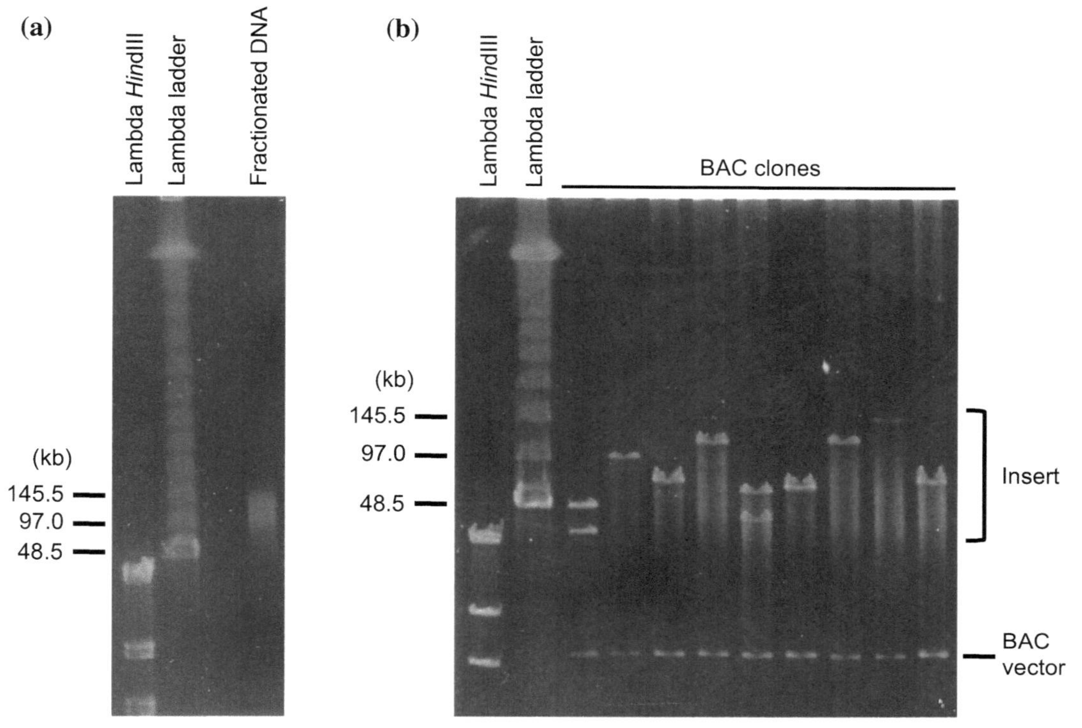

Fig. 11.1 BAC library construction in *Allium cepa*. **a** Size-fractionated HMW DNA (ca. 100–150 kb), which was partially digested with *Hin*dIII, was subjected to pulsed field gel electrophoresis (PFGE) analysis. The fractionated DNA was used as a BAC insert. **b** Insert check of the BAC clones. Randomly selected BAC clones were digested with *Not*I, and subjected to PFGE analysis

screening of alliinase BACs revealed that 4F4 and 5G5 pools were positive (Suzuki et al. 2002). Then, from the 2nd PCR screening of these pools, the positive BACs, 4F4-77 (86 kb) and 5G5-67 (105 kb), which overlapped each other, were successfully obtained (Do et al. 2004). This result suggested that the partial BAC library of *A. cepa* could be used for molecular cloning of bulb onion genes. Although, at that time, only a part of the 4F4-77 BAC (35 kb) could be sequenced by the conventional dideoxy chain termination method using model 310 DNA sequencer (Applied Biosystems), many repetitive sequences (including retrotransposons and MITE elements) were found in the 5'- and 3'-flanking regions of the *ALL1* gene (Table 11.1). Such highly repetitive nature of the *ALL1*-flanking regions was confirmed by genomic DNA gel blot analysis hybridized with the flanking subclones (Do et al. 2004). Thus, the *ALL1* BAC showed interesting structural genomic characteristics of *A. cepa*.

Similarly, the 2E8/10 and 4F10/155 BACs were found to contain the *LFS* genes (Masamura et al. 2012), and the 3B1-60 BAC (100 kb) containing the *AcRAD21-1* gene was obtained from the BAC screening in *A. cepa* (Suzuki et al. 2013). It is noteworthy that NGS technology, such as 454 GS FLX system (Roche), could be applied for sequencing of these BAC clones (Masamura et al. 2012; Suzuki et al. 2013). In the flanking regions of the *LFS* and *AcRAD21-1* genes, retrotransposon-related sequences were also found, indicating that such highly repetitive characteristics of intergenic regions are common in *A. cepa* (Table 11.1).

Jakse et al. (2008) also determined sequences of the 1G-12-89 BAC (108 kb) with degenerated sulfite reductase fragments and the S1-D12 BAC

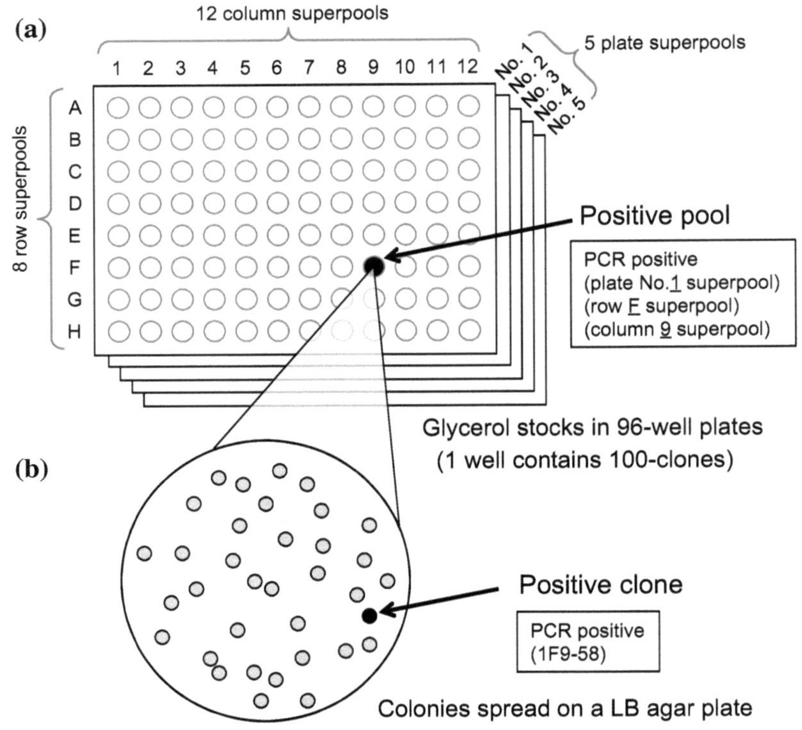

Fig. 11.2 PCR screening from the pooled BAC stocks. **a** One-hundred-clones-pooled BAC colonies were stocked in each well of the five 96-well plates with glycerol at −80 °C. PCR screening can be performed using the superpooled DNA isolated from these clones. For example, if positive amplified fragments are obtained from plate-No.1, row-F, and column-9 superpools, the 1F9 positive pool can be identified. **b** *E. coli* of the positive pools is then spread on a LB plate, and screened by colony PCR. As a result, a positive clone (e.g., 1F9-58) can be identified

(94 kb) without genes, in addition to 460 BAC-end sequences from 316 BACs. These BAC sequences revealed the estimated mean gene density of *A. cepa* as one gene per 168 kb (Jakse et al. 2008); this gene density is extremely low when compared to plants with small genome size, such as *Arabidopsis*. Highly repeated retrotransposon sequences between genes and the low gene density in *A. cepa* suggested that retroelements could expand genome and chromosome sizes in *Allium* species. As for other sequence characteristics of the *Allium* genome, average GC content in the sequenced regions of bulb onion BACs was 33–36% (Table 11.1), which is similar to *Arabidopsis* genome (The Arabidopsis Genome Initiative 2000). Thus, the BAC clones for sequencing are useful resources revealing genomic organization and sequence characteristics of *Allium* species.

11.2.3 Molecular Cytogenetic Use of BAC Clones in *A. Cepa*

Large chromosomes in plants generally contain a number of complex repetitive sequences, including retroelements, which influence their genome sizes and chromosome evolution (San-Miguel and Bennetzen 1998; Shirasu et al. 2000; Neumann et al. 2006; Suzuki et al. 2012; Zhang et al. 2004). Actually, chromosomes of *A. cepa* are very large in size (chromosome length of almost 20 μm observed in the mitotic metaphase), and it is consistent with the fact that BAC clones of *A. cepa* contain many repetitive sequences and retroelements as described above.

In the conventional cytogenetic analysis, plants with large chromosomes are advantageous for karyotyping and visualizing chromosomes' movement in the cell division, because the large

Table 11.1 Onion BAC clones sequenced to date

BAC clone no.	Insert length (kb)	Gene(s) contained	Repetitive element(s) contained	Average GC content (%) in the sequenced region	Reference
4F4-77	86	*ALL1*	retrotransposon, MITE	33.2*	Do et al. (2004)
5G5-67	105	*ALL1*	n.d.	n.d.	Do et al. (2004)
2E8/10	n.d.	*LFS*	retrotransposon-like	n.d.	Masamura et al. (2012)
4F10/155	n.d.	*LFS*	retrotransposon-like	n.d.	Masamura et al. (2012)
3B1-60	100	*AcRAD21-1*	retrotransposon-like	n.d.	Suzuki et al. (2013)
1G-12-89	108	*MYB-related*	transposon-like	34.2	Jakse et al. (2008)
S1-D12	94	no gene	retrovirus or transposon	36.2	Jakse et al. (2008)

*this GC content was calculated from the 35-kb sequence
n.d., not determined

chromosomes are easy to observe under microscopes. On the other hand, in the molecular cytogenetic analysis (e.g., FISH), complex repetitive elements make it difficult to detect the particular locus on the large chromosome(s). In the FISH analysis, longer DNA fragments for the probes are usually better; because the amount of fluorescence labeling depends on the fragment length, and signal intensity of the fluorescence is quite important for the detection. Therefore, BAC clones with large target DNAs are known to be efficient probes for FISH analysis. However, the BAC clones obtained from large plant chromosomes typically contain complex repetitive elements, which can hybridize with many dispersed genomic DNA to disturb the single-signal detection. Actually, in *A. cepa*, the FISH signals probed with specific BACs without competitive DNAs were frequently observed on entire chromosomes (Suzuki et al. 2001; 2012). It can be noted that appropriate usage of the BAC probes with competitive DNAs (e.g., C_0t-1) resulted in successful detection of the specific locus; FISH signals of the 2E8/10 and 4F10/155 BACs with LFS genes were detected on the proximal region of the long arm of chromosome 5 in *A. cepa* (Masamura et al. 2012).

The FISH analysis of randomly selected BAC clones without competitive DNAs (random BAC FISH analysis) can visually reveal the repetitive nature of the plant genome (Suzuki et al. 2001; 2012; Matsuba et al. 2015). The random BAC FISH of monocot plants revealed that less unique and denser dispersed signals of BAC FISH were observed in species with larger chromosomes in both the Poales and Asparagales species (Suzuki et al. 2012).

11.3 Functional Genomic Resources Developed

11.3.1 Transcriptome Analysis in Onion, Bunching Onion, and Garlic

Due to the fact that *Allium* species have huge genome size, transcript information has been considered as a practical functional genomic resource of *Allium* species. During the era of conventional Sanger sequencing technology, only medium-scale sequencing of ESTs was conducted in bulb onion (11,008 ESTs; Kuhl et al. 2004) and garlic (4752 ESTs; Kim et al. 2009). The development of next-generation sequencing technology changed the situation, and transcriptome level sequencing and assembly has been successfully applied to *Allium* species.

A number of transcriptome sequencing of bulb onion (*A. cepa* Common onion group) using NGS platforms have been carried out targeted to the identification of IgE epitopes, adaptation to cold stress, and analysis of sucrose metabolism (Rajkumar et al. 2015; Han et al. 2016; Zhang et al. 2016). Along with these target oriented transcriptome analyses, an approach for developing reference transcriptome catalog was also carried out (Khosa et al. 2016). As the highly heterozygous nature of bulb onion genome makes an analysis of sequencing data complicated, homozygous double haploid line "CUDH2107" was selected as the target of the transcriptome analysis aimed to reference catalog creation. A total of 271,665 transcript contigs were generated by assembling accumulated 396 million 100 bp paired-end Illumina reads, and the obtained contigs were analyzed with respect to gene ontology (GO). As a major research target of *Allium* species, the reference transcriptome data on doubled haploid bulb onion represent a valuable resource for genetic and genomic studies of *Allium* spp.

Shallot (*A. cepa* Aggregatum group) is recognized as a potential genetic resource for *Allium* crop improvement because of their adaptability to environmental stresses and thus it has been applied for several important material resources such as the monosomic addition lines (MALs) of the Japanese bunching onion (*Allium fistulosum*) with an extra chromosome from the shallot developed by Shigyo and colleagues (Shigyo et al. 1996). With the aim to make the best use of these resources, transcriptome analysis of doubled haploid shallot (accession "Chiang Mai") was carried out by applying Illumina sequence technology (Abdelrahman et al. 2017). Around 25 million qualified reads were accumulated from three organs (bulb, leaf, and root) and assembled into unigenes. The data set of 56,161 bulb unigenes was used as reference data set and annotated by similarity search against rice and Arabidopsis protein databases as well as nr database.

In garlic (*A. sativum*), the first NGS based transcriptome analysis was reported on RNA from the renewal buds, and it resulted in de novo assembly of 128,000 unigenes which were annotated and analyzed with respect to GO and metabolic pathways (Sun et al. 2012). In order to apply transcriptome information on genetic studies and breeding in garlic, the transcriptome analysis of fertile garlic was carried out. Over 32 million 250-bp paired-end reads were assembled into an extensive transcriptome of 240,000 contigs (Kamenetsky et al. 2015). Among these contigs, 102,000 highly expressed ones were annotated and analyzed for gene ontology and metabolic pathways. As containing a large number of specific reads in inflorescences and flowers, the transcriptome of fertile garlic represents a new resource for research and breeding of garlic as well as for the development of effective molecular markers for useful traits, including fertility and seed production.

In bunching onion (*A. fistulosum*), a large-scale transcript analysis was carried out by using two types of NGS platforms, GS-FLX and HiSeq 2000. By using GS-FLX platform, 42,511 contigs with an N50 of 1,114 bp were obtained. Then the unigene sequences obtained from HiSeq 2000 platform were mapped onto the 42,511 contigs to combine the two datasets, and finally, a total of 54,903 unigenes were obtained. The accumulated unigene sequences were applied for marker creation for development of genetic linkage map (Tsukazaki et al. 2015). Another large-scale transcriptome analysis of bunching onion was carried out by using Illumina HiSeq 2000 platform. A total of 53,837 unigenes with an average length of 619 bp were assembled, of which 35,250 were annotated in public protein databases. Based on the annotation, candidate unigenes involved in sulfur, selenium, and vitamin metabolism were revealed. A total of 2014 SSR markers were developed, which facilitated genetic linkage mapping and enabled the analysis of genetic diversity in *A. fistulosum* (Sun et al. 2016).

Reduction in the cost of sequencing by NGS technologies expanded the targets of transcriptome data collection. In this context, Zhou et al. (2015) reported the transcriptome data of Chinese chive (*Allium tuberosum*). By assembling 52 million reads obtained by Illumina HiSeq 2000 sequencing platform, a total of 60,031 unigenes were obtained, of which 35,648 were annotated in public protein databases. Using the Kyoto Encyclopedia of Genes and

Genomes (KEGG) pathway database, 21,361 unigenes were mapped onto 128 pathways. SSR sequences applicable for marker creation were identified on 2125 sequences. This dataset can be considered as the most comprehensive functional genomic resource that will facilitate further genetic research on Chinese chive and related species.

The NGS-based transcriptome analysis was also applied for comparative genomic analysis of *Allium* species. For the purpose of analysis of fistular leaf formation, the transcriptomes of nine economically important *Allium* species, including six species with fistular leaf (*A. cepa* L. var. cepa, *A. cepa* var. aggregatum, *A. fistulosum*, *A. ascalonicum*, *A. chinense* and *A. macrostemon*) and three species with solid-leaf (*A. sativum*, *A. porrum* and *A. tuberosum*), were accumulated (Zhu et al. 2017). By comparing the unigene sequences obtained by assembling 99.9–128.9 million Illumina paired-end sequencing of leaf transcriptomes of nine *Allium* species, it was found that many genes that were potentially related to programmed cell death either exhibited rapid diversification in fistular-leaved species, or were conserved in solid-leaved species in evolutionary history. This research indicated the practical approach for evolutional and phylogenetical study in *Allium* species based on comparative genomic analysis, and thus accumulated transcriptome sequence information will be used as valuable genetic resource in the *Allium* species with huge genome sizes.

11.3.2 Anchoring Transcriptome Information onto Genetic Map

Considering the application of accumulated transcriptome information as a functional genomic resource, its linkage to the genetic map information is one of the key issues. Genetic map development in onion and other *Allium* species has been limited by difficulty in developing and maintaining the mapping population as well as high degrees of heterozygosity. The first published genetic map of an *Allium* species was developed by King and colleagues (King et al. 1998) in the intraspecific onion cross "BYG15-23 x AC43" by using cDNA clone based RFLP markers. As the onion ESTs proved to be an excellent source of codominant PCR-based markers for genetic analyses in onion, this map was subsequently augmented with SNP and SSR markers derived from EST sequencing. Due to the transferable nature of EST-based markers, these markers were applied for map-based genetic analysis of a number of traits using intraspecific onion crosses (Gokce et al. 2002; Khar et al. 2008; McCallum et al. 2006, 2007). In order to increase the efficiency of PCR-based marker design from next-generation sequence data, a web-based toolkit enabling rapid development of sets of PCR assays targeting sequence variants identified from Illumina and 454 sequence data was developed (Baldwin et al. 2012).

Genetic map of bunching onion has been developed in several larger mapping pedigrees by using SSR and AFLP markers (Ohara et al. 2005; Tsukazaki et al. 2008), and applied for QTL analysis of seedling vigor (Ohara et al. 2009). NGS derived sequence information has been applied on bunching onion genetic map by anchoring SSR markers created from bunching onion genome sequences and onion EST sequences analyzed by NGS (Tsukazaki et al. 2010).

To increase the anchored markers on genetic maps of *Allium* species, mapping population of interspecific cross between onion and *Allium roylei*, which is the only *Allium* relative known to readily produce fertile hybrids with onion, was constructed (Meer and Vries 1990; van Heusden et al. 2000). Since *A. roylei* also crosses with bunching onion, trihybrid *A. cepa* x (*A. roylei* x *A. fistulosum*) populations were constructed for the purpose of introgression of *A. fistulosum* genetics into onion (Khrustaleva et al. 2005). By using reliable SNP information obtained by the assembly of transcriptome sequencing from the interspecific F_1 hybrid between *A. roylei* and *A. fistulosum* and nine onion cultivars, 667 SNP markers were anchored on the genetic map of trihybrid population (Scholten et al. 2016). These interspecies cross mapping populations have also been applied for the efforts on the integration of genetic map information generated by onion and bunching onion (McCallum et al. 2012).

Another valuable genetic resource that has enabled alignment of *Allium* genetic maps to physical chromosomes and facilitated comparison among species is the sets of bunching onion-shallot alien MALs developed by Shigyo and colleagues (Shigyo et al. 1996). By adapting the polymorphism between shallot and bunching onion, MALs can be used to allocate the markers onto physical chromosomes. A number of genetic linkage maps in shallot and bunching onion were aligned to physical chromosomes by taking this advantageous feature of MALs (van Heusden et al. 2000; Martin et al. 2005; Tsukazaki et al. 2010, 2015). Transcriptome data of MALs were accumulated by using NGS technology in a study aimed to identify the candidate genes related to physiological characters of each line. The obtained transcriptome sequence reads from each MAL were mapped onto doubled haploid shallot (accession "Chiang Mai") unigene sequence, and transcription level of each unigene was evaluated by RPKM (Reads Per Kilobase of exon per Million mapped reads) value (Abdelrahman et al. 2017). As an advanced application of this mapping data, SNP discovery and genotyping were carried out. By comparing the genotype call of parental bunching onion and shallot transcriptome mapping data, SNP sites with alternative homozygous call in bunching onion and reference homozygous call in shallot were selected. Anchored chromosome of corresponding unigenes was estimated based on the genotype calls of one of the MAL with heterozygous call, and more than 20,000 unigenes were anchored onto the chromosome (Manuscript in preparation). These anchoring markers would be useful for the proceeding genome sequencing project.

Along with the progress of NGS technologies, a new strategy named genotyping by sequencing (GBS) is becoming an increasingly important and cost-effective tool for the detection and scoring of SNPs in a large segregating population (Deschamps et al. 2012). Due to the enormous genome sizes, however, the genome DNA targeted GBS would be difficult in *Allium* species even in taking RAD sequencing approach. Thus, transcriptome-based approach for GBS was tested. Transcriptome data were accumulated from mapping population derived from F1 hybrids between doubled haploid of shallot and doubled haploid of bulb onion, and mapped onto shallot unigene sequences. Genotyping of the SNPs detected by mapping of transcriptome reads from doubled haploid of bulb onion were carried out based on the mapping results of transcriptome sequences from mapping population. By using genotype information on 98 lines of mapping population obtained by transcriptome based GBS, a high-resolution genetic linkage map with 610 genotype blocks composed of around 4400 unigenes was constructed (Manuscript in preparation). As shallot unigenes were also used as chromosome anchoring markers based on MALs transcriptome analysis, 8 linkage groups obtained by transcriptome based GBS were allocated to physical chromosomes (Fig. 11.3). With the sequence and annotation of the unigenes mapped on the linkage map, the linkage map will be a useful resource for genome sequencing and breeding of *Allium* species.

11.3.3 Databases Providing *Allium* Genomic Resource Information

The AlliumMap database provided the information on genetic map and marker data from multiple *Allium* species and populations. The integrated genetic map information constructed by using the mapping population of interspecific cross between onion and *A. roylei* provided a valuable bridge between the genetics of onion and *A. fistulosum* (McCallum et al. 2012). A large number of EST-SSR markers anchored onto the interspecific genetic linkage map are valuable resources for genetic and genomic studies of *Allium* species. This database is currently not available, but the contents of the database will be valuable to be integrated into other databases in the near future.

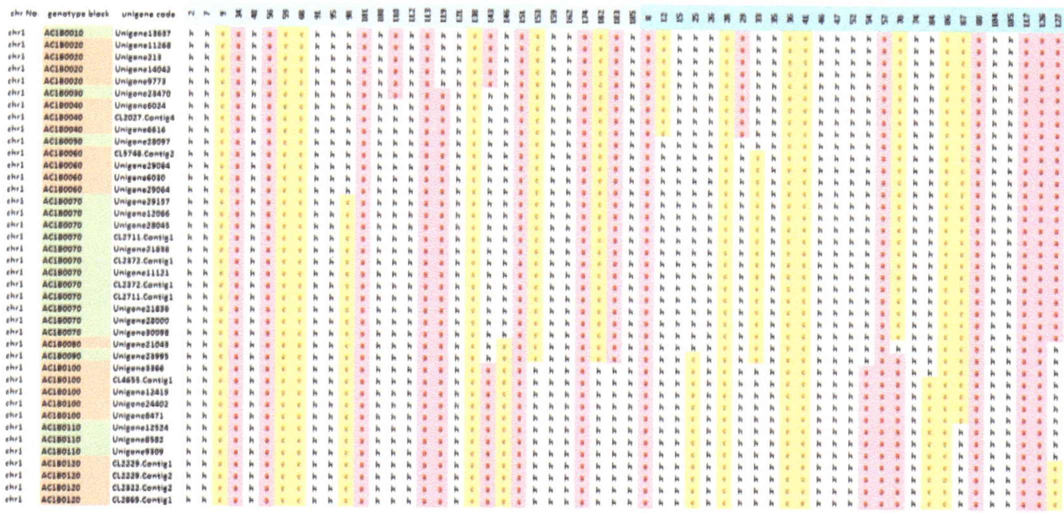

Fig. 11.3 Color map of RNA-seq based genotyping. An example of color map of RNA-seq based genotyping of the F2 mapping population derived from hybrids between doubled haploid of shallot and doubled haploid of bulb onion. Each row corresponds to a single unigene. The first three columns indicate chromosome number, genotype block, and unigene code, respectively. The remaining columns (from the fourth column) show the genotype of a single F2 line. Three different genotypes, homozygous shallot, homozygous bulb onion, and heterozygous are colored by pink, yellow, and white, respectively

VegMarks, a web database for genetic map and DNA markers for vegetable crops developed by National Institute of Vegetable and Tea Science (NIVTS), is available at [http://vegmarks.nivot.affrc.go.jp/]. This database contains detailed information on *A. fistulosum* marker characteristics such as ID number, map position, nucleotide sequence of the clones/PCR primers, and polymorphism data among varieties/accessions. This database is updated regularly and provides a link to other genome resource related sites that might be useful for *Allium* research.

A database for *Allium* EST-based information is available at [http://webtom.cabgrid.res.in/ogr/]. This database, named as the Onion Genomic Resource (OGR), contains the integrated information on the basis of 20,204 EST of bulb onion retrieved from NCBI. The accumulated ESTs were assembled into contigs that were used for gene prediction and annotation. Primers for amplification of SSR identified on the assembled contigs are available along with SSR markers in the published literature. With the aim to accelerate traditional improvement program by supplementing molecular data for precise and faster selection, the database and web interface have been designed in MySQL and PHP for user-friendly access.

The database for NGS-derived data-based information is also established. "Allium Transcriptome DataBase" (TDB) at [http://alliumtdb.kazusa.or.jp] provides the unigene sequences of doubled haploid of shallot and onion, *A. roylei*, and bunching onion with annotation (Abdelrahman et al. 2017). Target sequences can be accessed by BLAST searches (BLASTN, BLASTP, TBLASTN, TBLASTX) against the unigene sequences on the "BLAST" page and/or the keyword search against the results of the BLAST search against TAIR10 pep, IRGSP, and NR in each gene is available on the "KEYWORD" page. The detailed information including sequence, expression profile, BLAST search result against protein databases, and corresponding unigenes can be obtained at each page of DHA bulb unigene (Fig. 11.4). Expression profile contains RPKM data in bulb, root, and leaf of shallot and *A. roylei* as well as RPKM data of MAL. Information on the assigned

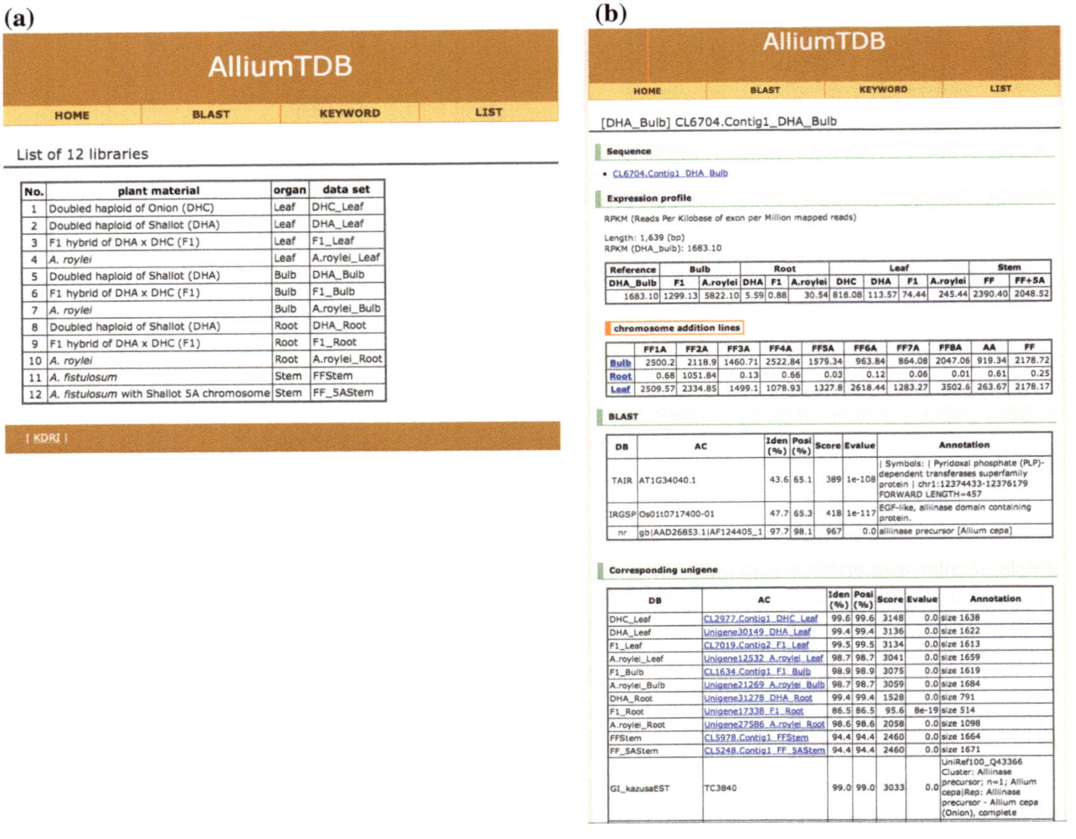

Fig. 11.4 Allium transcriptome database (AlliumTDB). **a** The top page of AlliumTDB. **b** An example of the detailed information on a unigene of DHA bulb

chromosome based on the genotype information on MAL will be presented in the near feature. By integrating the accumulated information on the basis of reference unigene, this database will provide the basis of *Allium* genomic resources.

11.4 Conclusion

Although the availability of the structural genomic resources in *Allium* species has been restricted to date, the recent rapid progress of NGS technology is making the whole genome sequencing extremely easier than before. Actually, genome sequencing of *A. cepa* is now ongoing as SEQUON project [http://www.oniongenome.net] by using HiSeq 2500 NGS sequencer (Illumina). On the other hand, many SSR markers have already been developed (Tsukazaki et al. 2007, 2015) and the integrated map database is available online (McCallum et al. 2012) in *Allium* species. In the post-genome era, the BAC resources and FISH techniques as described above are expected to connect the NGS whole genome sequence and the EST-derived marker-based map in *Allium* species. NGS technologies also have a sound applicability in functional genomics research of *Allium* species. As the NGS technologies become a routine part of biological research, studies focused on the functionality of the genomic sequences of *Allium* species are expected to increase in the near future. Along with the studies that are mentioned in this chapter, future research on the structural and functional genomic resources will provide substantial insight regarding the improvement of the yield of *Allium* species.

References

Abdelrahman M, El-Sayed M, Sato S, Hirakawa H, Ito S, Tanaka K, Mine Y, Sugiyama N, Suzuki Y, Yamauchi N, Shigyo M (2017) RNA-sequencing-based transcriptome and biochemical analyses of steroidal saponin pathway in a complete set of *Allium fistulosum—A. cepa* monosomic addition lines. PLoS ONE 12(8):e0181784. https://doi.org/10.1371/journal.pone.0181784

Arabidopsis Genome Initiative (2000) Analysis of the genome sequence of the flowering plant *Arabidopsis thaliana*. Nature 408:796–815

Arumuganathan K, Earle ED (1991) Nuclear DNA content of some important plant species. Plant Mol Biol Rep 9:208–218

Baldwin S, Revanna R, Thomson S, Pither-Joyce M, Wright K, Crowhurst R, Fiers M, Chen L, Macknight R, McCallum JA (2012) A toolkit for bulk PCR-based marker design from next-generation sequence data: application for development of a framework linkage map in bulb onion (*Allium cepa* L.). BMC Genom 13:637

Deschamps S, Llaca V, May GD (2012) Genotyping-by-sequencing in plants. Biology 1:460–483

Do GS, Suzuki G, Mukai Y (2004) Genomic organization of a novel root alliinase gene, *ALL1*, in onion. Gene 325:17–24

Gokce A, McCallum J, Sato Y, Havey M (2002) Molecular tagging of the *Ms* locus in onion. J Am Soc Hortic Sci 127:576–582

Han J, Thamilarasan SK, Natarajan S, Park JI, Chung MY, Nou IS (2016) De Novo assembly and transcriptome analysis of bulb onion (*Allium cepa* L.) during cold acclimation using contrasting genotypes. PLoS ONE 11:e0161987

Jakse J, Meyer JD, Suzuki G, McCallum J, Cheung F, Town CD, Havey MJ (2008) Pilot sequencing of onion genomic DNA reveals fragments of transposable elements, low gene densities, and significant gene enrichment after methyl filtration. Mol Genet Genomics 280:287–292

Kamenetsky R, Faigenboim A, Mayer ES, Michael TB, Gershberg C, Kimhi S, Esquira I, Shalom SR, Eshel D, Rabinowitch HD, Sherman A (2015) Integrated transcriptome catalogue and organ-specific profiling of gene expression in fertile garlic (*Allium sativum* L.). BMC Genom 2015(16):12

Khar A, Jakse J, Havey M (2008) Segregations for onion-bulb colors reveal that red is controlled by at least three loci. J Am Soc Hortic Sci 133:42–47

Khosa JS, Lee R, Bräuning S, Lord J, Pither-Joyce M, McCallum J, Macknight RC (2016) Doubled haploid 'CUDH2107'as a reference for bulb onion (*Allium cepa* L.) research: development of a transcriptome catalogue and identification of transcripts associated with male fertility. PLoS ONE 11:e0166568

Khrustaleva LI, De Melo PE, van Heusden AW, Kik C (2005) The integration of recombination and physical maps in a large-genome monocot using haploid genome analysis in a trihybrid Allium population. Genetics 169:1673–1685

Kim DW, Jung TS, Nam SH, Kwon HR, Kim A, Chae SH, Choi SH, Kim DQ, Kim RN, Park HS (2009) GarlicESTdb: an online database and mining tool for garlic EST sequences. BMC Plant Biol 9:61

Kim UJ, Birren BW, Slepak T, Mancino V, Boysen C, Kang HL, Simon MI, Shizuya H (1996) Construction and characterization of a human bacterial artificial chromosome library. Genomics 34:213–218

King J, Bradeen J, Bark O, McCallum J, Havey M (1998) A low-density genetic map of onion reveals a role for tandem duplication in the evolution of an extremely large diploid genome. Theor Appl Genet 96:52–62

Kuhl J, Cheung F, Yuan Q, Martin W, Zewdie Y, McCallum J, Catanach A, Rutherford P, Sink K, Jenderek M, Prince JP (2004) A unique set of 11,008 onion expressed sequence tags reveals expressed sequence and genomic differences between the monocot orders Asparagales and Poales. Plant Cell 16:114–125

Labani RM, Elkington TT (1987) Nuclear DNA variation in the genus *Allium* L. (Liliaceae). Heredity 59:119–128

Lee HR, Eom EM, Lim YP, Bang JW, Lee DH (2003) Construction of a garlic BAC library and chromosomal assignment of BAC clones using the FISH technique. Genome 46:514–520

Lysak M, Fransz P, Schubert I (1998) Cytogenetic analysis of *Arabidopsis*. In: Salinas J, Sanchez-Serrano JJ (eds) *Arabidopsis* protocols, 2nd edn. Humana Press, Totowa, New Jersey, pp 173–186

Martin W, McCallum J, Shigyo M, Jakse J, Kuhl J, Yamane N, Pither-Joyce M, Gokce A, Sink K, Town CD, Havey MJ (2005) Genetic mapping of expressed sequences in onion and in silico comparisons with rice show scant colinearity. Mol Genet Genom 274:197–204

Masamura N, McCallum J, Khrustaleva L, Kenel F, Pither-Joyce M, Shono J, Suzuki G, Mukai Y, Yamauchi N, Shigyo M (2012) Chromosomal organization and sequence diversity of genes encoding lachrymatory factor synthase in Allium cepa L. G3 (2):643–651

Matsuba A, Fujii M, Lee SS, Suzuki G, Yamamoto M, Mukai Y (2015) Molecular cytogenetic use of BAC clones in *Neofinetia falcata* and *Rhynchostylis coelestis*. The Nucleus 58:207–210

McCallum J, Baldwin S, Shigyo M, Deng Y, van Heusden S, Pither-Joyce M, Kenel F (2012) AlliumMap-A comparative genomics resource for cultivated *Allium vegetables*. BMC Genom 13:168

McCallum J, Clarke A, Pither-Joyce M, Shaw M, Butler R, Brash D, Scheffer J, Sims I, van Heusden S, Shigyo M, Havey MJ (2006) Genetic mapping of a major gene affecting onion bulb fructan content. Theor Appl Genet 112:958–967

McCallum J, Pither-Joyce M, Shaw M, Kenel F, Davis S, Butler R, Scheffer J, Jakse J, Havey MJ (2007) Genetic mapping of sulfur assimilation genes reveals a

QTL for onion bulb pungency. Theor Appl Genet 114:815–822
Meer Q, Vries J (1990) An interspecific cross between *Allium roylei* Stearn and *Allium cepa* L., and its backcross to *A. cepa*. Euphytica 47:29–31
Neumann P, Koblízková A, Navrátilová A, Macas J (2006) Significant expansion of *Vicia pannonica* genome size mediated by amplification of a single type of giant retroelement. Genetics 173:1047–1056
Ohara T, Song Y, Tsukazaki H, Wako T, Nunome T, Kojima A (2005) Genetic mapping of AFLP markers in Japanese bunching onion (*Allium fistulosum*). Euphytica 144(3):255–263
Ohara T, Tsukazaki H, YeonSang S, Wako T, Yamashita K, Kojima A (2009) Mapping of quantitative trait loci controlling seedling growth in bunching onion (*Allium fistulosum* L.). J Jpn Soc Horticultural Sci 78:436–442
Rajkumar H, Ramagoni RK, Anchoju VC, Vankudavath RN, Syed AU (2015) De novo transcriptome analysis of Allium cepa L. (Onion) bulb to identify allergens and epitopes. PLoS ONE 10:e0135387
Ricroch A, Yockteng R, Brown SC, Nadot S (2005) Evolution of genome size across some cultivated *Allium* species. Genome 48:511–520
SanMiguel P, Bennetzen JL (1998) Evidence that a recent increase in maize genome size was caused by the massive amplification of intergene retrotransposons. Ann Bot 82:37–44
Scholten OE, van Kaauwen MP, Shahin A, Hendrickx PM, Keizer LC, Burger K, van Heusden AW, van der Linden CG, Vosman B (2016) SNP-markers in *Allium* species to facilitate introgression breeding in onion. BMC Plant Biol 16:187
Shigyo M, Tashiro Y, Isshiki S, Miyazaki S (1996) Establishment of a series of alien monosomic addition lines of Japanese bunching onion (*Allium fistulosum* L.) with extra chromosomes from shallot (*A. cepa* L. Aggregatum group). Genes Genet Syst 71:363–371
Shirasu K, Schulman AH, Lahaye T, Schulze-Lefert P (2000) A contiguous 66-kb barley DNA sequence provides evidence for reversible genome expansion. Genome Res 10:908–915
Sun X, Zhou S, Meng F, Liu S (2012) *De novo* assembly and characterization of the garlic (*Allium sativum*) bud transcriptome by Illumina sequencing. Plant Cell Rep 31:1823–1828
Sun XD, Yu XH, Zhou SM, Liu SQ (2016) *De novo* assembly and characterization of the Welsh onion (*Allium fistulosum* L.) transcriptome using Illumina technology. Mol Genet Genomics 291:647–659
Suzuki G, Do GS, Mukai Y (2002) Efficient storage and screening system for onion BAC clones. Breeding Sci 52:157–159
Suzuki G, Nishiuchi C, Tsuru A, Kako E, Li J, Yamamoto M, Mukai Y (2013) Cellular localization of mitotic RAD21 with repetitive amino acid motifs in *Allium cepa*. Gene 514:75–81
Suzuki G, Ogaki Y, Hokimoto N, Xiao L, Kikuchi-Taura A, Harada C, Okayama R, Tsuru A, Onishi M, Saito N, Do GS, Lee SH, Ito T, Kanno A, Yamamoto M, Mukai Y (2012) Random BAC FISH of monocot plants reveals differential distribution of repetitive DNA elements in small and large chromosome species. Plant Cell Rep 31:621–628
Suzuki G, Ura A, Saito N, Do G, So BB, Yamamoto M, Mukai Y (2001) BAC FISH analysis in *Allium cepa*. Genes Genet Syst 76:251–255
Thomas JW, Summers TJ, Lee-Lin SQ, Maduro VV, Idol JR, Mastrian SD, Ryan JF, Jamison DC, Green ED (2000) Comparative genome mapping in the sequence-based era: early experience with human chromosome 7. Genome Res 10:624–633
Tsukazaki H, Yaguchi S, Shusei Sato S, Hirakawa H, Katayose Y, Kanamori H, Wako T (2015) Development of transcriptome shotgun assembly-derived markers in bunching onion (*Allium fistulosum*). Mol Breed 35:55
Tsukazaki H, Yamashita K, Yaguchi S, Masuzaki S, Fukuoka H, Yonemaru J, Kanamori H, Kono I, Hang T, Shigyo M, Kojima A (2008) Construction of SSR-based chromosome map in bunching onion (*Allium fistulosum*). Theor Appl Genet 2008 (117):1213–1223
Tsukazaki H, Yamashita K, Yaguchi S, Yamashita K, Hagihara T, Shigyo M, Kojima A, Wako T (2010) Direct determination of the chromosomal location of bunching onion and bulb onion markers using bunching onion-shallot monosomic additions and allotriploid-bunching onion single alien deletions. Theor Appl Genet 122:501–510
Tsukazaki H, Nunome T, Fukuoka H, Kanamori H, Kono I, Yamashita KI, Wako T, Kojima A (2007) Isolation of 1,796 SSR clones from SSR-enriched DNA libraries of bunching onion (*Allium fistulosum*). Euphytica 157:83–94
van Heusden A, van Ooijen J, Vrielink-van Ginkel R, Verbeek W, Wietsma W, Kik C (2000) A genetic map of an interspecic cross in *Allium* based on amplified fragment length polymorphism (AFLP™) markers. Theor Appl Genet 100:118–126
Zhang C, Zhang H, Zhan Z, Liu B, Chen Z, Liang Y (2016) Transcriptome analysis of sucrose metabolism during bulb swelling and development in onion (*Allium cepa* L.). Frontiers. Plant Sci 2016:7
Zhang HB, Wu C (2001) BAC as tools for genome sequencing. Plant Physiol Biochem 39:195–209
Zhang P, Li W, Fellers J, Friebe B, Gill BS (2004) BAC-FISH in wheat identifies chromosome landmarks consisting of different types of transposable elements. Chromosoma 112:288–299
Zhou SM, Chen LM, Liu SQ, Wang XF, Sun XD (2015) *De novo* assembly and annotation of the Chinese chive (*Allium tuberosum* Rottler ex Spr.) transcriptome using the Illumina platform. PLoS ONE 10(7):e0133312
Zhu S, Tang S, Tan Z, Yu Y, Dai Q, Liu T (2017) Comparative transcriptomics provide insight into the morphogenesis and evolution of fistular leaves in *Allium*. BMC Genom 18:60
Zonneveld BJ, Leitch IJ, Bennett MD (2005) First nuclear DNA amounts in more than 300 angiosperms. Ann Bot 96:229–244

Metabolomics of *Allium* Crops

12

Yuji Sawada

Abstract

Metabolomics is the broad study of small molecules and can involve either quantitative or comparative analysis of metabolic profiles of various biological groups, belonging to different genotypes, developmental stages, stress responses or tolerance, storage conditions, and methods of cooking or fermentation. *Allium* is a rich source of diverse metabolites, such as fructooligosaccharides, amino acids, peptides, and a variety of secondary metabolites. The *Allium* metabolites not only affect economically important traits such as color, flavor, and pungency but also play an important role in plant physiology and in promoting human health. Thus, *Allium* metabolomics may reveal the agricultural potential of *Allium*. This section will summarize plant metabolomics and case studies of *Allium* omics approaches.

12.1 Introduction

After the establishment of omics platforms, the study of plant metabolism has transitioned from the study of individual gene functions to metabolic system research. Advancement in nucleotide sequencing, for example, has provided omics information infrastructure for components of the central dogma: genome, transcriptome, and proteome (Mochida and Shinozaki 2011). Such nucleotide sequence-based omics approaches measure similar chemical properties, and common standardized analytical platforms are thus available for microbes, plants, and animals. However, the target molecules of metabolomics have diverse chemical properties, which make standardized comprehensive metabolomics challenging (Sumner et al. 2003; Fiehn et al. 2005; Fiehn et al. 2008; Manach et al. 2009; Kumar et al. 2017).

Mass spectrometry (MS) is a major instrument that has high sensitivity and broad metabolite detection capabilities. MS-based metabolomics approaches have facilitated the detection of hundreds of metabolites having diverse bioactivities in *Allium* crop, which has attracted the attention of many researchers (Morrissey and Osbourn 1999; Guillet et al. 2000; Oleszek and Marston 2000; Graf et al. 2005).

Liquid chromatography is a widely used separation technique, which is suitable for high to medium polar plant-specific metabolites, *e.g.*, flavonoids, terpenoids, and alkaloids (Cimpan and Gocan 2002; Graf et al. 2005; Negi et al. 2011). Liquid chromatography is coupled with mass spectrometry (LC–MS) to achieve broader detection and annotation of metabolites for plant metabolomics (Becker et al. 2012; Beltran et al.

Y. Sawada (✉)
RIKEN Center for Sustainable Resource Science, 1-7-22 Suehiro-cho, Tsurumi-ku, Yokohama, Kanagawa 230-0045, Japan
e-mail: yuji.sawada@riken.jp

M. Shigyo et al. (eds.), *The Allium Genomes*, Compendium of Plant Genomes,
https://doi.org/10.1007/978-3-319-95825-5_12

2012; Creek et al. 2012; Vinaixa et al. 2012; Zhou et al. 2012). Comprehensive analysis of diverse metabolites can be achieved by combining multiple instruments and methods (Naz et al. 2014). Gas chromatography and capillary electrophoresis techniques, for example, coupled with MS are used for the analysis of metabolites generated by central carbon metabolism, *e.g.*, amino acid, organic acid, and carbohydrate metabolism, that are common to all living organisms (Ramautar et al. 2009; Koek et al. 2011). In this section, we focus on LC–MS-based metabolomics for *Allium* using three types of methodologies: widely targeted metabolomics, un-targeted metabolomics, and elemental composition analysis (Fig. 12.1).

12.2 Widely Targeted Metabolomics

Selected reaction monitoring (SRM) using LC-tandem quadrupole (QqQ)-MS offers targeted metabolite analysis (Fig. 12.1a). More than a hundred SRM conditions have been used to conduct widely targeted metabolomics (Yost and Enke 1979; Sawada et al. 2009; Tsugawa et al. 2014a, b). In a previous study conducted in our laboratory, a small amount of sample (single grain (ca. 40 μg/ml extract of a seed) of model plant *Arabidopsis thaliana*) was successfully used for widely targeted analysis, and approximately 100 metabolite profiles were detected (Sawada et al. 2017). Using LC-QqQ-MS, more than hundreds of metabolites have been detected in *Allium*.

Environmental stress affects plant growth and yield, and identification of a practical biomarker is a major requirement for the breeding of stress-tolerant crops. In this regard, because of its high adaptability to tropical and subtropical environment, shallots (*Allium cepa* L. Aggregatum group) constitute an important genetic resource for the breeding of common onion (Phuong et al. 2006). Omics evaluation will allow the identification of such practical biological markers. Using LC-QqQ-MS, the common onion, shallot, and its F1 hybrid were evaluated and 113 targeted metabolites were detected using analytical criteria (Abdelrahman et al. 2015). The principal component analysis and volcano plot analysis clearly showed genotype-specific metabolites, which can be used as metabolic markers of environmental tolerance. Further integrated omics approaches, *e.g.*, linkage mapping and correlation analysis between transcriptome and metabolome, can elucidate the gene-to-metabolite networks in environmental responses or stress tolerance.

Allium crops have diverse sulfur-containing metabolites that are the source of their characteristic flavors. In *Allium* crops, *S*-alk(en)yl-L-cystein sulphoxides (ACSOs) are the precursors of both the flavors located in cytoplasm and the hydrolytic enzyme alliinase that is found in the vacuole. These ACSOs are hydrolyzed and result in the formation of volatile sulfides upon disruption of cells of *Allium* plants during cutting or crushing. *Trans-S*-1-propenyl-L-cysteine sulfoxide (PRENCSO) has been detected as the most abundant ACSO in onion by quantitative HPLC analysis (Yamazaki et al. 2011). The ACSOs of onion undergo changes during storage. The temperature of the stock room is an important factor for this metabolic change (Kopsell et al. 1999). Owing to the high analytical load of huge number of samples for such onion storage experiments (use of biological replicates, different temperature conditions, storage periods, and 2-year test), in one of our studies, we used LC-QqQ-MS due to its high sensitivity and high-throughput metabolomics (Kamata et al. 2016). In these onion storage tests, we also used classical quantitative HPLC for the detection of PRENCSO and related sulfur-containing metabolites, including *S*-2-carboxypropyl glutathione (2-CPGTH), gamma-glutamyl-*S*-1-propenyl-L-cysteine (GG-PRENCS), gamma-glutamyl-*S*-1-propenyl-L-cysteine sulfoxide (GG-PRENCSO), and 3-carboxy-5-methyl-1,4-thiazane sulfoxide (cycloalliin). In the validation, five-metabolite profiles identified using LC-QqQ-MS-based metabolomics were similar to those using quantitative HPLC analysis. At low storage temperature (0 and 5 °C), the five sulfur-containing metabolites had similar metabolic profiles. In the 2-year test, 2-CPGTH and GG-PRENCSO were able to

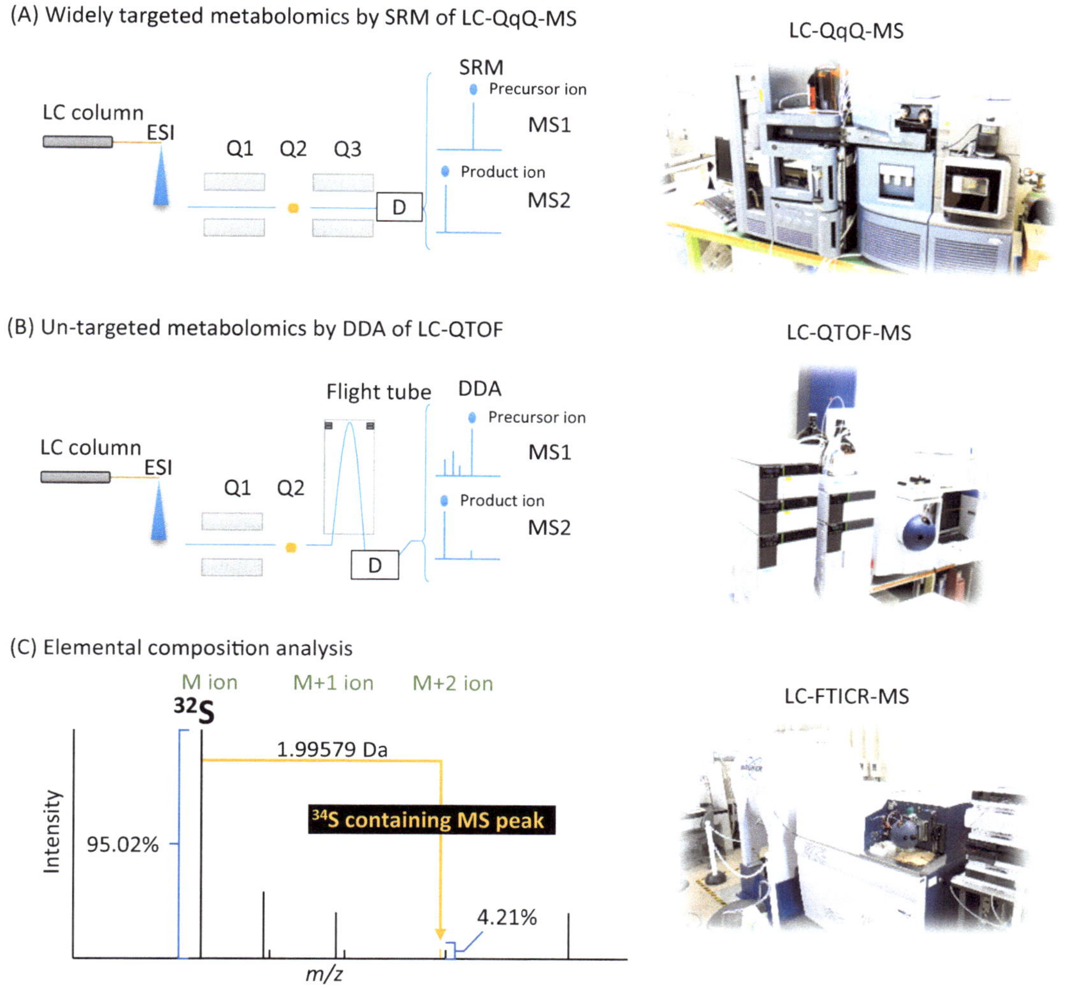

Fig. 12.1 Overview of plant metabolomics. **a** Widely targeted metabolomics of LC-QqQ-MS provides hundreds of metabolic profiles utilizing the selected reaction monitoring (SRM) conditions, which are optimized by using authentic standard compounds. In SRM mode, an ion with relevant molecular weight (precursor ion) is selected by first quadrupole (Q1), and the selected ion is fragmented by second quadrupole (Q2). Finally, the fragmented ion (product ion) is selected by third quadrupole (Q3). **b** Un-targeted metabolomics of LC-QTOF-MS provides data-dependent analysis (DDA) for almost all detectable precursor and product ions. Isolated precursor ion by Q1 is fragmented by Q2, and the product ions are measured by flight tube base on the time-of-flight. **c** Ultrahigh-resolution mass spectrometry (UHR-MS) of FTICR-MS is used for elemental composition analysis. In sulfur-containing metabolites, stable isotopes of ^{32}S and ^{34}S are separated and detected by UHR-MS, and the difference and natural abundance ratio are used for identification of sulfur ion

decrease an increase in PRENCSO and cyloalliin. A visible change in the phenotype was noted in which the leaves of some bulb onions grew after storage for 4 months at 5 °C, which suggested that the dormancy of the onion bulb was broken. At a storage temperature of 20 °C, the onion blub sprouted after 4 months. The metabolic profiles and visible phenotype support the FAO recommendation of 0 °C as the ideal long-term storage temperature for onions.

12.3 Un-targeted Metabolomics

LC-quadrupole time-of-flight (QTOF)-MS provides un-targeted metabolomics utilizing high-resolution MS and tandem MS (MS/MS). To illustrate, un-targeted MS/MS of data-dependent analysis (DDA) can detect almost all detectable metabolites in a sample matrix, followed by a public MS/MS database search that allows us to narrow down putative structures (Fig. 12.1b) (Smith et al. 2005; Sawada et al. 2012; Sakurai et al. 2013; Wishart et al. 2013; Tsugawa et al. 2015). At present, we have more than 1.6 million MS/MS tags in plant samples including *Arabidopsis*, rice, wheat, soybean, pear, and onion (Matsuda et al. 2009; Saito and Matsuda 2010; Matsuda et al. 2012). Reference MS, MS/MS, and category database have helped to improve the annotation and classification of un-targeted metabolomics (Nakayama et al. 2014; Tsugawa et al. 2015; Tsugawa 2018).

Toward comprehensive metabolic profiling of common onion bulb, LC-QTOF-MS coupled with two chromatographic methods assisted in detection and annotation of more than a hundred metabolites. This analytical workflow detected polar and semi-polar onion metabolites such as fructooligosaccharides, proteinogenic amino acids, peptides, sulfur-containing metabolites, flavonoids, and saponins (Bottcher et al. 2017). Comparison analysis between onion cultivars revealed cultivar-specific metabolite accumulation patterns. The association between metabolic profiles and their related loci will be elucidated by metabolome linkage mapping (Kumar et al. 2017).

Fermentation of *Allium* foods enhances its antibacterial, antigenotoxic, and antiproliferative activities (Millet et al. 2012), and use of the sequence-based omics approaches is challenging for the analysis of such fermented foods. Metabolic differences between fresh and black garlic have been assessed by LC-QTOF-MS (Molina-Calle et al. 2017). Ninety-five metabolites in polar extracts of fresh and black garlic were successfully annotated, and multivariate discriminant analysis suggested that the major change in metabolic profile occurred within the first 5 days. The fermentation of black garlic affected three metabolite species: organosulfur compounds, amino acids, and saccharides. This study can provide valuable information with regard to optimizing the fermentation process for the production of bioactive compounds.

12.4 Elemental Composition Analysis

Different plants have characteristic metabolites, and their structural diversity is not wholly elucidated. To explore these unidentified metabolites through classical approaches, the individual metabolites need to be isolated and characterized (Nakabayashi et al. 2009). Un-targeted metabolomics has been used for the detection of unknown metabolites, followed by the annotation utilizing similarity search with reference MS and MS/MS databases (Tsugawa 2018).

Ultrahigh resolution (>250,000 full width at half-maximum) and mass accuracy (<1 mDa) of LC-FTICR-MS can provide the precise elemental composition of the detectable ions (Nakabayashi and Saito 2013, 2017). As a case study, the sulfur-containing metabolites of *Allium* crops were analyzed by LC-FTICR-MS (Fig. 12.1c) using a ^{13}C-labeled onion bulb, and elemental compositions of ACSOs and glutathione derivatives were successfully ascribed (Nakabayashi et al. 2013). Using the same technique, a novel S-containing metabolite "asparaptine" was identified in Asparagus, which is an inhibitor of angiotensin-converting enzyme that plays a major role in hypertension (Nakabayashi et al. 2015). Sulfur-containing metabolites were also identified in garlic using LC-FTICR-MS (Nakabayashi et al. 2016). Exploring novel sulfur-containing metabolites will contribute to the advancement of cost-effective metabolic breeding and study of biologically active sulfur-containing metabolites present in *Allium* crop.

12.5 Future Aspects for *Allium* Metabolomics

Allium metabolomics will elucidate characteristic metabolites and their accumulation patterns among accessions, landraces, cultivars, and storage and fermentation conditions. The individual bioresource-specific metabolic patterns can be used for molecular breeding of *Allium* crops while the broad metabolic profiles of *Allium* bioresources can be used for integrated omics approaches. The integration of metabolomics and transcriptomics will provide insight into the molecular mechanism of *Allium* metabolite biosynthesis. The metabolic profiles can also be used for elucidation of the molecular architecture underlying environmental responses and stress tolerance in *Allium*.

12.6 Conclusions

This chapter introduced recent advances in *Allium* metabolomics. These advancements will contribute to the improvement of *Allium* crops via metabolic and genetic marker-assisted breeding. *Allium* metabolomics also provides opportunities to identify novel metabolites with innovative biological activities that may allow improved environmental adaptation or that can be utilized to promote human health.

Acknowledgements This work was supported by the RIKEN Plant Science Center, Japan Science and Grant-in-Aid for Scientific Research (B). We thank RIKEN members: Dr. Ryo Nakabayashi, Mrs. Makoto Suzuki, Akane Sakata, Yutaka Yamada, Tetsuya Mori, and Muneo Sato for their useful comments and LC–MS/MS analysis of the metabolomics technologies.

References

Abdelrahman M et al (2015) Integrating transcriptome and target metabolome variability in doubled haploids of Allium cepa for abiotic stress protection. Mol Breed 35:195. https://doi.org/10.1007/S11032-015-0378-2

Becker S, Kortz L, Helmschrodt C, Thiery J, Ceglarek U (2012) LC-MS-based metabolomics in the clinical laboratory. J Chromatogr B, Anal Technol Biomed Life Sci 883–884:68–75. https://doi.org/10.1016/j.jchromb.2011.10.018

Beltran A et al (2012) Assessment of compatibility between extraction methods for NMR- and LC/MS-based metabolomics. Anal Chem. 84:5838–5844. https://doi.org/10.1021/ac3005567

Bottcher C, Krahmer A, Sturtz M, Widder S, Schulz H (2017) Comprehensive metabolite profiling of onion bulbs (*Allium cepa*) using liquid chromatography coupled with electrospray ionization quadrupole time-of-flight mass spectrometry. Metabolomics 13:35. https://doi.org/10.1007/s11306-017-1175-1

Cimpan G, Gocan S (2002) Analysis of medicinal plants by HPLC: recent approaches. J Liq Chromatogr Relat Technol 25:2225–2292. https://doi.org/10.1081/Jlc-120014003

Creek DJ, Jankevics A, Burgess KE, Breitling R, Barrett MP (2012) IDEOM: an excel interface for analysis of LC-MS-based metabolomics data. Bioinformatics 28:1048–1049. https://doi.org/10.1093/bioinformatics/bts069

Fiehn O, Wohlgemuth G, Scholz M (2005) Setup and annotation of metabolomic experiments by integrating biological and mass spectrometric metadata. Data Integr Life Sci 3615:224–239

Fiehn O et al (2008) Quality control for plant metabolomics: reporting MSI-compliant studies. Plant J 53:691–704. https://doi.org/10.1111/j.1365-313X.2007.03387.x

Graf BA, Milbury PE, Blumberg JB (2005) Flavonols, flavones, flavanones, and human health: epidemiological evidence. J Med Food 8:281–290. https://doi.org/10.1089/jmf.2005.8.281

Guillet G, Harmatha J, Waddell TG, Philogene BJ, Arnason JT (2000) Synergistic insecticidal mode of action between sesquiterpene lactones and a phototoxin, alpha-terthienyl. Photochem Photobiol 71:111–115

Kamata Y et al (2016) Changes in trans-S-1-propenyl-L-cysteine sulfoxide and related sulfur-containing amino acids during Onion storage. J Agric Food Chem 64:9063–9071. https://doi.org/10.1021/acs.jafc.6b04073

Koek MM, Jellema RH, van der Greef J, Tas AC, Hankemeier T (2011) Quantitative metabolomics based on gas chromatography mass spectrometry: status and perspectives. Metab Off J Metabol Soc 7:307–328. https://doi.org/10.1007/s11306-010-0254-3

Kopsell DE, Randle WM, Eiteman MA (1999) Changes in the S-alk(en)yl cysteine sulfoxides and their biosynthetic intermediates during onion storage. J Am Soc Hortic Sci 124:177–183

Kumar R, Bohra A, Pandey AK, Pandey MK, Kumar A (2017) Metabolomics for plant improvement: status and prospects. Front Plant Sci 8:1302. https://doi.org/10.3389/fpls.2017.01302

Manach C, Hubert J, Llorach R, Scalbert A (2009) The complex links between dietary phytochemicals and human health deciphered by metabolomics. Mol Nutr Food Res 53:1303–1315. https://doi.org/10.1002/mnfr.200800516

Matsuda F et al (2012) Dissection of genotype-phenotype associations in rice grains using metabolome quantitative trait loci analysis. Plant J 70:624–636. https://doi.org/10.1111/j.1365-313X.2012.04903.x

Matsuda F, Yonekura-Sakakibara K, Niida R, Kuromori T, Shinozaki K, Saito K (2009) MS/MS spectral tag-based annotation of non-targeted profile of plant secondary metabolites. Plant J 57:555–577. https://doi.org/10.1111/j.1365-313X.2008.03705.x

Millet A, Lamy E, Jonas D, Stintzing F, Mersch-Sundermann V, Merfort I (2012) Fermentation Enhances the Biological Activity of Allium cepa Bulb Extracts. J Agric Food Chem 60:2148–2156. https://doi.org/10.1021/jf2041643

Mochida K, Shinozaki K (2011) Advances in omics and bioinformatics tools for systems analyses of plant functions. Plant and Cell Physiol 52:2017–2038. https://doi.org/10.1093/pcp/pcr153

Molina-Calle M, de Medina VS, Calderon-Santiago M, Priego-Capote F, de Castro MDL (2017) Untargeted analysis to monitor metabolic changes of garlic along heat treatment by LC-QTOF MS/MS. Electrophoresis 38:2349–2360. https://doi.org/10.1002/elps.201700062

Morrissey JP, Osbourn AE (1999) Fungal resistance to plant antibiotics as a mechanism of pathogenesis. Microbiol Mol Biol Rev: MMBR 63:708–724

Nakabayashi R et al (2009) Metabolomics-oriented isolation and structure elucidation of 37 compounds including two anthocyanins from Arabidopsis thaliana. Phytochemistry 70:1017–1029. https://doi.org/10.1016/j.phytochem.2009.03.021

Nakabayashi R, Saito K (2013) Metabolomics for unknown plant metabolites. Anal Bioanal Chem 405:5005–5011. https://doi.org/10.1007/s00216-013-6869-2

Nakabayashi R, Saito K (2017) Ultrahigh resolution metabolomics for S-containing metabolites. Curr Opin Biotechnol 43:8–16. https://doi.org/10.1016/j.copbio.2016.07.003

Nakabayashi R et al (2016) Chemical assignment of structural isomers of sulfur containing metabolites in garlic by liquid chromatography fourier transform ion cyclotron resonance mass spectrometry. J Nutr 146: 397s–402s. https://doi.org/10.3945/jn.114.202317

Nakabayashi R et al (2013) Combination of liquid chromatography-fourier transform ion cyclotron resonance-mass spectrometry with 13C-labeling for chemical assignment of sulfur-containing metabolites in onion bulbs. Anal Chem 85:1310–1315. https://doi.org/10.1021/ac302733c

Nakabayashi R, Yang ZG, Nishizawa T, Mori T, Saito K (2015) Top-down targeted metabolomics reveals a sulfur-containing metabolite with inhibitory activity against angiotensin-converting enzyme in asparagus officinalis. J Nat Prod 78:1179–1183. https://doi.org/10.1021/acs.jnatprod.5b00092

Nakayama Y, Tamada Y, Tsugawa H, Bamba T, Fukusaki E (2014) Novel strategy for non-targeted isotope-assisted metabolomics by means of metabolic turnover and multivariate analysis. Metabolites 4:722–739. https://doi.org/10.3390/metabo4030722

Naz S, Moreira dos Santos DC, Garcia A, Barbas C (2014) Analytical protocols based on LC-MS, GC-MS and CE-MS for nontargeted metabolomics of biological tissues. Bioanalysis 6:1657–1677. https://doi.org/10.4155/bio.14.119

Negi JS, Singh P, Pant GJ, Rawat MS (2011) High-performance liquid chromatography analysis of plant saponins: an update 2005–2010. Pharmacog Rev 5:155–158. https://doi.org/10.4103/0973-7847.91109

Oleszek W, Marston A (2000) Saponins in food, feedstuffs, and medicinal plants. Kluwer Academic Publishers, Dordrecht and Boston, MA

Phuong PTM, Isshiki S, Tashiro Y (2006) Genetic variation of shallot (Allium cepa L. aggregatum group) in Vietnam. J Japan Soc Hortic Sci 75:236–242. https://doi.org/10.2503/Jjshs.75.236

Ramautar R, Somsen GW, de Jong GJ (2009) CE-MS in metabolomics. Electrophoresis 30:276–291. https://doi.org/10.1002/elps.200800512

Saito K, Matsuda F (2010) Metabolomics for functional genomics, systems biology, and biotechnology. Annu Rev Plant Biol 61:463–489. https://doi.org/10.1146/annurev.arplant.043008.092035

Sakurai T et al (2013) PRIMe update: innovative content for plant metabolomics and integration of gene expression and metabolite accumulation. Plant Cell Physiol 54:e5. https://doi.org/10.1093/pcp/pcs184

Sawada Y et al (2009) Widely targeted metabolomics based on large-scale MS/MS data for elucidating metabolite accumulation patterns in plants. Plant Cell Physiol 50:37–47. https://doi.org/10.1093/pcp/pcn183

Sawada Y et al (2012) RIKEN tandem mass spectral database (ReSpect) for phytochemicals: a plant-specific MS/MS-based data resource and database. Phytochemistry 82:38–45. https://doi.org/10.1016/j.phytochem.2012.07.007

Sawada Y, Tsukaya H, Li Y, Sato M, Kawade K, Hirai MY (2017) A novel method for single-grain-based metabolic profiling of arabidopsis seed. Metabolomics 13:75. https://doi.org/10.1007/s11306-017-1211-1

Smith CA et al (2005) METLIN—A metabolite mass spectral database. Ther Drug Monit 27:747–751. https://doi.org/10.1097/01.ftd.0000179845.53213.39

Sumner LW, Mendes P, Dixon RA (2003) Plant metabolomics: large-scale phytochemistry in the functional genomics era. Phytochemistry 62:817–836. https://doi.org/10.1016/S0031-9422(02)00708-2

Tsugawa H (2018) Advances in computational metabolomics and databases deepen the understanding of metabolisms. Curr Opin Biotechnol 54:10–17. https://doi.org/10.1016/j.copbio.2018.01.008

Tsugawa H, et al (2015) MS-DIAL: data-independent MS/MS deconvolution for comprehensive metabolome analysis. Nat Methods 12(6):523. https://doi.org/10.1038/nmeth.3393

Tsugawa H, Kanazawa M, Ogiwara A, Arita M (2014a) MRMPROBS suite for metabolomics using

large-scale MRM assays. Bioinformatics 30:2379–2380. https://doi.org/10.1093/bioinformatics/btu203
Tsugawa H et al (2014b) MRM-DIFF: data processing strategy for differential analysis in large scale MRM-based lipidomics studies. Front Genet 5:471. https://doi.org/10.3389/fgene.2014.00471
Vinaixa M, Samino S, Saez I, Duran J, Guinovart JJ, Yanes O (2012) A guideline to univariate statistical analysis for LC/MS-Based untargeted metabolomics-derived data. Metabolites 2:775–795. https://doi.org/10.3390/metabo2040775
Wishart DS et al (2013) HMDB 3.0-The Human Metabolome Database in 2013. Nucleic Acids Res 41:D801–D807. https://doi.org/10.1093/nar/gks1065
Yamazaki Y, Iwasaki K, Mikami M, Yagihashi A (2011) Distribution of eleven flavor precursors, S-Alk(en)yl-L-cysteine derivatives, in seven Allium vegetables. Food Sci Technol Res 17:55–62
Yost RA, Enke CG (1979) Triple quadrupole mass-spectrometry for direct mixture analysis and structure elucidation. Anal Chem 51:1251–1264. https://doi.org/10.1021/Ac50048a002
Zhou B, Xiao JF, Tuli L, Ressom HW (2012) LC-MS-based metabolomics. Mol Biosyst 8:470–481. https://doi.org/10.1039/c1mb05350g

Molecular Mapping of Genes and QTL: Progress to Date and Development of New Population Resources for NGS Genetics

13

John McCallum, Samantha Baldwin, Richard Macknight, Jiffinvir Khosa and Martin Shaw

Abstract

Ability to conduct molecular analyses in *Allium* species was previously limited by the power of available technologies to sample very large heterozygous genomes in a cost-effective manner. The emergence of third-generation long-read sequencing technologies, coupled with maturing strategies for population genomics in non-model systems provides exciting opportunities for *Allium* crops. Now that these technical advances have removed major barriers to progress, it is necessary to consider the horticultural resources to which we can apply such technologies. This will hasten delivery of genetics tools that will ultimately benefit breeders, farmers, and consumers of *Allium* vegetables. Here, we describe progress to date in *Allium* mapping, with an emphasis on onion, and review approaches used on other non-model systems that could be applied. We also present strategies we have developed over the past decade to establish genetic stocks for onion suited to modern population genomics analyses.

J. McCallum · R. Macknight · J. Khosa
Biochemistry Dept, University of Otago, Dunedin, New Zealand

J. McCallum (✉) · S. Baldwin · R. Macknight · M. Shaw
New Zealand Institute for Plant and Food Research, Private Bag 4704, Christchurch, New Zealand
e-mail: john.mccallum@plantandfood.co.nz

13.1 Introduction: Status, Challenges, and Forward Strategies For Allium Mapping

Owing to the large genome sizes of *Allium* species (1 N > 10^{10} bp) molecular mapping has been based on sampling polymorphisms in the gene space with evolving technologies. The current reference map of onion is based on the map of USDA-ARS family BYG15-23 × AC43, built initially with cDNA RFLP markers (King et al. 1998), and then augmented with PCR-based markers (Martin et al. 2005), and then second-generation transcriptome sequencing (Duangjit et al. 2013). This map was anchored to physical chromosomes using a set of *A. cepa - A.fistulosum* monosomic alien addition lines originally developed by Shigyo et al. (1997). These lines have similarly been used to anchor linkage maps in *A. fistulosum* based on genomic SSR and transcriptome-derived markers (Tsukazaki et al. 2008, 2015). Although the PCR-based SSR and SNP marker assays used to date for these studies have been the standard diagnostics employed over the past two decades, they pose challenges for new entrants. These include reproducibility (especially for genomic SSRs), portability (especially for SNPs), scalability (SSR and SNP), and accessibility (proprietary platforms such as LGC Group Kasp https://www.lgcgroup.com/kasp/). The challenge for researchers without large in-house data or marker

M. Shigyo et al. (eds.), *The Allium Genomes*, Compendium of Plant Genomes,
https://doi.org/10.1007/978-3-319-95825-5_13

libraries is to link genes and/or polymorphisms with anchored features on maps. Although the studies to date have employed manual markers or relatively low-plex genotyping strategies, future strategies must accommodate targeted or global genotyping by sequencing data and bulk assignment to maps or assemblies.

Although it is likely that first generation genome assemblies of onion may be developed in the near future, the usefulness of these to the community will depend on how scaffolds can be ordered in relation to the gene space and aligned with genetic knowledge. We suggest that the major challenge facing Allium genome research is not technical but rather how to achieve better connectivity across studies based on some common stocks, coordinates, diagnostics, and community data curation.

Our previous review (McCallum 2007) provides a summary of progress in onion genetic mapping just prior to the availability of second-generation sequencing. Despite the remarkable explosion in genome analysis technology and bioinformatics over the past decade, progress in gene and QTL mapping in *Allium* species, as indicated by published literature, has been remarkably limited. Despite the vast economic and cultural value of *Allium* crops such as onion and garlic, recent progress is insufficient compared to the remarkable progress made in numerous non-model species without economic value (Ellegren 2014). A major factor that may be limiting public R&D in *Allium* genetics is the ongoing consolidation of the global seed industry, most notably acquisitions and mergers by biotech companies (Dunwell 2016; Maisashvili et al. 2016; Bonny 2017). This trend has been accompanied by a corresponding decline in public vegetable genetics R&D and a smaller research community. Modern applied genomic research requires access not only to analytical services but also specialist bioinformatics and appropriate genetic stocks. The trends of the past 10 years suggest that progress in *Allium* genetics is currently limited not by technology so much as horticultural capability and genetic resources.

This review considers research progress and the question of what resources are needed for *Allium* mapping to develop stronger collaborations, better science, and outcomes for business and consumers alike. To provide some context for discussion of strategies for applying next-generation sequencing for *Allium* genetics we describe our research experience using doubled-haploid onion materials, emphasizing practical aspects of population development and plant husbandry. These details are not typically captured in scientific publication but are a significant barrier to entry for new researchers.

13.2 Progress in Genetic Mapping

13.2.1 Cytoplasmic Male Sterility (CMS)

Because the use of CMS underpins most commercial *Allium* breeding, improved genetic tools for selecting cytotypes and nuclear restorer loci continue to be key targets for *Allium* genetics. The detailed high-resolution mapping of restorer loci is an exception to the generally low resolution and power of most *Allium* mapping. Earlier genetic mapping studies of the *Ms* locus, which conditions restoration of onion S-cytoplasmic male sterility, placed it on chromosome 2 but these RFLP-derived markers proved to be in linkage equilibrium with Ms (Gökçe and Havey 2002). The set of SNP markers developed by Duangjit et al. (2013) was, however, shown to provide three SNP markers in linkage disequilibrium with *Ms*. Using markers from these maps to generate intron-spanning probes for a sensitive tyramide-FISH method allowed Khrustaleva et al. (2016) to construct a physical map of the *Ms* region. Importantly, this showed that the locus lies in a region of lower recombination. Additional codominant PCR-based markers were developed by Bang et al. (2011) from RFLP markers on the USDA-ARS SNP map.

Recent transcriptome analyses in onion (Khosa et al. 2016) and bunching onion (Liu et al. 2016) have provided some putative candidates for mitochondrial and nuclear genes underpinning CMS and nuclear restorer genetics, but these have not directly contributed useful targets for mapping. By contrast, Korean researchers using variations of bulked segregant analysis (BSA; Michelmore et al. 1991) have provided very useful anonymous genomic and candidate gene markers for onion CMS fertility restoration. By using classic RAPD-BSA, Park et al. (2012) identified RAPD markers distinguishing male-sterile and -fertile bulks and converted these to more portable codominant markers (Bang et al. 2013), shown to be useful for both S and T cytoplasms (Kim 2014). They subsequently used RNA-SEQ in a BSA approach to identify 30 contigs that they confirmed to be in complete LD with *Ms* by association analysis (Kim et al. 2015). This kind of approach employing a combination of large segregating populations, association panels, and pool sequencing is a very powerful strategy for major genes segregating in non-model organisms such as onion. The other groups who have developed Ms-linked markers are Huo and collaborators (Huo et al. 2015) who reported that the AcSKP1 gene is very tightly linked. These researchers also identified that pectin methylesterase (PME) expression in flower buds is Ms-linked by cDNA-SRAP (Huo et al. 2012).

Validation of these markers by Havey and von Kohn (2017) in North American onion germplasm confirmed tight linkage of the marker Jnurf13 (Kim 2014) with *Ms* and complete linkage of AcPMS1. However, a similar study in Indian germplasm by Khar and Saini (2016) found that none of the markers evaluated were in complete disequilibrium with *Ms*. This may reflect the greater diversity of Indian SD/ID germplasm and highlights the challenges of transferring markers in an out-breeding crop with a poorly characterized genome.

13.2.2 Genetic Mapping of Color

Among the earliest morphological markers used in onion are skin color phenotypes due to flavonoid pigmentation (Cramer and Havey 1999). Classic genetic studies demonstrated that at least 3 loci (*I, C and R)* conditioned onion color (Clarke et al. 1944). Molecular studies of expression and allelic variation in genes encoding anthocyanin biosynthesis led to identification of structural or trans-regulator variants of *chalcone synthase* (*CHS*), *dihydroflavonol 4-reductase* (*DFR*) and *anthocyanidin synthase* (*ANS*) as putative candidates for the *C* (Kim et al. 2005a, b, c, d), *R* (Kim et al. 2004a, b, 2005a, b, c, d) and *L* loci (Kim et al. 2005a, b, c, d), respectively. Khar et al. (2008) studied co-segregation of color phenotypes and SNPs in a larger family than these earlier studies, confirming that red color was conditioned by the *DFR* gene and a locus more loosely linked to the *ANS* gene (*L2*). Assignment of genes encoding flavonoid biosynthesis in onion by Masuzaki et al. (2006a, b) revealed the chromosomal assignment of the major flavonoid biosynthetic genes, as well as locating the 3-glycosyltransferase activity responsible for anthocyanin glycosylation to chromosome 4. Khar et al. (2008) reported that markers reported by Kim et al. were not informative in their families and suggested that absence of red coloration is due to multiple mutations, selected at multiple occasions in the domestication and breeding of onion. This was confirmed by subsequent studies of Kim and coworkers who showed that flesh color phenotypes might be conditioned by multiple distinct mutations in *DFR* and *ANS* genes (Song et al. 2014; Kim et al. 2009; Park et al. 2013; Kim et al. 2017; Kim et al. 2014a, b). Experience by breeding companies also suggests that developing universal markers to loss-of-function alleles, such as brown alleles of *DFR*, will be challenging until reference sequence and detailed re-sequencing data become available. In an attempt to identify the *C* locus Baek et al. (2017) used BSA of RNA from pools of yellow or white bulbs. The data were used for differential expression analysis to determine regulatory and biosynthetic genes but also to design markers to the *C* locus based on polymorphisms that fitted the expected allele heterozygosity status for *C*. A completely linked marker in the GST1 gene was identified but not expected to be the functional gene for *C* locus.

Global genetic mapping studies of pigmentation have been carried out on onion bulb colors by Duangjit et al. (2014) and on Japanese

Bunching Onion (JBO; *Allium fistulosum)* pseudostem color by Tsukazaki et al. (2017). JBO pigmentation QTL were detected on Chromosomes 4 and 5, with the latter (*qPig5a-2*) being a major gene. Binary quantitative genetic analysis for red versus yellow color in the OH1 × 5225 onion family showed that this could be modeled by interactions between a locus on chromosome 7 mapping to the *DFR* gene and two linked loci on chromosome 4 that likely correspond to the *L1* and *L2* loci.

13.2.3 Flavor

The most distinctive chemical attribute of *Allium* vegetables are the organosulfur compounds which impart flavor and other bioactivities (Block 2010). Discovery of the onion Lachrymatory *Factor Synthase* (*LFS*) gene by Imai et al. (2002) established that tearing arises due to a specific enzyme activity. Based on evidence from the second-generation sequencing of a DH onion, BAC sequencing, interspecific linkage mapping and FISH it was possible to show that this is encoded by duplicated genes on onion Chromosome 5 (Masamura et al. 2012). By using the same BAC and mapping resources, it was also possible to characterize the genome organization and map locations of onion *sulfite reductase* gene and its expressed pseudogene (McManus et al. 2012). Tsukazaki et al. (2012) mapped a major pungency QTL in *A. fistulosum* to a region of chromosome 2 and developed a useful marker from this. The relationship of this QTL to onion pungency QTL is not clear.

The market classes, culinary properties, and *Allium* bulb and stem tissues are in major part determined by the balance of fructo-oligosaccharides to hexoses. Carbon physiology is central to growth and productivity, and although there have been no major advances in this area, this area seems ripe for integrative genetic and physiological studies. Two recent studies by Zhang et al. (2016) and Han et al. (2016) have provided quite large and highly relevant transcriptome datasets from developing and stored bulbs, respectively. There is a tight connection between carbohydrate metabolism and flowering physiology, and it was not surprising that genetic mapping of onion- bolting revealed a putative association between bulb sucrose and bolting linked to a region on Chromosome 3 containing an invertase gene (Baldwin et al. 2013a, b). Integrative global transcriptome-based genetics now provides the potential to tease out the genetic and physiological factors determining variability in *Allium* carbon metabolism.

13.2.4 Adaptive and Morphological Loci

The adaptive phenotypes and appearance of *Alliums* are central to breeding goals and may often constrain the ability to access genetic diversity. Some of the key genes involved in regulation of bulb and flower formation were identified by traditional methods (Taylor et al. 2010) but RNA-SEQ methods have enabled a rapid expansion in the knowledge of key gene families and variation underlying development and adaptation in garlic (Kamenetsky et al. 2015) and onion (Lee et al. 2013; Khosa et al. 2016; Manoharan et al. 2016). In contrast to these transcriptome approaches within a single species, Zhu et al. (2017) used a comparative approach to identify candidate genes for the fistular leaf phenotype. With the growing availability of such catalogs in *Allium* species such as those reported by Han et al. (2016) and Sun et al. (2015), and bioinformatics tools to support comparative transcriptomics, this kind of evolutionary approach deserves wider usage.

One of the key adaptation traits in *Allium* is variation in vernalization requirements for flowering (Brewster 2008). Genetic mapping in both onion and *A. fistulosum* has located QTL associated with bolting (premature flowering). Using a wide cross between a long day (LD) doubled-haploid (DH) onion "CUDH2150" and a highly adaptable short day/intermediate day (SD/ID) Indian landrace "Nasik Red" we were able to map and validate a major bolting QTL *AcBlt1* to chromosome 1 (Baldwin et al. 2013a, b). These studies were conducted in uncontrolled field conditions on large population sizes.

By contrast, Wako et al. (2016) used a combination of field and heated glasshouse evaluations to map bolting time QTL in two *A. fistulosum* families. In addition to utilizing the SaT03 family employed earlier for QTL mapping (Tsukazaki et al. 2012) they used a cross between an ever-flowering line "Ki" and late-flowering "C". Reproducible QTL for bolting time were detected on chromosome 2 in both "Ki x C" and SaT03 families. In addition, a reproducible QTL on Chromosome 1 was detected in "Ki x C" in the open field but not in a heated glasshouse. We hope that the use of controlled phenotyping and orthologous gene markers may permit testing the homology of this QTL with onion *AcBlt1* in the near future. Although a number of key flowering genes have been placed on the *Allium* genetic maps in these two studies, there is currently no strong evidence for linkage of candidates to QTL.

A conspicuous form of morphological variation in onion is seen in the leaf wax variants originally reported in classical studies by Jones et al. (reviewed by Cramer and Havey 1999). Glossy and semi-glossy types with lower epicuticular waxes show reduced susceptibility to onion thrips (Jones et al. 1935), and the more practical semi-glossy types are, therefore, attractive as a genetic strategy to mitigate impacts of this serious pest. Damon et al. (2014) confirmed that glossy and semi-glossy types contained lower leaf waxes andthese phenotypes were associated to reduce thrips load. They then conducted QTL mapping of the leaf wax chromatographic phenotypes, identifying two major QTL on chromosomes 2 and 5 associated with different pathways (Damon and Havey 2014). Similar phenotypes have been reported in *A. fistulosum* (Yang et al. 2017) and RNA-SEQ comparisons performed between glossy and wild-types (Liu et al. 2014), but no mapping has yet been reported.

Considerable progress has been reported in interspecific disease resistance, most notably the dominant downy mildew resistance from *A. roylei* backcrossed into onion (Van der Meer and De Vries 1990). Observations of skewed segregation in F_2 populations using the original introgressions motivated analysis by FISH and AFLP which revealed that the introgression contained a lethal gene (Scholten et al. 2007). The markers enabled identification of rare crossovers and selection of a homozygous introgression which was deployed to commercial hybrid varieties. Kim et al. (2016) reported a simple PCR marker that can be used for tagging this introgression based on linkage to SNPs in the map of Duangjit et al. (2013). Resistance to *Botrytis* leaf blight has been introgressed into onion from *A. roylei* and genetically-characterized but not mapped (Hyde et al. 2015). However Scholten and colleagues (Scholten et al. 2016) were able to map this resistance in a hybrid population to A. *roylei* chromosome 6 using a set of SNP assays developed by RNASEQ of onion genotypes and an A. *fistulosum x* A. *roylei* F1 hybrid. We hope that wider access to genotyping and data platforms will enable more work like this that can enhance *Allium* sustainability and quality in the future.

13.3 Future Strategies and Resources

The choices for conducting powerful and economical genome analysis in *Allium* are growing as service costs fall and bioinformatics tools expand. Recently Jo et al (2017) used a double-digest GBS method to develop a reference-free genetic map of an F_2 onion population, demonstrating that this widely used method can be employed in large *Allium* genomes. This study did not assign linkage groups to the physical map but indicates the potential for using modern reduced-representation methods to conduct genetic analyses. Currently, one of the most fertile areas in genetics is employing creative NGS strategies to conduct genome scans or population genetic studies (Matz 2017). There is great scope for relatively economical *Allium* population genetic analysis using mature technologies such as pool sequencing (Schlötterer et al. 2014). The limiting factor for applying such technologies is the horticultural research context.

We broadly categorize the critical areas as materials, information and horticultural strategy.

13.4 References and Control Materials for *Allium* Research

A hallmark of good science is the use and choice of controls. Model systems research clearly demonstrates that selecting and distributing common reference stocks is the key to building research capability and community. Maize is a good example where the choice of key reference lines and mapping populations has enabled deep basic and commercial research (Schnable et al. 2009; Tian et al. 2011; Coe et al. 2002). The most critical barrier to raising the standard of *Allium* research is an absence of reference lines that can be used to develop integrated data sets among research groups. Onion poses many challenges for the choice and exchange of reference materials. The default option for a reference or control line in most agronomic and genetic studies is usually a local landrace or hybrid favored by local industry. This approach ensures a perception of local relevance for funding bodies but limits global relevance and applicability. Seedlots for such "control" lines may be unavailable or may be agronomically inappropriate, e.g., adapted to different photoperiod. An obvious alternative to variable open-pollinated standards is the use of commercial hybrids or publicly available inbreds released from US public breeding programmes (Goldman et al. 2001). However, the maintenance of inbreds especially involving CMS is not straightforward and onion "inbreds" can still harbor significant genetic variability (McCallum et al. 2008).

Although the exchange of DNA samples has enabled some collaborations among research groups, the usefulness of this approach has been limited by the practicalities of producing sufficient DNA and exchanging mapping data. By contrast, the best example of the successful exchange of genetic resources in *Allium* genetics has been the use *A. cepa* - *A. fistulosum* monosomic alien addition lines developed by Shigyo et al. (1997). Although these stocks have only permitted chromosome-level mapping of genes and phenotypes, they have contributed significantly to anchoring and sharing of genetic marker data in *A. cepa* and *A. fistulosum.* The key to success of these stocks has been the simplicity of exchange (only 10–12 samples), ease of genotyping (interspecific polymorphisms are easily identified or dominant) and trivial data analysis. The availability of RNA-SEQ from these lines (Abdelrahman et al. 2017) now offers the potential for performing this assignment in silico. We expect that these resources will provide the foundation of a more global *Allium* framework gene map.

13.5 Information

The establishment of RNA-SEQ as a routine typing methodology has led to numerous publications, providing opportunities as well as bioinformatics challenges for would-be Allium genetics researchers. These studies have employed a variety of genotypes at different physiological stages and have distributed data as supplementary data, bespoke databases, and/or public repositories, or have not released such data. To date, there does not appear to have been any efforts to construct consensus gene indices or agree on common stocks. Among studies in onion, that of Kim et al. (2014a, b) is notable in that it employed a DH reference line "H6" (Walker 2010), as we did in both mapping (Baldwin et al. 2012a, b) and more recent transcriptome studies (Khosa et al. 2016). Excellent transcriptome catalogs have been developed for garlic, most notably that of Kamenetsky et al. (2015) but applications to date appear to be limited to marker design for fingerprinting genetic resource collections (Liu et al. 2015) due to lack of fertile accessions for genetic stock development. Similarly, Zhou et al. (2015) have generated a seedling transcriptome catalog for *A. tubersosum*, opening the way for genetic studies of this widely cultivated vegetable.

We previously published a database compiling mapping and marker data from *Allium* genetic mapping experiments (McCallum et al. 2012). This is still available at http://alliumgenetics.org/ but has proved challenging to migrate and maintain since it is based on CMAP (Youens-Clark et al. 2009) which uses deprecated web technologies. This curation and technology problem has also affected other initiatives, notable RobustDb (Bhasi et al. 2010). A more recent publication (Shukla et al. 2016) describes an online database based on data extracted from Genbank. These resources are at great risk of "URL decay" (Wren et al. 2017) since they are reliant on usage, citation and potentially unreliable funding and institutional support. There is a clear need for community-curated resources. Submission of curated, integrated data sets into Genbank is challenging for projects with limited bioinformatics resources. The use of generic collaborative platforms like Cyverse (http://www.cyverse.org/; Merchant et al. 2016) that support community collaborated data and workflows (Grover et al. 2017) may provide a more future-proofed solution that could encourage greater participation in *Allium* genetics. Rather than focus on providing database functionality and interactive graphical features, it may be more important to focus on curated, accessible source-controlled tabular data that could serve multiple analytical uses, including reproducible analyses and web applications. The use of a source-control model such as Github could also provide a simple basis for managing tabular community data. We have provided SQL dumps of the AlliumMap (McCallum et al. 2012) CMAP database at https://github.com/PlantandFoodResearch/AlliumMap/ to assist access to this data.

13.6 Doubled-Haploid Onions as Reference Materials

Although there has been a long history of research concerning the development of doubled haploids for onion breeding (Martínez 2003; Bohanec 2002; Campion et al. 1995), this has rarely provided genotypes with fitness and fecundity suitable for breeding and genetics. Use of DH lines has been key to more recent transcriptome (Duangjit et al. 2013; Kim et al. 2014a, b; Khosa et al. 2016) and metabolic analyses (Abdelrahman et al. 2015; Joshi 2014). An exception to the generally low fitness of DH onion genotypes is a series of DH lines extracted for the synthetic "YIX-A" which were released by Cornell University (Hyde et al. 2012; Alan et al. 2004). These have proven to be very practical in our hands and one selection is currently the subject of a whole genome sequencing initiative (http://www.oniongenome.net/). The DH lines referred were provided under MTA from the Cornell University Plant Breeding Department and we can provide these direct to researchers who have signed the same agreement. In future, these may be available from gene banks (Martha Mutschler *pers comm.*). The use of DH lines extracted from an F_1 plant to provide full-sib haploid mapping families as practiced by Duangjit et al. (2014) is a highly attractive but technically challenging strategy.

13.7 Development and Propagation of Large F_2 Stocks for Molecular Mapping

In order to encourage greater experimentation by existing and new entrants into the research community, we now summarize our current practices for onion population development, developed over the past 20 years. Although these methods have been influenced by commercial breeding practices, they are focused on experimental stock development. This work was carried out in a temperate intermediate/long day environment in Lincoln New Zealand (43.6 °S, 172.4 °E). We focus on field horticultural practices employed during research on adaptive and carbon physiology of onion (Baldwin et al. 2013a, b; Baldwin et al. 2012a, b; Revanna et al. 2013). In a companion article (Khosa et al. 2018), we describe methodologies for propagation under controlled environments.

The germplasm used in this study included seed lots obtained by the former DSIR (Grant and Carter 1986) and a series of diverse landrace accessions from the EU Allium collection (previously at HRI Wellesbourne; http://ealldb.ipk-gatersleben.de/861/). Because such seed lots are usually of highly variable age and quality, and in very limited quantities we germinate on moist paper towels at 25 °C and transfer viable seedlings (< 10 mm) to Jiffy-7 Peat Pellets (http://www.jiffypot.com/en/products/jiffy7.html). These can be conveniently managed in cell trays and then transplanted to the field or glasshouse once they reach the 3–4 leaf stage. These plugs minimize dehydration and transplant shock. For direct sowing of uncoated seed, we use Earthway 1001-B garden seeders (https://earthway.com) fitted with a custom Blank Seed Plate (Cat # 18106) drilled with twelve 4 mm holes at 30 mm spacing 4 mm from the plate edge. Alternatively, a Radish plate (Cat # 18101) may be used after filing the lugs off flush with the plate.

Although we typically propagate our experimental populations in sections of commercial production fields provided by growers or breeding companies, special care is needed when first propagating such mal-adapted, valuable, and limited materials with lower fitness and seed quality compared to commercial cultivars and breeding lines. Commercial herbicide regimes may impact on growth and weak stands may require manual weeding. Other options we have used include propagating sets of 8–12 plants in 20L tubs with trickle irrigation in a tunnel house and propagating small trials transplanted through synthetic mulch cloth under drip irrigation (Fig. 13.1).

Fig. 13.1 Detail of a small experimental plot of a DH x landrace F_2 population transplanted in synthetic mulch cloth under drip irrigation

During propagation of doubled-haploid stocks these should be visually inspected for evidence of phenotypic variation that might suggest contamination before or during sowing. The Cornell DH lines are conspicuously uniform in plant form and bulb shape, though these may vary considerably year-to-year (Figs. 13.2 and 13.3). Once plants are mature, as judged by leaf collapse in ~ 50% of plants ("top-down"), we lift all the plots gently and allow plants to cure in the field for about 1 week. Severely mal-adapted materials such as SD fresh market types or very LD types may require pre-emptive removal and

Fig. 13.2 Bulbs of Cornell University DH Onion genotype "CUDH2107"

Fig. 13.3 Bulbs of Cornell University DH Onion genotype "CUDH2122"

other special treatment. If conditions are cool and/or damp it is preferable to remove sensitive and valuable materials to a warm, dry, and well-ventilated environment for curing. After removal from the field in shallow slotted produce crates we will typically cure bulbs for a further 1–2 weeks in a well-ventilated tunnel or glasshouse prior to clipping off leaves and placing in storage. We store onions in a small cool store at ~ 5°C with humidity maintained at 65–70% RH by a domestic dehumidifier.

If we need to sample bulbs for chemical or molecular analysis, they can be bisected and the bases saved for flowering in the spring. We usually place cut bases in net fruit bags with labels and then allow them to dry in a glasshouse for 3 days before dipping in a 5% solution of Carbendazim. Dipped or painted bulbs are best planted rather than stored to minimize losses. Bulb tissues are a good source of DNA or RNA, provided that high-salt precipitations are used to reduce carbohydrate coprecipitation.

After storage for 4–5 months, we select bulbs for crossing or seed production and plant into a soil/peat mix in 18–20 L bins with drainage holes in an unheated tunnel house. Initially, the soil mix is maintained just sufficiently moist to mitigate peat dust risks. When bulbs have rooted and sprouted we install trickle irrigation on a timer set to provide modest watering twice daily. Watering duration is increased once plants are flowering, especially in periods of greater summer heat. For pair crossing or self-pollination, we bag the umbels in a perforated bread bag prior to flower opening and seal the base with electrical tape. Just prior to flowers opening, we cut the bag corner and introduce pupae and/or flies of the species *Lucilia sericata* (common green bottle fly; usually sourced from pet food suppliers), resealing with a paper clip. It is important to shade umbels bagged in this manner to minimize heat stress on flies and flowers alike. We do not normally emasculate since the F_1 hybrid progeny of the wide crosses we are concerned with are conspicuous by their size, vigor, and bulb color phenotypes compared to DH parents. Molecular marker tests may be used to confirm hybridity. For limited-mass-pollinations, we cage a single tub containing 6–12 bulbs or 2–3 such tubs where we wish to ensure larger useful population sizes. When inflorescences emerge, we place a simple frame constructed from 1.8–2.1 m × 16 mm plastic-coated garden stakes as uprights, inserted into a frame constructed with dowels inserted into 19 mm irrigation corner elbows. It is desirable to tie inflorescences together at this stage so that umbels do not lie against the cage material later. Prior to splitting the leafy spathe that encloses the umbel we cover the frame with 2 mm mesh cover sewn in a tube and close this at the top with plastic twine (Fig. 13.4). As flowers open we introduce blowflies that we have previously hatched from pupae in a warm dry environment. Pupae may be introduced directly into cages but emergence may be less. To prolong the working life of pollinators, especially where nectar or flower numbers are limiting we provide stale beer soaked into paper towels in a cup. Blowflies do not work as effectively in larger cages over 1 m^3. In such cases, small bumblebee boxes used for glasshouse pollinations are more effective. An alternative means to

Fig. 13.4 Limited mass cages at peak flowering

cross-pollinate pair crosses or small cages is to do this manually using instruments such as a makeup brush or feather duster (dedicated to a cross or cage). This requires enclosures to have zips or other means of access.

When flowering has ended irrigation is maintained until seed development is complete as evidenced by black appearance. At this stage, irrigation is turned off and plants allowed to dry until the black seed is evident on the top of the umbel. Umbels are clipped off and placed in a heavy paper or cloth bags and dried in cool, low humidity conditions. Cleaned seed should be stored in cool, low humidity conditions.

Following pollination, especially if this is limited, some genotypes may spontaneously form bulbils or "topsets" in the inflorescences (Brewster 2008). These enable clonal multiplication of F_1 plants (Fig. 13.5). Topset production may be forced by trimming unopened flowers from the umbel base after splitting of the leafy spathe and painting with dilute benzyladenine (Dunstan and Short 1979) in a wetting agent. Dry topsets may be stored for several months, planted in the spring and mass-pollinated, enabling large-scale production of F_2 seed. Alternatively, many wide hybrids are highly multicentred and produce lateral "cloves" that that can be split before or after flowering to enable mass pollinations of clonal F_1 plants over one or more seasons.

Fig. 13.5 Topset production an F_1 umbel from the cross "CUDH2107 x Southport White Globe"

We refined these strategies during the development of the "Nasik Red × CUDH2150" F_2 families used to map bulb and adaptive traits (Baldwin et al. 2013a, b; Revanna et al. 2013; Baldwin et al. 2012a, b). Since the low-solids DH line "CUDH2150" proved very hard to grow and store we conducted pair crossing between the much better-storing genotypes "CUDH2122" (used for skim genomic sequencing and SSR development in Baldwin et al. 2012a, b) and "CUDH2170" (used for RNASEQ in Khosa et al. 2016) and landrace and other selections used in these studies. We confirmed the observations of Hyde et al. (2012) that these DH lines form vigorous F_1 hybrids which are very practical to handle. Segregation of bulb phenotypes in one such F_2 family is shown in Fig. 13.6. This population segregates strongly for a multicentred trait that may be homologous with the tillering trait analysed by QTL mapping in *A. fistulosum* (Tsukazaki et al. 2017).

13.8 Conclusion

In times of rapid and ongoing change in genetic technologies, it can be very challenging for horticultural researchers to choose among these and access informatics required to exploit them fully. Realizing the promises of *Allium* QTL and gene mapping will require developing a new generation of shared resources that extend beyond national and institutional boundaries. Although funding is frequently tied to local business needs, it is essential that wider initiatives are developed that can raise the quality of *Allium* genetics for academic publishing and business application. We encourage new and existing *Allium* genetics researchers to design their research to both leverage and contribute to community germplasm and data sources.

Fig. 13.6 Segregation for multicentred ("tillering") phenotype in F_2 family from the cross CUDH2122 (Cornell University doubled-haploid onion line "DH2122 E21A9M-9C 2 N") x Pakistani Landrace ("HRIGRU 3493")

References

Abdelrahman M, El-Sayed M, Sato S, Hirakawa H, Ito S, Tanaka K, Mine Y, Sugiyama N, Suzuki Y, Yamauchi N, Shigyo M (2017) RNA-sequencing-based transcriptome and biochemical analyses of steroidal saponin pathway in a complete set of *Allium fistulosum*—*A. cepa* monosomic addition lines. PLoS ONE 12:e0181784. https://doi.org/10.1371/journal.pone.0181784

Abdelrahman M, Sawada Y, Nakabayashi R, Sato S, Hirakawa H, El-Sayed M, Hirai MY, Saito K, Yamauchi N, Shigyo M (2015) Integrating transcriptome and target metabolome variability in doubled haploids of *Allium cepa* for abiotic stress protection. Mol Breeding 35:195. https://doi.org/10.1007/s11032-015-0378-2

Alan AR, Brants A, Cobb E, Goldschmied PA, Mutschler MA, Earle ED (2004) Fecund gynogenic lines from onion (*Allium cepa* L.) breeding materials. Plant Sci 167:1055–1066

Baek G, Kim C-W, Kim S (2017) Development of a molecular marker tightly linked to the *C* locus conferring a white bulb color in onion (*Allium cepa* L.) using bulked segregant analysis and RNA-Seq. Mol Breeding. https://doi.org/10.1007/s11032-017-0697-6

Baldwin S, Pither-Joyce M, Wright K, Chen L, McCallum J (2012a) Development of robust genomic simple sequence repeat markers for estimation of genetic diversity within and among bulb onion (*Allium cepa* L.) populations. Mol Breeding 30:1401–1411. https://doi.org/10.1007/s11032-012-9727-6

Baldwin S, Revanna R, Pither-Joyce M, Shaw M, Wright K, Thomson S, Moya L, Lee R, Macknight R, McCallum J (2013a) Genetic analyses of bolting in bulb onion (*Allium cepa* L.). Theor Appl Genet 127:535–547. https://doi.org/10.1007/s00122-013-2232-4

Baldwin S, Revanna R, Pither-Joyce M, Shaw M, Wright K, Thomson S, Moya L, Lee R, Macknight R, McCallum J (2013b) Genetic analyses of bolting in bulb onion (*Allium cepa* L.). Theor Appl Genet 127:535–547. https://doi.org/10.1007/s00122-013-2232-4

Baldwin S, Revanna R, Thomson S, Pither-Joyce M, Wright K, Crowhurst R, Fiers M, Chen L, Macknight R, McCallum JA (2012b) A toolkit for bulk PCR-based marker design from next-generation sequence data: application for development of a framework linkage map in bulb onion (*Allium cepa* L.). BMC Genom 13:637. https://doi.org/10.1186/1471-2164-13-637

Bang H, Cho DY, Yoo K-S, Yoon M-K, Patil BS, Kim S (2011) Development of simple PCR-based markers linked to the *Ms* locus a restorer-of-fertility gene in onion (*Allium cepa* L.). Euphytica 179:439–449. https://doi.org/10.1007/s10681-010-0342-5

Bang H, Kim S, Park SO, Yoo K-S, Patila BS (2013) Development of a codominant CAPS marker linked to the *Ms* locus controlling fertility restoration in onion (*Allium cepa* L.). Sci Hortic 153:42–49. https://doi.org/10.1016/j.scienta.2013.01.020

Bhasi A, Senalik D, Simon PW, Kumar B, Manikandan V, Philip P, Senapathy P (2010) RoBuST: an integrated genomics resource for the root and bulb crop families Apiaceae and Alliaceae. BMC Plant Biol 10:161. https://doi.org/10.1186/1471-2229-10-161

Block E (2010) Garlic and other Alliums: the lore and the science. Royal society of Chemistry

Bohanec B (2002) Doubled-haploid onions. In: Allium crop science: recent advances. CABI, pp. 145–157

Bonny S (2017) Corporate concentration and technological change in the global seed industry. Sustainability 9:1632. https://doi.org/10.3390/su9091632

Brewster JL (2008) Physiology of crop growth. In: Onions and other vegetable alliums. CABI Press

Campion B, Bohanec B, Javornik B (1995) Gynogenic lines of onion (*Allium cepa* L.): evidence of their homozygosity. Theor Appl Genet 91:598–602. https://doi.org/10.1007/bf00223285

Clarke AE, Jones HA, Little TM (1944) Inheritance of bulb color in the onion. Genetics 29:569

Coe E, Cone K, McMullen M, Chen S-S, Davis G, Gardiner J, Liscum E, Polacco M, Paterson A, Sanchez-Villeda H, Soderlund C, Wing R (2002) Access to the maize genome: an integrated physical and genetic map. Plant Physiol 128:9–12

Cramer CS, Havey MJ (1999) Morphological, biochemical, and molecular markers in onion. HortScience 34:589–593

Damon SJ, Groves RL, Havey MJ (2014) Variation for epicuticular waxes on onion foliage and impacts on numbers of onion thrips. J Am Soc Hortic Sci 139:495–501

Damon SJ, Havey MJ (2014) Quantitative trait loci controlling amounts and types of epicuticular waxes in onion. J Am Soc Hortic Sci 139:597–602

Duangjit J, Bohanec B, Chan AP, Town CD, Havey MJ (2013) Transcriptome sequencing to produce SNP-based genetic maps of onion. Theor Appl Genet 126:2093–2101. https://doi.org/10.1007/s00122-013-2121-x

Duangjit J, Welsh K, Wise ML, Bohanec B, Havey MJ (2014) Genetic analyses of anthocyanin concentrations and intensity of red bulb color among segregating haploid progenies of onion. Mol Breeding 34:75–85. https://doi.org/10.1007/s11032-014-0018-2

Dunstan DI, Short KC (1979) Shoot production from the flower head of *Allium cepa* L. Sci Hortic 10:345–356

Dunwell JM (2016) Mergers and acquisitions in global ag-biotech. In: Biotechnology of major cereals. CABI, pp 206–222

Ellegren H (2014) Genome sequencing and population genomics in non-model organisms. Trends Ecol Evol 29:51–63. https://doi.org/10.1016/j.tree.2013.09.008

Goldman IL, Schroeck G, Harvey MJ (2001) History of public onion breeding programs in the United States. Plant Breeding Rev 20:67–104

Grant D, Carter B (1986) Onion breeding in the DSIR. In: Proceedings of the Agronomy Society of New Zealand, pp 115–117

Grover JW, Bomhoff M, Davey S, Gregory BD, Mosher RA, Lyons E (2017) CoGe LoadExp+: a web-based suite that integrates next-gen sequencing data analysis workflows and visualization. bioRXiv https://doi.org/10.1101/118802

Gökçe AF, Havey MJ (2002) Linkage equilibrium among tightly linked RFLPs and the *Ms* locus in open-pollinated onion populations. J Am Soc Hortic Sci 127:944–946

Han J, Thamilarasan SK, Natarajan S, Park J-I, Chung M-Y, Nou I-S (2016) De novo assembly and transcriptome analysis of bulb onion (*Allium cepa* L.) during cold acclimation using contrasting genotypes. PLoS ONE 11:e0161987. https://doi.org/10.1371/journal.pone.0161987

Havey MJ, von Kohn C (2017) Efficacy of molecular markers Jnurf13 and AcPms1 for prediction of genotypes at the nuclear *Ms* locus in three open-pollinated populations of onion from North America. HortScience 52:1052–1053. https://doi.org/10.21273/hortsci11954-17

Huo Y, Miao J, Liu B, Yang Y, Zhang Y, Tahara Y, Meng Q, He Q, Kitano H, Wu X (2012) The expression of pectin methylesterase in onion flower buds is associated with the dominant male-fertility restoration allele. Plant Breeding 131:211–216

Huo YM, Liu BJ, Yang YY, Miao J, Gao LM, Kong SP, Wang ZB, Kitano H, Wu X (2015) $AcSKP_1$ a multiplex PCR-based co-dominant marker in complete linkage disequilibrium with the male-fertility restoration (*Ms*) locus, and its application in open-pollinated populations of onion. Euphytica 204:711–722. https://doi.org/10.1007/s10681-015-1374-7

Hyde PT, Earle ED, Mutschler MA (2012) Doubled haploid onion (*Allium cepa* L.) lines and their impact on hybrid performance. HortScience 47:1690–1695

Hyde PT, Leckie BM, Carpenter SCD, Lorbeerb JW, Hoeptingc CA, Ullrichd MR, Mutschler MA (2015) Evaluation of onion (Allium cepa L.) lines and hybrids possessing *Bs1* for resistance against multiple isolates of *Botrytis squamosa*. Crop Science Abstract 55:136. https://doi.org/10.2135/cropsci2014.03.0255

Imai S, Tsuge N, Tomotake M, Nagatome Y, Sawada H, Nagata T, Kumagai H (2002) Plant biochemistry: an onion enzyme that makes the eyes water. Nature 419:685

Jo J, Purushotham PM, Han K, Lee H-R, Nah G, Kang B-C (2017) Development of a genetic map for onion (*Allium cepa* L.) using reference-free genotyping-by-sequencing and SNP assays. Frontiers in Plant Sci. https://doi.org/10.3389/fpls.2017.01606

Jones HA, Bailey SF, Emsweller SL (1935) Field studies of *Thrips tabaci* Lind. with especial reference to resistance in onions. J Econ Entomol 28:678–680. https://doi.org/10.1093/jee/28.4.678

Joshi S (2014) Interaction between sulfur (S) and nitrogen (N) assimilation pathways in response to S and N supply in onion (*Allium cepa* L.) Ph.D. thesis, Plant Molecular Biology at Massey University, Palmerston North, New Zealand

Kamenetsky R, Faigenboim A, Mayer ES, Michael TB, Gershberg C, Kimhi S, Esquira I, Shalom SR, Eshel D, Rabinowitch HD, Sherman A (2015) Integrated transcriptome catalogue and organ-specific profiling of gene expression in fertile garlic (*Allium sativum* L.). BMC Genomics. https://doi.org/10.1186/s12864-015-1212-2

Khar A, Jakse J, Havey MJ (2008) Segregations for onion bulb colors reveal that red is controlled by at least three loci. J Am Soc Hortic Sci 133:42–47

Khar A, Saini N (2016) Limitations of PCR-based molecular markers to identify male-sterile and maintainer plants from Indian onion (*Allium cepa* L.) populations. Plant Breeding 135:519–524. https://doi.org/10.1111/pbr.12373

Khosa JS, Lee R, Bräuning S, Lord J, Pither-Joyce M, McCallum J, Macknight RC (2016) Doubled Haploid 'CUDH2107' as a reference for bulb onion (*Allium cepa* L.) research: development of a transcriptome catalogue and identification of transcripts associated with male fertility. PLoS ONE 11(11):e0166568. https://doi.org/10.1371/journal.pone.0166568

Khosa J, Lee R, Joshi S, Shaw M ,McCallum J, Macknight R (2018) A Guide for the Cultivation of Onion (Allium cepa L.) under Controlled-Environment Conditions. Submitted

Khrustaleva L, Jiang J, Havey MJ (2016) High-resolution tyramide-FISH mapping of markers tightly linked to the male-fertility restoration (*Ms*) locus of onion. Theor Appl Genet 129:535–545

Kim B, Cho Y, Kim S (2017) Identification of a novel *DFR-A* mutant allele determining the bulb color difference between red and yellow onions (*Allium cepa* L.). Plant Breeding Biotechnol 5:45–53. https://doi.org/10.9787/pbb.2017.5.1.045

Kim S (2014) A codominant molecular marker in linkage disequilibrium with a restorer-of-fertility gene (*Ms*) and its application in reevaluation of inheritance of fertility restoration in onions. Mol Breeding 34:769–778. https://doi.org/10.1007/s11032-014-0073-8

Kim S, Baek D, Cho DY, Lee E-T, Yoon M-K (2009) Identification of two novel inactive *DFR-A* alleles responsible for failure to produce anthocyanin and development of a simple PCR-based molecular marker for bulb color selection in onion (*Allium cepa* L.). Theor Appl Genet 118:1391–1399. https://doi.org/10.1007/s00122-009-0989-2

Kim S, Binzel ML, Park S, Yoo K-S, Pike L (2004a) Inactivation of DFR (Dihydroflavonol 4-reductase) gene transcription results in blockage of anthocyanin production in yellow onions (*Allium cepa*). HortScience 39:889

Kim S, Binzel ML, Yoo KS, Park S, Pike LM (2004b) *Pink* (*P*), a new locus responsible for a pink trait in onions (*Allium cepa*) resulting from natural mutations of anthocyanidin synthase. Mol Genet Genomics 272:18–27. https://doi.org/10.1007/s00438-004-1041-5

Kim S, Jones R, Yoo K-S, Pike LM (2005a) The *L* locus one of complementary genes required for anthocyanin production in onions (*Allium cepa*), encodes anthocyanidin synthase. Theor Appl Genet 111:120–127. https://doi.org/10.1007/s00122-005-2000-1

Kim S, Kim C-W, Choi M-S, Kim S (2016) Development of a simple PCR marker tagging the *Allium roylei* fragment harboring resistance to downy mildew (*Peronospora destructor*) in onion (*Allium cepa* L.). Euphytica 208:561–569

Kim S, Kim C-W, Park M, Choi D (2015) Identification of candidate genes associated with fertility restoration of cytoplasmic male-sterility in onion (*Allium cepa* L.) using a combination of bulked segregant analysis and RNA-seq. Theor Appl Genet 128:2289–2299. https://doi.org/10.1007/s00122-015-2584-z

Kim S, Kim M-S, Kim Y-M, Yeom S-I, Cheong K, Kim K-T, Jeon J, Kim S, Kim D-S, Sohn S-H, Lee Y-H, Choi D (2014a) Integrative structural annotation of de novo RNA-Seq provides an accurate reference gene set of the enormous genome of the onion (*Allium cepa* L.). DNA Res 22:19–27. https://doi.org/10.1093/dnares/dsu035

Kim S, Park JY, Yang T-J (2014b) Characterization of three active transposable elements recently inserted in three independent *DFR-A* alleles and one high-copy DNA transposon isolated from the *Pink* allele of the *ANS* gene in onion (*Allium cepa* L.). Mol Genet Genomics 290:1027–1037. https://doi.org/10.1007/s00438-014-0973-7

Kim S, Yoo K-S, Pike LM (2005b) The basic color factor the *C* locus, encodes a regulatory gene controlling transcription of chalcone synthase genes in onions (*Allium cepa*). Euphytica 142:273–282. https://doi.org/10.1007/s10681-005-2239-2

Kim S, Yoo K-S, Pike LM (2005c) Development of a co-dominant PCR-based marker for allelic selection of the pink trait in onions (*Allium cepa*), based on the insertion mutation in the promoter of the anthocyanidin synthase gene. Theor Appl Genet 110:573–578. https://doi.org/10.1007/s00122-004-1875-6

Kim S, Yoo KS, Pike LM (2005d) Development of a PCR-based marker utilizing a deletion mutation in the dihydroflavonol 4-reductase (DFR) gene responsible for the lack of anthocyanin production in yellow onions (*Allium cepa*). Theor Appl Genet 110:588–595. https://doi.org/10.1007/s00122-004-1882-7

King JJ, Bradeen JM, Bark O, McCallum JA, Havey MJ (1998) A low-density genetic map of onion reveals a role for tandem duplication in the evolution of an extremely large diploid genome. Theor Appl Genet 96:52–62. https://doi.org/10.1007/s001220050708

Lee R, Baldwin S, Kenel F, McCallum J, Macknight R (2013) FLOWERING LOCUS T genes control onion bulb formation and flowering. Nat Commun 4:2884

Liu Q, Lan Y, Wen C, Zhao H, Wang J, Wang Y (2016) Transcriptome sequencing analyses between the cytoplasmic male sterile line and its maintainer line in welsh onion (*Allium fistulosum* L.). Int J Mol Sci 17:1058. https://doi.org/10.3390/ijms17071058

Liu Q, Wen C, Zhao H, Zhang L, Wang J, Wang Y (2014) RNA-Seq reveals leaf cuticular wax-related genes in Welsh onion. PLoS ONE 11:e113290. https://doi.org/10.1371/journal.pone.0113290

Liu T, Zeng L, Zhu S, Chen X, Tang Q, Mei S, Tang S (2015) Large-scale development of expressed sequence tag-derived simple sequence repeat markers by deep transcriptome sequencing in garlic (*Allium sativum* L.). Mol Breed. https://doi.org/10.1007/s11032-015-0399-x

Maisashvili A, Bryant H, Raulston JM, Knapek G, Outlaw J, Richard J (2016) Seed prices, proposed mergers and acquisitions among biotech firms. Choices 31:1–10

Manoharan RK, Han JSH, Vijayakumar H, Subramani B, Thamilarasan SK, Park J-I, Nou I-S (2016) Molecular and functional characterization of *FLOWERING LOCUS T* homologs in *Allium cepa*. Molecules 21:217. https://doi.org/10.3390/molecules21020217

Martin WJ, McCallum J, Shigyo M, Jakse J, Kuhl JC, Yamane N, Pither-Joyce M, Gokce AF, Sink KC, Town CD, Havey MJ (2005) Genetic mapping of expressed sequences in onion and in silico comparisons with rice show scant colinearity. Mol Genet Genomics 274:197–204. https://doi.org/10.1007/s00438-005-0007-6

Martínez L (2003) *In vitro* gynogenesis induction and doubled haploid production in onion (*Allium cepa* L.). In: Doubled haploid production in crop plants. Springer Netherlands, pp 275–279

Masamura N, McCallum J, Khrustaleva L, Kenel F, Pither-Joyce M, Shono J, Suzuki G, Mukai Y, Yamauchi N, Shigyo M (2012) Chromosomal organization and sequence diversity of genes encoding lachrymatory factor synthase in Allium cepa L. 3G: Genes Genomes. Genetics 2:643–651. https://doi.org/10.1534/g3.112.002592

Masuzaki S, Shigyo M, Yamauchi N (2006a) Direct comparison between genomic constitution and flavonoid contents in *Allium* multiple alien addition lines reveals chromosomal locations of genes related to biosynthesis from dihydrokaempferol to quercetin glucosides in scaly leaf of shallot (*Allium cepa* L.). Theor Appl Genet 112:607–617. https://doi.org/10.1007/s00122-005-0157-2

Masuzaki S-I, Shigyo M, Yamauchi N (2006b) Complete assignment of structural genes involved in flavonoid biosynthesis influencing bulb color to individual chromosomes of the shallot (*Allium cepa* L.). Genes & Genetic Systems 81:255–263. https://doi.org/10.1266/ggs.81.255

Matz MV (2017) Fantastic beasts and how to sequence them: genomic approaches for obscure model organisms. BioRxiv. https://doi.org/10.1101/165928

McCallum J (2007) Onion. In: Kole C (eds.) Vegetables. Genome mapping and molecular breeding in plants, vol 5. Springer, Berlin, Heidelberg, pp 331–347

McCallum J, Baldwin S, Shigyo M, Deng Y, van Heusden S, Pither-Joyce M, Kenel F (2012) AlliumMap-A comparative genomics resource for cultivated *Allium* vegetables. BMC Genom 13:168. https://doi.org/10.1186/1471-2164-13-168

McCallum J, Thomson S, Pither-Joyce M, Kenel F, Clarke A, Havey MJ (2008) Genetic diversity analysis and single-nucleotide polymorphism marker development in cultivated bulb onion based on expressed sequence tag–simple sequence repeat markers. J Am Soc Hortic Sci 133:810–818

McManus MT, Joshi S, Searle B, Pither-Joyce M, Shaw M, Leung S, Albert N, Shigyo M, Jakse J, Havey MJ, McCallum J (2012) Genotypic variation in sulfur assimilation and metabolism of onion (*Allium cepa* L.) III. Characterization of sulfite reductase. Phytochemistry 83:34–42. https://doi.org/10.1016/j.phytochem.2012.07.028

Merchant N, Lyons E, Goff S, Vaughn M, Ware D, Micklos D, Antin P (2016) The iPlant collaborative: cyberinfrastructure for enabling data to discovery for the life sciences. PLoS Biol 14:e1002342. https://doi.org/10.1371/journal.pbio.1002342

Michelmore RW, Paran I, Kesseli RV (1991) Identification of markers linked to disease-resistance genes by bulked segregant analysis: a rapid method to detect markers in specific genomic regions by using segregating populations. Proc Natl Acad Sci 88:9828–9832

Park J, Bang H, Cho DY, Yoon M-K, Patil BS, Kim S (2012) Construction of high-resolution linkage map of the *Ms* locus, a restorer-of-fertility gene in onion (*Allium cepa* L.). Euphytica 192:267–278. https://doi.org/10.1007/s10681-012-0851-5

Park J, Cho DY, Moon JS, Yoon M-K, Kim S (2013) Development of functional markers for detection of inactive *DFR-A* alleles responsible for failure of anthocyanin production in onions (*Allium cepa* L.). Korean J Hortic Sci Technol 31:72–79. https://doi.org/10.7235/hort.2013.12131

Revanna R, Turnbull MH, Shaw ML, Wright KM, Butler RC, Jameson PE, McCallum JA (2013) Measurement of the distribution of non-structural carbohydrate composition in onion populations by a high-throughput microplate enzymatic assay. J Sci Food Agric 93:2470–2477. https://doi.org/10.1002/jsfa.6062

Schlötterer C, Tobler R, Kofler R, Nolte V (2014) Sequencing pools of individuals mining genome-wide polymorphism data without big funding. Nat Rev Genet 15:749–763. https://doi.org/10.1038/nrg3803

Schnable PS, Ware D, Fulton RS et al (2009) The B73 maize genome: complexity diversity, and dynamics. Science 326:1112–1115. https://doi.org/10.1126/science.1178534

Scholten OE, van Heusden AW, Khrustaleva LI, Burger-Meijer K, Mank RA, Antonise RGC, Harrewijn JL, Van haecke W, Oost EH, Peters RJ, Kik C (2007) The long and winding road leading to the successful introgression of downy mildew resistance into onion. Euphytica 156:345–353

Scholten OE, van Kaauwen MPW, Shahin A, Hendrickx PM, Paul Keizer LC, Burger K, van Heusden AW, van der Linden CG, Vosman B (2016) SNP-markers in *Allium* species to facilitate introgression breeding in onion. BMC Plant Biol 16:187. https://doi.org/10.1186/s12870-016-0879-0

Shigyo M, Iino M, Isshiki S, Tashiro Y (1997) Morphological characteristics of a series of alien monosomic addition lines of Japanese bunching onion (*Allium fistulosum* L.) with extra chromosomes from shallot (*A. cepa* L. Aggregatum group). Genes & Genetic Systems 72:181–186. https://doi.org/10.1266/ggs.72.181

Shukla S, Iquebal MA, Jaiswal S, Angadi UB, Fatma S, Kumar N, Jasrotia RS, Fatima Y, Rai A, Kumar D (2016) The onion genomic resource: A genomics and bioinformatics driven resource for onion breeding. Plant Gene 8:9–15. https://doi.org/10.1016/j.plgene.2016.09.003

Song S, Kim C-W, Moon JS, Kim S (2014) At least nine independent natural mutations of the DFR-A gene are responsible for appearance of yellow onions (*Allium cepa* L.) from red progenitors. Mol Breeding 33:173–186. https://doi.org/10.1007/s11032-013-9942-9

Sun X-D, Yu X-H, Zhou S-M, Liu S-Q (2015) De novo assembly and characterization of the Welsh onion (*Allium fistulosum* L.) transcriptome using Illumina technology. Mol Genet Genomics 291:647–659. https://doi.org/10.1007/s00438-015-1131-6

Taylor A, Massiah AJ, Thomas B (2010) Conservation of *Arabidopsis thaliana* photoperiodic flowering time genes in onion (*Allium cepa* L.). Plant Cell Physiol 51:1638–1647. https://doi.org/10.1093/pcp/pcq120

Tian F, Bradbury PJ, Brown PJ, Hung H, Sun Q, Flint-Garcia S, Rocheford TR, McMullen MD, Holland JB, Buckler ES (2011) Genome-wide association study of leaf architecture in the maize nested association mapping population. Nat Genet 43:159–162

Tsukazaki H, Yaguchi S, Sato S, Hirakawa H, Katayose Y, Kanamori H, Kurita K, Itoh T, Kumagai M, Mizuno S, Hamada M, Fukuoka H, Yamashita K, McCallum JA, Shigyo M, Wako T (2015) Development of transcriptome shotgun assembly-derived markers in bunching onion (*Allium fistulosum*). Mol Breeding 35:55. https://doi.org/10.1007/s11032-015-0265-x

Tsukazaki H, Yaguchi S, Yamashita K-I, Shigyo M, Kojima A, Wako T (2012) QTL analysis for pseudostem pungency in bunching onion (*Allium fistulosum*). Mol Breeding 30:1689–1698. https://doi.org/10.1007/s11032-012-9752-5

Tsukazaki H, Yaguchi S, Yamashita K-I, Wako T (2017) QTL analysis of morphological traits and pseudostem pigmentation in bunching onion (*Allium fistulosum*). Euphytica 213:152. https://doi.org/10.1007/s10681-017-1944-y

Tsukazaki H, Yamashita K-I, Yaguchi S, Masuzaki S, Fukuoka H, Yonemaru J, Kanamori H, Kono I, Hang TTM, Shigyo M, Kojima A, Wako T (2008) Construction of SSR-based chromosome map in bunching onion (*Allium fistulosum*). Theor Appl Genet 117:1213–1223. https://doi.org/10.1007/s00122-008-0849-5

Van der Meer QP, De Vries JN (1990) An interspecific cross between *Allium roylei* Stearn and *Allium cepa* L., and its backcross to *A. cepa*. Euphytica 47:29–31

Wako T, Tsukazaki H, Yaguchi S, Yamashita K-I, Ito S-I, Shigyo M (2016) Mapping of quantitative trait loci for bolting time in bunching onion (*Allium fistulosum* L.). Euphytica 209:537–546. https://doi.org/10.1007/s10681-016-1686-2

Walker RL (2010) Evaluation of short-day onion doubled haploid lines. Ph.D. thesis, Texas A & M University

Wren JD, Georgescu C, Giles CB, Hennessey J (2017) Use it or lose it: citations predict the continued online availability of published bioinformatics resources. Nucleic Acids Res 45:3627–3633. https://doi.org/10.1093/nar/gkx182

Yang L, Liu Q, Wang Y, Liu L (2017) Identification and characterization of a glossy mutant in Welsh onion (*Allium fistulosum* L.). Sci Hortic 225:122–127. https://doi.org/10.1016/j.scienta.2017.05.014

Youens-Clark K, Faga B, Yap IV, Stein L, Ware D (2009) CMap 1.01: a comparative mapping application for the Internet. Bioinformatics 25:3040–3042. https://doi.org/10.1093/bioinformatics/btp458

Zhang C, Zhang H, Zhan Z, Liu B, Chen Z, Liang Y (2016) Transcriptome analysis of sucrose metabolism during bulb swelling and development in onion (*Allium cepa* L.). Frontiers in Plant Science. https://doi.org/10.3389/fpls.2016.01425

Zhou S-M, Chen L-M, Liu S-Q, Wang X-F, Sun X-D (2015) De Novo assembly and annotation of the Chinese chive (*Allium tuberosum* Rottler ex Spr.) transcriptome using the Illumina platform. PLoS ONE 10:e0133312. https://doi.org/10.1371/journal.pone.0133312

Zhu S, Tang S, Tan Z, Yu Y, Dai Q, Liu T (2017) Comparative transcriptomics provide insight into the morphogenesis and evolution of fistular leaves in *Allium*. BMC Genom 18:60. https://doi.org/10.1186/s12864-016-3474-8

Impact of the *Allium* Genomes on Plant Breeding

14

Michael J. Havey

Abstract
A better understanding of the structures and characteristics of the chloroplast, mitochondrial, and nuclear genomes have played significant roles in the genetic improvement of *Allium* crops. In this chapter, I will reflect upon the practical use of this genomic information for genetic improvement of the Alliums.

14.1 The Chloroplast Genome

The chloroplast DNAs of the Alliums are similar in sizes and structures to that of other angiosperms. In onion, the chloroplast DNA is a circular mapping molecule with two inverted repeats that separate large (LSC) and small (SSC) single copy regions, as is true for most other angiosperms (Havey 1991b; Katayama et al. 1991). Its size is approximately 153 kilobases with a GC content of 36.8% (von Kohn et al. 2013; Kim et al. 2015c). Slight differences reported for the estimated sizes of the chloroplast DNAs of onion cytoplasms are likely due to small indels across intergenic regions. The overall gene content is similar to other members of the monocot order Asparagales, except for numerous differences with the cpDNAs of the orchids (Von Kohn et al. 2013).

Polymorphisms in the *Allium* chloroplast DNAs have proven useful for phylogenetic analyses. Havey (1991b) identified polymorphic restriction enzyme sites in the chloroplast DNAs among major cultivated Alliums and observed that maternal phylogenies agreed with morphological classifications. Havey (1992b) then used chloroplast polymorphisms to assess maternal phylogenies among plants in *Allium* sections *Cepa* and *Phyllodolon*, revealing that *Allium vavilovii* and *Allium altaicum* were the likely progenitors of bulb (*Allium cepa*) and bunching (*Allium fistulosum*) onions, respectively, and that related *Allium galanthum, Allium pskemense,* and *Allium oschaninii* were placed more basal to the cultivated forms. A similar phylogeny was reported by Yamashita et al. (2001) using chloroplast and mitochondrial polymorphisms. Maternal phylogenies of *Allium* sections *Cepa* and *Phyllodolon* were consistent with crossabilities among the species (Havey 1992b; Raamsdonk et al. 1992).

Polymorphisms in specific coding and noncoding regions of the chloroplast DNAs have also been used for phylogenetic analyses of specific groups within the genus *Allium*. For example, Li et al. (2010) used polymorphisms in the chloroplast rps16 gene and Huang et al. (2014) in two noncoding regions of the chloroplast DNAs

M. J. Havey (✉)
USDA-ARS and Department of Horticulture,
University of Wisconsin, 1575 Linden Drive,
Madison, WI 53706, USA
e-mail: mjhavey@wisc.edu

M. Shigyo et al. (eds.), *The Allium Genomes*, Compendium of Plant Genomes,
https://doi.org/10.1007/978-3-319-95825-5_14

(trnL-F and rpl32-trnL) to estimate phylogenies in *Allium* subgenus *Cyathophora*. Li et al. (2016) used the matK, trnH-psbA, and rps16 regions to demonstrate the monophyletic origin of species *in Allium* subgen. *Anguinum*.

Although cytoplasmic male sterility (CMS) is likely conditioned by the mitochondrial DNA, numerous chloroplast polymorphisms distinguish the normal (N) male-fertile and male-sterile (S and T) cytoplasms of onion. Courcel et al. (1989) and Holford et al. (1991) used using gel electrophoresis of restriction enzyme digestions of N- and S-cytoplasmic chloroplast DNAs and revealed numerous banding differences. Subsequent studies using DNA-gel-blot hybridizations (Havey 1993, 2000; Satoh et al. 1993), indels or polymorphic restriction enzyme sits in PCR amplicons (Havey 1995; Sato 1998; Kim and Yoon 2010), and single nucleotide polymorphisms (SNPs) (Von Kohn et al. 2013) revealed numerous polymorphisms distinguishing these two cytoplasms. This relatively large number of polymorphisms led to the recognition that S cytoplasm of onion is an alien cytoplasm introgressed into bulb onion populations in antiquity (Havey 1993, discussed below). Havey (2000), Engelke et al. (2003), and Kim et al. (2009b) identified mitochondrial polymorphisms between N and T cytoplasms, which are much more closely related to each other relative to S cytoplasm. Even though CMS in onion is likely associated with the mitochondrial DNA, maternal transmission of both the mitochondrial and chloroplast DNAs allows for polymorphisms in either genome to be used to classify onion cytoplasms.

Other alloplasmic sources of CMS exist for the Alliums. The cytoplasm of *A. galanthum* has been introgressed into bulb onion (Havey 1999), shallot (Yamashita and Tashiro 1999, 2004), and bunching onion (Yamashita et al. 1999) to generate new sources of CMS for hybrid seed production (discussed below). No nuclear restorers of male fertility were detected in onion for the galanthum CMS (Havey 1999). The cytoplasm of *Allium rolyei* has also been introgressed into *A. cepa* and conditioned male sterility (Vu et al. 2011). These interspecific sources of CMS offer the opportunity to diversify cytoplasms used for hybrid seed production and reduce the genetic vulnerability due to predominant use of S cytoplasm (Havey 2000).

14.2 The Mitochondrial Genome

The mitochondrial (mt) DNAs of the *Alliums* are not well characterized. Kim et al. (2016a) sequenced, assembled, and annotated the mt DNA of S-cytoplasmic onion. As expected, it maps as a circular molecule of approximately 316 kb and carries the coding regions for major genes for oxidative respiration, transfer and ribosomal RNAs, and significant amounts of chloroplast DNA from intergenomic transfers (Kim et al. 2016a). A chimeric gene (*orf25*) comprised of part of the cox1 gene and another sequence has been proposed as the candidate for the male sterility conditioned by S cytoplasm (Kim and Yoon 2010; Kim et al. 2016a). Sequencing of other sources of CMS from onion [such as T and T-like cytoplasms (Havey 2000)], bunching onion (Moue and Uehara 1985), and chive (*Allium schoenoprasum*) (Tatlioglu 1982; Potz and Tatlioglu 1993; Engelke and Tatlioglu 2000) should provide insights into unique and potentially common mechanisms of CMS among the Alliums, as well as clarify mechanisms of nuclear male fertility restoration. Although the mt DNA has not been widely used to estimate phylogenies due to structural variation among closely related plants, there are examples of its use in the Alliums. Yamashita et al. (2001) reported that similar phylogenies were estimated using polymorphisms in the cp and mt DNAs, and Kik et al. (1997) used mt DNA variation to estimate relationships among species within *Allium ampeloprasum* complex and found that accessions within a mitochondrially defined clade were crossable.

14.3 The Nuclear Genome

Most of the cultivated Alliums are diploid ($2n=2x=16$) plants and show disomic inheritance ratios, including garlic (*Allium sativum*),

Japanese bunching onion, bulb onion, and chive (Eickmeyer et al. 1990; Cramer and Havey 1999; Ipek et al. 2005; Zewdie et al. 2005;Tsukazaki et al. 2008). Leek (*A. ampeloprasum*) is an autotetraploid ($2n=4x=32$) and shows tetrasomic inheritance (Berninger and Buret 1967). The nuclear genomes of the Alliums have numerous unique characteristics including enormous sizes (Ohri et al. 1998; Ricroch et al. 2005) with little evidence of recent polyploidization event(s), relatively low GC contents (Kirk et al. 1970; Kuhl et al. 2004), and large stretches of the DNA with degraded (Pearce et al. 1996; Jakse et al. 2008) and intact (Vitte et al. 2013; Kim et al. 2015d) transposable elements. Onion is often used in the classroom for cytogenetic analyses because it possesses relatively few, huge chromosomes. The nuclear DNA of onion is one of the largest among cultivated plants at approximately 15,290 megabase pairs (Arumuganathan and Earle 1991) per 1C nucleus, and on average each onion chromosome carries an amount of DNA equal to 75% of the 1C content of the maize nuclear genome (Bennett and Smith 1976). Molecular studies have revealed unique characteristics of this extremely large genome. The GC content of onion DNA is 32%, one of the lowest known for any angiosperm (Kirk et al. 1970; Stack and Comings 1979). Density gradient centrifugation revealed a 375 base pair (bp) telomeric sequence representing 4% of the genome (Barnes et al. 1988). Stack and Comings (1979) used Cot reassociation kinetics to reveal three repetitive fractions in the bulb onion genome. The first fraction represents approximately 41% of the genome and is repeated approximately 21,600 times, fraction two comprises about 36% of the genome and is repeated approximately 225 times, and fraction three comprises about 6% of the genome and consists of low-copy DNAs. Approximately 10% of the DNA was not detectable by reassociation kinetics (Stack and Comings 1979). The results of these studies indicate that the onion genome likely consists of middle-repetitive sequences occurring in short-period interspersions among low-copy regions (Stack and Comings 1979).

Enormous genome size differences exist among closely related Alliums (Ohri et al. 1998). For example, bulb onion has 28% more nuclear DNA than *A. fistulosum* (Labani and Elkington 1987; Jones and Rees 1968); this difference of 5.4 pg per 1C nucleus is approximately equal to the total 1C DNA content of barley (*Hordeum vulgare*), pepper (*Capsicum annuum*), or radish (*Raphanus sativus*) (Bennett and Smith 1976). This increase in DNA content cannot be attributed to duplication of one or a few chromosomes, and Jones and Rees (1968) and Ranjekar et al. (1978) proposed that intrachromosomal duplications contributed to increased chromosome sizes in onion. Ohri and Pistrick (2001) observed no relationship between genome size and phenology for *Allium* species across six subgenera, indicating that life history was not a significant factor contributing to genome size differences. Jones and Rees (1968) and Narayan (1988) studied interspecific hybrids between *A. cepa* and *A. fistulosum* and observed that all eight bivalents were asymmetric, indicating that DNA differences were spread across all eight chromosomes. However, size differences among individual bivalents varied from a maximum of 60% to a minimum of 20% (Jones and Rees 1968). Pachytene pairing in interspecific hybrids between the bulb and bunching onions revealed loops and overlaps, which is the evidence for the accumulation of repetitive sequences or tandem duplication of chromosome segments (Jones and Rees 1968). King et al. (1998) observed relatively high frequencies of dominant restriction fragment length polymorphisms (RFLPs), which could have resulted from duplication and deletion events after unequal crossing over. Masamura et al. (2012) studied the genomic region carrying the tearing factor, lachrymatory factor synthase (LFS), and demonstrated that the region carries at least two tandemly arranged copies of LFS. The origin of tandemly repeated regions could also be due to retrotransposition events, in which a transcribed RNA is converted back to DNA and gets inserted into the genome at a nearby location. The prevalence of localized tandem duplications will become apparent once we have annotated sequences of the *Allium* genomes. Doubled-haploid (DH) lines of onion and other Alliums are useful for molecular analyses because any polymorphisms within a single DH must be among paralogous regions (Baldwin et al. 2012).

Telomeres are located at the end of chromosomes and are required for stable maintenance and transmission of chromosomes. Almost all plants possess the *Arabidopsis*-type telomere as repeats of the TTTAGGG motif (Fuchs et al. 1995). However, some species of the Alliaceae and Liliaceae are unique because they do not possess this widely conserved telomeric repeat (Fuchs et al. 1995; Pich et al. 1996). The previously identified 375 base pair GC-rich satellite DNA (Barnes et al. 1988) was proposed to have replaced the Arabidopsis-type telomere in *A. cepa* (Pich et al. 1996; Pich and Schubert 1998); however subsequent research revealed that the telomeres of *A. cepa* are repeats of the unusual sequence CTCGGTTATGGG (Fajkus et al. 2016).

14.4 Nuclear 45s Ribosomal (R) DNA

The nuclear genome carries highly repeated sequences encoding the 5 and 45s rDNAs. The 5s region exists as tandemly arrayed repeats of 120-bp coding sequence with a non-transcribed spacer of variable sizes (Garcia et al. 2014). In chive and onion, two 5s clusters exist in each species located on chromosomes 6 and 7, respectively (Shibata and Hizume 2002a). The 45s coding region exists as multiple copies of tandemly arrays of three conserved regions encoding the 5.8, 18, and 26s rRNAs (Garcia et al. 2014). Two internal transcribed spacers (ITS) separate the three rRNA-coding regions, and an intergenic spacer (IGS) separates the rRNA-coding and ITS regions. Differences have been observed in the Alliums for the relative proportion of rDNA and number of nucleolus organizer regions (NOR) per cell (Maggini and Garbari 1977; Maggini et al. 1978). In situ hybridizations of the nuclear 45s rDNA revealed the locations and numbers of NORs in bulb and bunching onions (Schubert and Wobus 1985; Ricroch et al. 1992). In bulb onion, NORs were located on chromosomes 6 and 8; bunching onion possessed NORs on chromosomes 5 and 8. For both species, the smallest chromosome (#8) carried a NOR. Schubert and Wobus (1985) observed that recombination can occur between chromosomes carrying NOR regions, resulting in exchanges of terminal regions termed "NOR jumping". Schubert et al. (1983) used differences in the NORs between *A. cepa* and *A. fistulosum* to establish the interspecific origin of diploid viviparous onion.

The sizes of the 45s rDNA repeats are approximately 11.8, 13.1, 11.5, 10.4, and 12.3 kb for bulb onion, bunching onion, garlic, leek, and chive, respectively, due primarily to length differences in the IGS (Havey 1992c). Variation in the ITS regions has been widely used alone and with chloroplast sequences to estimate phylogenies in the Alliums (Havey 1992c; Li et al. 2010; Huang et al. 2014). Friesen et al. (2006) reported ITS sizes between 612 and 661 bp among Alliums and polymorphisms in the ITS have been used to clarify relationships among Alliums and related genera (Hirschegger et al. 2010). Li et al. (2016) used in both regions to estimate phylogenies among 12 taxa in *Allium* subgenus *Anguinum*, and reported that this group is monophyletic even though species are indigenous to disjoined regions in the high mountains of southwestern Europe to eastern Asia and in northeastern North America. Huang et al. (2014) reported that phylogenies estimated using both ITS and chloroplast sequences were better resolved than those produced with only one of the regions and that major clades were consistent with earlier estimations reported by Friesen et al. (2006).

14.5 Interspecific Hybridizations

There are numerous examples of asexually and sexually propagated Alliums that resulted from interspecific hybridizations. Top-setting (viviparous) onions [*Allium x proliferum* (Moench) Schrad. syn. *A. cepa* L. var. *viviparum* (Metzger)

Alefeld] exist that are interspecific hybrids with 16 chromosomes, eight from bulb onion or shallot and eight from bunching onion (Havey 1991a; Hizume 1994; Shibata and Hizume 2002a, b). Analyses of the chloroplast and mitochondrial DNAs of top-setting interspecifics revealed that *A. fistulosum* was most commonly the female parent (Havey 1991a; Yamashita and Tashiro 2001). Top-setting onions are widely grown [e.g., *A. wakegi* Araki in Japan or TS6007 in North America (Havey 1991a)], and may have originated from independent interspecific hybridizations. Amphidiploids from interspecific hybrids between bunching and bulb onions can be seed propagated and commercially grown as green bunching onions [an example is the cultivar "Beltsville Bunching" (Jones and Clarke 1942)]. "Delta Giant" is grown as an asexually propagated shallot and resulted from an interspecific hybrid between *A. fistulosum* and shallot (Perkins et al. 1958). There also exist viviparous forms of Japanese bunching onion (Havey 1992a) and presumably bulb onion that are not the result of interspecific hybrids. These are likely homeotic variants in which normal flower production has been replaced with vegetative top sets, and growers or gardeners noticed these unique plants and propagated them asexually.

Two examples of triploid ($2n=3x=24$) top-setting onions are "Ludujka" from the Adriatic region (Puizina and Papes 1996) and "Pran" from India (Koul and Gohil 1971). Puizina et al. (1999) reported that both plants had identical banding patterns for a set of RAPD markers and may have a common origin. Although *A. cepa* is clearly one of the parents of most diploid and triploid top-setting onions, *A. fistulosum* may not always be a second parent (Havey 1991a, b; Maaβ 1997; Puizina et al. 1999). Puizina and Papes (1996) studied top-setting onions from the Adriatic region and reported that bulb and bunching onion were parents of both diploid and triploid types, indicating that multiple interspecific hybridizations may have occurred giving rise to these asexually propagated Alliums. Puizina et al. (1999) undertook GISH analyses and concluded that one parent of triploid top-setting onions may be *A. roylei*, disagreeing with results of other researchers (Havey 1991a; Maaβ 1997; Klaas 1998). Therefore, one or two other parental species of these triploid top-setting onions remains unknown. "Pran" is especially important because it may have been the donor of male-sterile (S) cytoplasm widely used to produce hybrid onion seed (Havey 1993).

Unique interspecific hybrids have been produced between garlic and bulb onion (Ohsumi et al. 1993) and between bunching onion and chive (Umehara et al. 2006). This latter interspecific hybrid produced more carotene than either parental species (Umehara et al. 2006). Although there are no reports of subsequent commercial use, these interspecific hybrids can be asexually propagated and may become unique vegetable products. Elephant garlic (*A. ampeloprasum* var. *giganteum*) is mostly an asexually propagated hexa- ($2n=6x=48$) or octoploid ($2n=8x=64$) that resulted from interspecific hybridization(s) of uncertain origin(s) (Hirschegger et al. 2006, 2010).

14.6 Introgression Research

There are relatively few examples of successful introgression of beneficial traits from one *Allium* species into another. One could consider pink root resistance in the amphidiploid "Beltsville Bunching" to be an example of successful utilization of a disease resistance from *A. fistulosum* in a cultivar grown and consumed as a green onion. The most successful interspecific introgression is downy mildew resistance from *A. roylei* to the bulb onion. Kofoet and Zinkernagel (1990) identified *A. roylei* as a source of downy mildew resistance. The interspecific hybrid and backcrosses to bulb onion were completed at Wageningen in Holland (de Vries et al. 1992). Introgression of this resistance proved to be difficult because the resistance allele was linked in coupling phase to a lethal factor, and only after isolating a recombination event was it possible to produced plants homozygous for the resistance allele (Scholten et al. 2007). Downy mildew resistance from *A. roylei* has been successfully commercialized and widely accepted by growers, providing the best example of the successful interspecific transfer of a major disease resistance in the Alliums.

Many researchers have produced interspecific hybrids between bunching and bulb onions and completed numerous backcrosses to *A. cepa* (Emsweller and Jones 1935; Maeda 1937; Peffley and Mangum 1990; Peffley and Hou 2000), and there is strong evidence of recombination between chromosomes of the two species (Stevenson et al. 1998; Hou and Peffley 2000; Budylina et al. 2014). Nevertheless, there are no examples of successful transfer and deployment of a beneficial trait from *A. fistulosum* to bulb onion. The reason for this may be the location of genes in the two species. Localization of chiasmata near the centromere has been reported for numerous *Alliums*, including *A. ampeloprasum, A. fistulosum, Allium kochii,* and *Allium cyathophorum* (Koul and Gohil 1970; Stack and Roelofs 1996; Maeda 1937; Ved Brat 1965). For autopolyploids such as leek (*A. ampeloprasum*), localized chiasmata would avoid the formation of multivalents among large metacentric chromosomes and increase the likelihood of balanced gametes. However, it is not clear what advantage results from chiasmata localized near the centromere of a diploid species such as bunching onion. Fiskesjo (1975) observed that the terminal ends of bunching onion chromosomes were largely heterochromatic. If gene-rich (euchromatic) regions are concentrated near the centromeres of bunching onion, successful transfer of desirable traits to bulb onion may be difficult because the interspecific hybrids show more randomly located as opposed to localized chiasmata (Maeda 1937; Stevenson et al. 1998; Khrustaleva and Kik 1998), which could transfer largely heterochromatic regions from bunching onion to bulb onion.

14.7 Genomic in Situ Hybridization (GISH)

GISH analyses have proven to be very useful to document origins of interspecific hybrids, establish chromosome compositions of alien addition lines, and reveal recombination among chromosomes for interspecific hybrids. The technique involves labeling of the DNA of one species with a fluorescent dye and blocking with unlabeled DNA from the other species to reduce any cross hybridization with the labeled DNA. GISH analyses have revealed little cross hybridization among chromosomes of closely related Alliums, indicating that the genomes have accumulated enough sequence diversity to the point of becoming unique. GISH analyses have revealed the interspecific origins of top-setting (Puizina et al. 1999; Shibata and Hizume 2002b) and ornamental Alliums (Friesen and Klaas 1998), as well as recombination among chromosomes in progenies from interspecific hybrids. As mentioned above, there is little to no evidence of successful introgression and commercialization of a beneficial trait from *A. fistulosum* to *A. cepa*, in spite of many instances of interspecific hybridizations and backcrossing (Emsweller and Jones 1935; Maeda 1937; Peffley 1986; Peffley and Hou 2000). Initially, there were questions regarding the extent of recombination between chromosomes of bunching and bulb onions due to incomplete synapsis and synaptic irregularities (Albini and Jones 1990). Because there is a little cross hybridization between the repetitive DNAs of most closely related Alliums, chromosome sets can often be clearly differentiated. GISH was used by Stevenson et al. (1998) to reveal recombinant chromosomes after meiosis in an interspecific hybrid between bunching and bulb onions. Hou and Peffley (2000) observed telomeric regions from bunching onion in advanced backcrosses to bulb onion, demonstrating that recombination among chromosomes had occurred. Budylina et al. (2014) analyzed progenies advanced backcross or selfing generations of interspecific hybrids between bunching and bulb onions and conclusively revealed recombination between chromosomes 7 from the two Alliums. This study also demonstrated that restitution gametes must have occurred during self-pollinations or backcrossing to produce progenies at 3x or higher ploidies with complete sets of chromosomes from both parental species. Extensive recombination between the chromosomes of *A. fistulosum* and *A. roylei* (Khrustaleva and Kik 1998; Khrustaleva et al. 2005) was revealed by GISH, providing a potentially useful approach to enhance introgression of genomic regions from *A. fistulosum* into bulb onion (Khrustaleva and Kik 2000). GISH was used to evaluate the prevalence of bulb onion chromosomes during backcrossing of an interspecific hybrid between *A. galanthum* as the seed parent to bulb onion, and by the BC3, there was little evidence of genomic regions from *A. galanthum* in progenies (Yamashita et al. 2000). In another study of progenies from backcrossing of an *A. galanthum-fistulosum* interspecific hybrid to *A. fistulousm*, GISH revealed that a nuclear male fertility restoration locus for the galanthum cytoplasm had been introgressed from *A. galanthum* into the nuclear genome *A. fistulosum* (Yamashita et al. 2005).

GISH has been used successfully to monitor the chromosome complement of progenies for the development of alien monosomic addition lines of the *Alliums*. Complete sets of monosomic addition lines have been developed for shallot (*A. cepa*) carrying individual chromosomes from bunching onion (Hang et al. 2014), bunching onion carrying individual chromosomes from shallot (Shigyo et al. 1996), and shallot and bulb onion with individual chromosomes from *A. roylei* (Vu et al. 2012). GISH clearly established the alien origin of additional chromosomes towards the development of monosomic lines (Shigyo et al. 1998; Vu et al. 2012).

Vu et al. (2012) analyzed the monosomic addition lines of *A. fistulosum* carrying single chromosomes from shallot (*A. cepa* var. *aggregatum*) for resistance to *Fusarium oxysporum* f. sp. *cepae* (FOC). The monosomic line carrying shallot chromosome 2 showed the highest resistance to FOC, indicating that this chromosome carries genes conferring FOC resistance from *A. cepa*. In another study, Galván Vivero (2009) evaluated a mapping population from *A. cepa* x (*A. roylei* x *A. fistulosum*) for resistance to Fusarium basal rot, and was able to assign resistance from *A. roylei* to chromosome 2 and from *A. fistulosum* to chromosome 8. In both of these cases, resistances to FOC could be potentially introgressed into bulb or bunching onions and given their interspecific origins (Galván et al. 2008), it should be relatively easy to use GISH or identify molecular markers across the alien chromosome segment to monitor introgression during backcrossing.

14.8 Fluorescent in Situ Hybridization (FISH)

FISH is a useful technique to visualize the locations of specific DNA clones on chromosomes. Initially, large DNA fragments were required such as those cloned into bacterial artificial chromosomes (BACs). However, larger clones may carry commonly repetitive sequences which can hybridize to many locations across chromosomes. FISH of random onion BAC clones of ~100 kb revealed that approximately 75% produced signals across chromosomes, 11–15% hybridized to discrete regions such as centromeres or telomeres, and relatively few (2–4%) hybridized to discrete chromosome regions (Suzuki et al. 2001, 2012). This indicates that repetitive sequences on the BACs commonly exist across the onion chromosomes. FISH signals can be augmented by hybridizing tandem repeats (Kirov et al. 2017) or approaches such as tyramide (t)-FISH which amplify signals (Khrustaleva and Kik 1998). Romanov et al. (2015) used t-FISH to visualize the locations of genetically linked cDNA clones along chromosome 5 of onion and revealed that tightly linked markers can be physically distant. In onion, recombination tends to occur interstitially with centromeric or telomeric regions experiencing significantly fewer chiasmata (Albini and Jones 1988). Molecular markers tightly linked to the nuclear male fertility restoration locus of onion were localized near the centromere of chromosome 2, in a region of relatively low recombination (Khrustaleva et al. 2016). This observation explained the relatively large numbers of markers tagging the *Ms* locus (Gökçe et al. 2002; Bang et al. 2011; Yang et al. 2012; Bang et al. 2013; Havey 2013; Park et al. 2013; Kim 2014; Kim et al. 2015a), even though in some cases, few polymorphisms were evaluated or relatively small family sizes at maximum linkage disequilibrium were used.

14.9 Nuclear Polymorphisms

The enormous sizes of the nuclear genomes of the Alliums have little to no bearing on genetic analyses or improvement. Of greater effect is the biennial generation time which significantly slows progress, as well as the relatively high costs of storing bulbs or plants for vernalization and crossing with insects. Prior to 1980, there were only 22 genetic loci described for onion (Cramer and Havey 1999) and only one linkage reported between yellow lethal (*yl*) seedlings and leaf glossiness (*gy*) (Jones et al. 1944). Relatively few loci were described in the other cultivated Alliums as well, and no reported linkages. The large sizes of the *Allium* genomes have no effect on genetic linkages, which are determined by the numbers and positions of chiasmata along the chromosomes. Onion has a mean number of chiasmata of 19.2 per meiosis (Albini and Jones 1988), indicating that there would be, on average, one crossover per arm with approximately three additional crossovers distributed across the eight bivalents at meiosis. This average number of crossovers is similar to other cross-pollinated diploid plants and therefore the construction of a genetic map of onion (and the other diploid Alliums) should not require any more genetic loci than other plants. However, the huge nuclear genomes of the Alliums result in enormous physical distances (in kilo- or mega-basepairs of DNA) between tightly linked markers, which negatively impacts map-based cloning of important genes.

A plethora of molecular markers have been developed for onion and to a lesser extent the other cultivated Alliums. Restriction fragment length polymorphisms (RFLPs) were the first molecular marker used to develop a genetic map of onion (King et al. 1998). RFLPs were costly and labor intensive, requiring isolation of large amounts of genomic DNA, digestion with a restriction enzyme, gel electrophoresis and blotting, and probe hybridization. The enormous nuclear genome of onion made detection and mapping of RFLPs challenging, requiring radioactive labeling of DNA probes to high specific activities (Bark and Havey 1995). Nevertheless, RFLPs provided robust estimates of the phylogenies of important onion populations (Bark and Havey 1995), introgression of S cytoplasm into elite onion populations (Havey and Bark 1994), and development of the first genetic map of onion (King et al. 1998; Martin et al. 2005). These markers were successfully used to identify chromosome regions controlling male fertility restoration (Gökçe et al. 2002), as well as pungency, soluble solids content, and health-enhancing antiplatelet activity (Galmarini et al. 2001).

After about 1995, development of molecular markers in the Alliums focused on polymorphisms detected using the polymerase chain reaction (PCR), including randomly amplified polymorphic DNAs (RAPDs), amplified fragment length polymorphisms (AFLPs), simple sequence repeats (SSRs), and single nucleotide polymorphisms (SNPs). There have also been numerous PCR-based markers developed using cleaved amplified polymorphisms (CAPs) or similar approaches for specific RFLPs or candidate genes.

RAPDs were a popular marker in the 1990s because they required little DNA, no cloning, and no prior sequence knowledge (Wilkie et al. 1993; Bradeen and Havey 1995). Single decamer primers were most commonly used added to a PCR reaction with genomic DNA; if sequences homologous to the decamer existed in both forward and reverse orientations and positions were relatively close to each other, an amplicon was produced. Polymorphisms in the priming sites would result in dominant markers appearing as presence/versus absence of amplicons, or if an insertion or deletion (indel) existed between the sites, a codominant marker could occur. Although initially regarded as both plentiful and easy to detect, RAPDs proved to be inconsistent with DNA isolations and PCR conditions and were especially challenging when amplifying from the huge genome of the Alliums (Bradeen and Havey 1995). RAPDs are primarily a dominant marker, which is less useful in highly heterozygous crops such as the Alliums, in which inbreds can retain relatively high levels of heterozygosity.

AFLPs are dominant and anonymous markers that result from restriction enzyme digestion of the DNA, ligation of specific primers onto the restricted sites, and PCR amplification. They are anonymous because no sequence information is produced, and the presence of an amplicon is dominant over its absence. An advantage of AFLPs for plants such as the Alliums is that relatively rare cutting enzymes can be used to reduce the numbers of amplicons produced. For example, digestion with an enzyme recognizing relative rare six or eight base pair sequences, coupled with a more frequently cutting enzyme recognizing a four-bp motif has been used to reduce the number of amplicons and enhance their resolution (Heusden et al. 2000). Genetic maps based primarily on AFLPs have been produced for bunching onion (Ohara et al. 2005) and from interspecific progenies (Heusden et al. 2000). AFLPs have been successfully used to assess genetic diversity among garlic accessions and elimination of duplications in germplasm collections (Ipek et al. 2003; Volk et al. 2004).

SSRs were the first codominant, PCR-based molecular marker that is both highly repeatable and commonly polymorphic among *Allium* populations and among individual plants within a cultivar. Sequencing of genomic fragments (Song et al. 2004; Tsukazaki et al. 2007; Baldwin et al. 2012) and cDNAs (Kuhl et al. 2004; McCallum et al. 2001, 2008; Jakše et al. 2005; Tsukazaki et al. 2015) produced high numbers of sequences that have been mined for SSR motifs. Unique priming sites on both sides of the putative SSR were identified and polymorphic amplicons produced. SSRs identified from the transcriptome often correspond to different numbers of amino acids in membrane-spanning regions that may not negatively affect protein function (Jakše et al. 2005). SSRs have been used for genetic mapping (Martin et al. 2005; Masuzaki et al. 2006; Tsukazaki et al. 2008), estimation of nuclear phylogenies among onion populations and related Alliums (Fischer and Bachmann 2000; McCallum et al. 2008; Araki et al. 2010; Khar et al. 2011; Baldwin et al. 2012), and assessment of genetic diversity within germplasm collections (Zhao et al. 2011). Eventual sequencing of the *Allium* nuclear genomes should reveal large numbers of SSRs for genetic and phylogenetic studies.

SNPs are the most useful molecular markers because of their abundance and ease of genotyping using high throughput platforms. SNP identification has been primarily focused on the transcriptome by sequencing highly inbred or doubled-haploid lines in order to avoid polymorphisms among paralogous sequences (Jakše et al. 2005; McCallum et al. 2008; Duangjit et al. 2013; Tsukazaki et al. 2015). Large numbers of SNPs have been identified and genotyped using platforms such as KASPar (Duangjit et al. 2013; Damon et al. 2014), as well as gel resolution of insertion–deletion (indel) events (Tsukazaki et al. 2015). However, segregation analyses to establish alleles at loci have been completed for only about 600 onion SNPs (Duangjit et al. 2013; Damon et al. 2014).

14.10 Marker–Phenotype Associations

Although the genetics of many morphological characteristics have been characterized and used for *Allium* improvement (Cramer and Havey 1999), there are relatively few examples of specific associations of traits with markers or candidate genes. Identification of DNA polymorphisms within candidate genes or the development of molecular markers tightly linked to important traits offer great potential for the genetic improvement of the Alliums, especially given the relatively high cost of vernalizing bulbs and crossing with insects. The genetics of onion bulb colors are the best example of classical genetic analyses followed by candidate gene identification. Rieman (1931) named the codominant inhibitor (*I*) locus which when homozygous dominant conditions white bulbs over yellow or red, and that red, yellow, or white bulbs were conditioned by an allelic series at the

W locus. Jones et al. (1944) built upon Rieman's (1931) work and presented the naming of loci in use today. Their results agreed with Rieman's observation that the *I* locus is codominant, in that plants homozygous dominant at I have white bulbs, and plants heterozygous at I show a light yellow or "buff" color. Jones et al. (1944) recognized that in addition to the *I* locus, a basic color factor (*C*) locus requires a dominant allele to produce red or yellow bulb colors; the homozygous recessive genotype (*cc*) conditions white bulbs (termed recessive white). Jones et al. (1944) identified the *R* locus at which a dominant allele conditions red bulb color and the homozygous recessive genotype yellow bulbs. Results did not support the allelic series at Rieman's *W* locus, and segregations could be explained by three loci (*I*, *C*, and *R*; Jones et al. 1944). Jones and Peterson (1952) observed that crosses between yellow onions from North America and Brazil occasionally produced red bulbs, which can be explained by the interaction of two complementary loci. El-Shafie and Davis (1967) worked out the genetics of bulb color using North American populations and proposed that red bulbs are conditioned by dominant alleles at both of two complementary loci (Jones' *R* locus and the *L* locus). Their results agreed with Rieman (1931) and Jones et al. (1944) that white bulbs can be conditioned either by a dominant allele at the *I* locus, or the homozygous recessive genotype at the *C* locus. El-Shafie and Davis (1967) named the golden (*G*) locus, at which dominant alleles condition a golden yellow bulb color and the homozygous recessive genotype produces chartreuse bulb color. Khar et al. (2008) and Duangjit et al. (2014) demonstrated that an additional locus (*L2*) also interacts with R to condition red bulb color, and is linked to *L* at approximately 6 cM on chromosome 3. Recessively inherited gold-colored bulbs have significantly lower amounts of the flavonoid quercetin and co-segregated with a premature stop codon in the chalcone isomerase (*CHI*) gene (Kim et al. 2004c).

The *L* and *R* loci map to chromosomes 4 and 7, respectively (Masuzaki et al. 2006), and the *C* locus to chromosome 6 (Khar et al. 2008). Mapping of polymorphisms in candidate genes in the biosynthetic pathway for anthocyanins demonstrated that the *L* locus encodes anthocyanidin synthase (*ANS*; Kim et al. 2004b, 2005a) and *R* locus Dihydroflavonol 4-reductase (*DFR*; Kim et al. 2004a). Kim et al. (2005c, 2006, 2009a, 2016b) and Song et al. (2014) have described many polymorphisms, indicating that significant variation has accumulated among alleles at these two loci or potentially among alleles across paralogous loci. However, there may still be uncharacterized loci controlling bulb colors in onion, which have not been studied simply because no one has undertaken the research. It may be that mutations at trans-acting factors or different structural genes in the anthocyanin biosynthetic pathway have occurred and are waiting to be discovered and described. It is important that researchers carefully determine if a unique bulb color phenotype is a new locus, or potentially an allele at a previously described locus. Cases in point are the naming of the *crb-1* locus controlling red bulb color (King et al. 1998) or the recessive *p* locus putatively controlling pink bulb color (Kim et al. 2004b), which have been subsequently recognized as alleles at the *L* or *R* loci (Khar et al. 2008; Kim et al. 2005b).

14.11 CMS and Production of Hybrid Seed

The *Allium* inflorescence is an umbel with many perfect flowers. Outcrossing is encouraged by protandry (Currah and Ockendon 1978), in which pollen is shed before the stigma elongates and becomes receptive within an individual flower. However flowers in the umbel mature at different times, first at the top of the umbel and subsequently downward across the umbel (Currah and Ockendon 1978). Therefore in a single umbel, there will be both mature pollen and receptive stigmas at the same time, and self-pollination can occur (Jones 1923). This allows for inbreeding to develop more uniform inbred lines for hybrid development (Jones and Davis 1944). However, the production of hybrid seed requires systems of male sterility to ensure hybridity. For the onion, systems of cytoplasmic-genic male sterility (CMS) are used to produce hybrid seed. The most widely used source of CMS was discovered by Dr. Henry Jones and colleagues and is conditioned by the interaction of male-sterile (S) cytoplasm and one nuclear male fertility restoration (*Ms*) locus (Jones and Clarke 1943). The first onion hybrid (Calred) was produced by asexually propagating a single S-cytoplasmic plant (Italian Red 13–53) as the female parent. Afterwards, Jones and Davis (1944) developed the approach to seed propagate male-sterile plants (S *msms*) by crossing with maintainer plants that are N cytoplasmic and homozygous recessive at *Ms*. S cytoplasm was backcrossed into numerous market classes of onion to develop male-sterile lines for hybrid production.

S cytoplasm is a divergent cytoplasm that was introgressed into onion population(s) sometime in antiquity, as evidenced by the plethora of polymorphisms in both the chloroplast and mitochondrial DNAs (Havey 1993, 2000; Kim et al. 2009b; Kim and Yoon 2010). A likely scenario was that an interspecific hybrid, such as a top-setting onion such as "Pran", grew sympatrically with diploid bulb onion, and occasional crossing occurred and viable seed was produced (Havey 1993). These progenies would possess the cytoplasm of the interspecific hybrid, and after multiple generations of backcrossing to onion, the cytoplasm would be introgressed into the onion population.

Berninger (1965) identified a second source of onion CMS in the French population "Juane Pais der Vertus"; this cytoplasm is closely related to N cytoplasm and likely arose as a variant of the wild-type onion cytoplasm. T cytoplasm has been used to produce hybrids in Europe and possibly other markets. Schweisguth (1973) proposed that male fertility restoration was conditioned by dominant alleles at two to three complementary loci. Independent sources of CMS different from S cytoplasm and more closely related to T cytoplasm have been used to produce hybrid onion seed (Havey 2000).

CMS has also been genetically characterized in chive [Tatlioglu 1982, 1994 (cited in Engelke and Tatlioglu 2000)] and Japanese bunching onion (Moue and Uehara 1985). In chive, a series of detailed studies revealed that male fertility restoration for CMS in chive is conditioned by three loci, one of which stably restores male fertility (*X* locus), one that restores only at higher temperatures (*T* locus), and a third that restores only after tetracycline treatment (a locus) (Tatlioglu 1987; Tatlioglu and Wricke 1988; Engelke and Tatlioglu 2004). Potz and Tatlioglu (1993) demonstrated that chive CMS was associated with DNA polymorphisms in the mt, but not cp, DNA. Gel blot hybridizations using mitochondrial genes revealed numerous differences between N and S cytoplasms of chive, and presences of an 18 kilo-Dalton protein were associated with CMS (Potz and Tatlioglu 1993). Subsequently, Engelke and Tatlioglu (2000) studied numerous sources of the male sterility phenotype and revealed extensive mitochondrial diversity within chive, and identified molecular markers distinguishing chive cytoplasms (Engelke and Tatlioglu 2003). In Japanese bunching onion, CMS was originally described by Moue and Uehara (1985), and additional sources of CMS were subsequently identified in two ("Zhangqiu" and "Guangzhou") populations of bunching onion (Yamashita et al. 2010). CMS in both chive and Japanese bunching onion are used commercially to produce hybrid seed.

14.12 Future Prospects

Transcriptome sequencing is an efficient approach to generate sequence information from expressed regions of the genome and has been widely used in the *Alliums*. Kuhl et al. (2004) and Baldwin et al. (2012) undertook sequencing of normalized cDNA libraries from onion populations. Subsequent transcriptome sequencing has been completed for garlic (Kamenetsky et al. 2015, Havey and Ahn 2016), Japanese bunching onion (Liu et al. 2014; Tsukazaki et al. 2015), and bulb onion (Duangjit et al. 2013; Kim et al.

2015b; Sohn et al. 2016). These efforts have produced reference transcriptomes for these *Alliums*, however, there is not a comprehensive assembly of sequences into a single database for individual species or across the *Alliums*. Some transcriptome sequencing has targeted specific germplasm or traits, such as fertile garlic (Kamenetsky et al. 2015), reduced epicuticular waxes (Liu et al. 2014), bulb storage (Chope et al. 2012), or anthocyanin biosynthesis (Schwinn et al. 2016). These studies provide a snapshot into gene expression across the major *Alliums*; however, a coordinated effort is necessary to sequence cDNAs from a variety of tissues or treatments so that a more global view of gene expression can be developed.

A major resource for *Allium* improvement will be the sequencing of the nuclear genomes from all of the cultivated *Alliums*. The enormous sizes of these *Allium* nuclear DNAs represent significant challenges for sequencing, assembly, and annotation. Historically short read technologies have been combined with end sequencing and fingerprinting of large insert libraries (such as BACs) to build sequence contigs and scaffolds towards building a genome assembly. This has been cost prohibitive for large genomes such as those found in the *Alliums*. However, with the development of longer read technologies, it is now feasible to generate large sequence reads for assembly and annotation. Pilot sequencing of onion genomic regions revealed very low gene densities and long stretches of degenerated transposable elements (Jakše et al. 2008); however, Vitte et al. (2013) identified intact retroelements in onion sequences. It will be interesting to determine to what extent the onion genome is comprised of relatively young, intact retrotransposons versus more ancient elements that have significantly degraded. Nevertheless, sequencing indicates that gene densities are likely very low in the *Alliums*, with individual genes separated by long tracts of sequences with few definitive characteristics.

Future sequencing efforts should concentrate on doubled-haploid lines to ensure complete genetic uniformity of the lines, and for mining of the sequences for SSRs and SNPs. Our community must work together to ensure that independent sequencing efforts are coordinated and result available to the community at large. These sequence resources will enable researchers to identify causal mutations for major phenotypes, allowing breeders to select desirable phenotypes before flowering to greatly increase selection efficiency, towards the production of new, value-added cultivars for growers, processors, and consumers.

References

Albini SM, Jones GH (1988) Synaptonemal complex spreading in *Allium cepa* and *Allium fistulosum*. II. Pachytene observations: The SC karyotype and the correspondence of late recombination nodules and chiasmata. Genome 30:399–410

Albini SM, Jones GH (1990) Synaptonemal complex spreading in *Allium cepa* and *Allium fistulosum*. III. The F_1 hybrid. Genome 33:854–866

Araki N, Masuzaki S, Tsukazaki H, Yaguchi S, Wako T, Tashiro Y, Yamauchi N, Shigyo M (2010) Development of microsatellite markers in cultivated and wild species of sections *Cepa* and *Phyllodolon* in *Allium*. Euphytica 173:321–328

Arumuganathan K, Earle ED (1991) Estimation of nuclear DNA content of plants by flow cytometry. Plant Mol Biol Rep 9:229–241

Baldwin S, Pither-Joyce M, Wright K, Chen L, McCallum J (2012) Development of robust genomic simple sequence repeat markers for estimation of genetic diversity within and among bulb onion (*Allium cepa* L.) populations. Mol Breed 30:1401–1411

Bang H, Cho DY, Yoo K, Yoon M, Patil BS, Kim S (2011) Development of simple PCR-based markers linked to the *Ms* locus, a restorer-of-fertility gene in onion (*Allium cepa* L.). Euphytica 179:439–449

Bang H, Kim S, Park SO, Yoo KS, Patil BS (2013) Development of a codominant CAPS marker linked to the *Ms* locus controlling fertility restoration in onion (*Allium cepa* L.). Sci Hort 153:42–49

Bark OH, Havey MJ (1995) Similarities and relationships among open-pollinated populations of the bulb onion as estimated by nuclear RFLPs. Theor Appl Genet 90:607–614

Barnes SR, James AM, Jamieson G (1988) The organization, nucleotide sequence, and chromosomal distribution of a satellite DNA from *Allium cepa*. Chromosoma 92:185–192

Bennett MD, Smith JB (1976) Nuclear DNA amounts in angiosperms. Phil Trans R Soc Lond B 274:227–274

Berninger E (1965) Contribution a l'etude de la sterilite male de l'oignon (*Allium cepa* L.). Ann Amelior Plant 15:183–199

Berninger E, Buret P (1967) Etudes des deficients chlorophyllines chez deux especes cultivees du genre *Allium*: l'oignon (*A. cepa* L.) et le poireau (*A. porrum* L.). Ann Amelior Plant 17:175–194

Bradeen JM, Havey MJ (1995) Randomly amplified polymorphic DNA in bulb onion and its use to assess inbred integrity. J Amer Soc Hort Sci 120:752–758

Budylina MV, Kan LY, Romanov VS, Khrustaleva LI (2014) GISH study of advanced generation of the interspecific hybrids between *Allium cepa* L. and *Allium fistulosum* L. with relative resistance to downy mildew. Russian J Genet 50:387–394

Chope GA, Cools K, Hammond JP, Thompson AJ, Terry LA (2012) Physiological, biochemical and transcriptional analysis of onion bulbs during storage. Ann Bot 109:819–831

Courcel AGL, Vedel F, Boussac JM (1989) DNA polymorphism in *Allium cepa* cytoplasms and its implications concerning the origin of onions. Theor Appl Genet 77:793–798

Cramer CS, Havey MJ (1999) Morphological, biochemical, and molecular markers in onion. HortScience 34:589–593

Currah L, Ockendon DJ (1978) Protandry and the sequence of flower opening in the onion (*Allium cepa* L.). New Phytol 81:419–428

Damon S, Groves R, Havey MJ (2014) Quantitative trait loci controlling amounts and types of epicuticular waxes in onion. J Amer Soc Hort Sci 139:597–602

de Vries JN, Wietsma WA, Jongerius MC (1992) Introgression of characters from *Allium roylei* to *A. cepa* L. In: Hanelt P, Hammer K, Knupffer H (eds) The genus *Allium*—taxonomic problems and genetic resources, Gatersleben, Germany, pp 321–325

Duangjit J, Bohanec B, Chan AP, Town CT, Havey MJ (2013) Transcriptome sequencing to produce SNP-based genetic maps of onion. Theor Appl Genet 126:2093–2101

Duangjit J, Welsh K, Wise M, Bohanec B, Havey MJ (2014) Genetic analyses of anthocyanin concentrations and intensity of red-bulb color among segregating haploid progenies of onion. Mol Breed 34:75–85

Eickmeyer F, Primus-Kopper S, Wricke G (1990) Genetic analysis of isozyme loci in chives (*Allium schoenoprasum* L.). Z Pflanzenzüchtg (Plant Breed) 105:327–331

El-Shafie M, Davis G (1967) Inheritance of bulb color in *Allium cepa*. Hilgardia 9:607–622

Emsweller SL, Jones HA (1935) An interspecifc hybrid in *Allium*. Hilgardia 9:265–273

Engelke T, Tatlioglu T (2000) Mitochondrial genome diversity in connection with male sterility in *Allium schoenoprasum* L. Theor Appl Genet 100:942–948

Engelke T, Tatlioglu T (2003) A PCR marker system for monitoring the frequencies of normal and sterility-inducing cytoplasm types in German chive varieties. Z Pflanzenzüchtg (Plant Breed) 122:467–472

Engelke T, Tatlioglu T (2004) The fertility restorer genes X and T alter the transcripts of a novel mitochondrial gene implicated in CMS1 in chives (*Allium schoenoprasum* L.). Mol Genet Genomics 271:150–160

Engelke T, Terefe D, Tatlioglu T (2003) A PCR-based marker system monitoring CMS-(S), CMS-(T) and (N)-cytoplasm in the onion (*Allium cepa* L.). Theor Appl Genet 107:162–167

Fajkus P, Peška V, Sitová Z, Fulnečková J, Dvořáčková M, Gogela R, Sýkorová E, Hapala J, Fajkus J (2016) *Allium* telomeres unmasked: the unusual telomeric sequence (CTCGGTTATGGG)n is synthesized by telomerase. Plant J 85:337–347

Fischer D, Bachmann K (2000) Onion microsatellites for germplasm analysis and their use in assessing intra- and interspecific relatedness within the subgenus *Rhizirideum*. Theor Appl Genet 101:153–164

Fiskesjo G (1975) Chromosomal relationships between three species of *Allium* as revealed by C-banding. Hereditas 81:23–32

Friesen N, Klaas M (1998) Origin of some minor vegetatively propagated *Allium* crops studies with RAPD and GISH. Genet Res Crop Evol 45:511–523

Friesen N, Fritsch RM, Blattner FR (2006) Phylogeny and new intrageneric classification of *Allium* (Alliaceae) based on nuclear rDNA ITS sequences. In: Columbus JT, Friar EA, Hamilton CW, Porter JM, Prince LM, Simpson MG (eds) Monocots: comparative biology and evolution I. Aliso, vol 22, pp 372–395

Fuchs J, Brandes A, Schubert I (1995) Telomere sequence localization and karyotype evolution in higher plants. Plant Syst Evol 196:227–241

Galmarini CR, Goldman IL, Havey MJ (2001) Genetic analyses of correlated solids, flavor, and health-enhancing traits in onion (*Allium cepa* L.). Mol Gen Genomics 265:543–551

Galván Vivero GA (2009) Resistance to *Fusarium* basal rot and response to arbuscular mycorrhizal fungi in *Allium*. PhD thesis, Wageningen University, Holland, p 157

Galván GA, Koning-Boucoiran C, Koopman WJ, Burger-Meijer K, González PH, Waalwijk C, Kik C, Scholten OE (2008) Genetic variation among *Fusarium* isolates from onion, and resistance to *Fusarium* basal rot in related *Allium* species. Eur J Plant Path 121:499–512

Garcia S, Gálvez F, Gras A, Kovařík A, Garnatje T (2014) Plant rDNA database: update and new features. Database bau063

Gökçe AF, McCallum J, Sato Y, Havey MJ (2002) Molecular tagging of the *Ms* locus in onion. J Amer Soc Hort Sci 127:576–582

Hang T, Shigyo M, Yamauchi N, Tashiro Y (2014) Production and characterization of alien chromosome additions in shallot carrying extra chromosome(s) of Japanese bunching onion. Genes Genet Syst 79: 263–269

Havey MJ (1991a) Molecular characterization of the interspecific origin of viviparous onion. J Hered 82:501–503

Havey MJ (1991b) Phylogenetic relationships among cultivated *Allium* species from restriction enzyme analysis of the chloroplast genome. Theor Appl Genet 81:752–757

Havey MJ (1992a) A viviparous *Allium fistulosum*. Allium Improv Newsl 2:13–14

Havey MJ (1992b) Restriction enzyme analysis of the chloroplast and nuclear 45s ribosomal DNA of *Allium* sections *Cepa* and *Phyllodolon*. Plant Syst Evol 183:17–31

Havey MJ (1992c) Restriction enzyme analysis of the nuclear 45s ribosomal DNA of six cultivated Alliums. Plant Syst Evol 181:45–55

Havey MJ (1993) A putative donor of S-cytoplasm and its distribution among open-pollinated populations of onion. Theor Appl Genet 86:128–134

Havey MJ (1995) Identification of cytoplasms using the polymerase chain reaction to aid in the extraction of maintainer lines from open-pollinated populations of onion. Theor Appl Genet 90:263–268

Havey MJ (1999) Seed yield, floral morphology, and lack of male-fertility restoration of male-sterile onion (*Allium cepa*) populations possessing the cytoplasm of *Allium galanthum*. J Amer Soc Hort Sci 124:626–629

Havey MJ (2000) Diversity among male-sterility-inducing and male-fertile cytoplasms of onion. Theor Appl Genet 101:778–782

Havey MJ (2013) Single nucleotide polymorphisms in linkage disequilibrium with the male-fertility restoration (*Ms*) locus of onion. J Amer Soc Hort Sci 138:306–309

Havey MJ, Ahn YK (2016) Single nucleotide polymorphisms and indel markers from the transcriptome of garlic. J Amer Soc Hort Sci 141:62–65

Havey MJ, Bark OH (1994) Molecular confirmation that sterile cytoplasm has been introduced into open-pollinated Grano onion cultivars. J Amer Soc Hort Sci 119:90–93

Hirschegger P, Galmarini C, Bohanec B (2006) Characterization of a novel form of fertile great headed garlic (*Allium* sp.). Plant Breed 125:635–637

Hirschegger P, Jakse J, Trontelj P, Bohanec B (2010) Origins of *Allium ampeloprasum* horticultural groups and a molecular phylogeny of the section *Allium* (*Allium*: Alliaceae). Mol Phylog Evol 54:488–497

Hizume M (1994) Allodiploid nature of *Allium wakegi* Araki revealed by genomic in situ hybridization and localization of 5S and 18S rDNAs. Jpn J Genet 69:407–415

Holford P, Croft JH, Newbury HJ (1991) Differences between, and possible origins of, the cytoplasms found in fertile and male-sterile onions (*Allium cepa* L.). Theor Appl Genet 82:737–744

Hou A, Peffley EB (2000) Recombinant chromosomes of advanced backcross plants between *Allium cepa* L. and *A. fistulosum* L. revealed by *in situ* hybridization. Theor Appl Genet 100:1190–1196

Huang DQ, Yang J-T, Zhou C-J, Zhou S-D, He XJ (2014) Phylogenetic reappraisal of *Allium* subgenus *Cyathophora* (Amaryllidaceae) and related taxa, with a proposal of two new sections. J Plant Res 127:275–286

Ipek M, Ipek A, Simon PW (2003) Comparison of AFLPs, RAPD markers, and isozymes for diversity assessment of garlic and detection of putative duplicates in germplasm collections. J Amer Soc Hort Sci 128:246–252

Ipek M, Ipek A, Almquist SG, Simon PW (2005) Demonstration of linkage and development of the first low-density genetic map of garlic based on AFLP markers. Theor Appl Genet 110:228–236

Jakše J, Martin W, McCallum J, Havey MJ (2005) Single nucleotide polymorphisms, indels, and simple sequence repeats for onion cultivar identification. J Amer Soc Hort Sci 130:912–917

Jakše J, Meyer JDF, Suzuki G, McCallum J, Cheung F, Town CD, Havey MJ (2008) Pilot sequencing of onion genomic DNA reveals fragmented transposable elements, low gene densities, and significant gene enrichment after methyl filtration. Mol Genet Genomics 280:287–292

Jones HA (1923) Pollination and self-fertility in the onion. Proc Amer Soc Hort Sci 20:191–197

Jones HA, Clarke AE (1942) A natural amphidiploid from an onion species hybrid (*Allium cepa* L. x *Allium fistulosum* L.). J Hered 23:25–32

Jones HA, Clarke AE (1943) Inheritance of male sterility in the onion and the production of hybrid seed. Proc Amer Soc Hort Sci 43:189–194

Jones HA, Davis G (1944) Inbreeding and heterosis and their relation to the development of new varieties of onions. USDA Tech Bull No 874

Jones HA, Peterson C (1952) Complementary factors for light-red bulb color in onions. Proc Amer Soc Hort Sci 59:457

Jones RN, Rees H (1968) Nuclear DNA variation in *Allium*. Heredity 23:591–605

Jones HA, Clarke AE, Stevenson F (1944) Studies in the genetics of the onion (*Allium cepa* L.). Proc Amer Soc Hort Sci 44:479–484

Kamenetsky R, Faigenboim A, Mayer ES, Michael TB, Gershberg C, Kimhi S, Esquira I, Shalom SR, Eshel D, Rabinowitch HD, Sherman A (2015) Integrated transcriptome catalogue and organ-specific profiling of gene expression in fertile garlic (*Allium sativum* L.). BMC Genom 16:12

Katayama H, Sasakuma T, Ogihara Y (1991) Physical map of chloroplast DNA of the onion *Allium cepa* L., showing the location of photosynthesis-related genes. Jap J Genet 66:421–431

Khar A, Jakše J, Havey MJ (2008) Segregations for onion-bulb colors reveal that red is controlled by at least three loci. J Amer Soc Hort Sci 133:42–47

Khar A, Lawande KE, Negi KS (2011) Microsatellite marker based analysis of genetic diversity in short day tropical Indian onion and cross amplification in related *Allium* spp. Genet Res Crop Evol 58:741–752

Khrustaleva LI, Kik C (1998) Cytogenetical studies in the bridge cross *Allium cepa* × (*A. fistulosum* × *A. roylei*). Theor Appl Genet 96:8–14

Khrustaleva LI, Kik C (2000) Introgression of *Allium fistulosum* into *A. cepa* mediated by *A. roylei*. Theor Appl Genet 100:17–26

Khrustaleva L, de Melo PE, van Heusden AW, Kik C (2005) The integration of recombination and physical maps in a large-genome monocot using haploid genome analysis in a trihybrid *Allium* population. Genetics 169:1673–1685

Khrustaleva L, Jiang J, Havey MJ (2016) High-resolution tyramide-FISH mapping of markers tightly linked to the male-fertility restoration (*Ms*) locus of onion. Theor Appl Genet 129:535–545

Kik C, Samoylov AM, Verbeek WHJ, Raamsdonk LWD (1997) Mitochondrial DNA variation and crossability of leek (*Allium porrum*) and its wild relatives from the *Allium ampeloprasum* complex. Theor Appl Genet 94:465–471

Kim S (2014) A codominant molecular marker in linkage disequilibrium with a restorer-of-fertility gene (*Ms*) and its application in reevaluation of inheritance of fertility restoration in onions. Mol Breed 34:769–778

Kim S, Yoon M (2010) Comparison of mitochondrial and chloroplast genome segments from three onion (*Allium cepa* L.) cytoplasm types and identification of a trans-splicing intron of *cox2*. Curr Genet 56:177–188

Kim S, Binzel ML, Park S, Yoo KS, Pike LM (2004a) Inactivation of DFR (Dihydroflavonol 4-reductase) gene transcription results in blockage of anthocyanin production in yellow onions (*Allium cepa*). Mol Breed 14:253–263

Kim S, Binzel ML, Yoo KS, Park S, Pike LM (2004b) *Pink* (*P*), a new locus responsible for a pink trait in onions (*Allium cepa*) resulting from natural mutations of anthocyanidin synthase. Mol Genet Genomics 272:18–27

Kim S, Jones R, Yoo K, Pike LM (2004c) Gold color in onions (*Allium cepa*), a natural mutation of the chalcone isomerase gene harboring a pre-mature stop codon. Mol Gen Genomics 272:411–419

Kim S, Jones R, Yoo KS, Pike LM (2005a) The *L* locus, one of complementary genes required for anthocyanin production in onions (*Allium cepa*), encodes anthocyanidin synthase. Theor Appl Genet 111:120–127

Kim S, Yoo KS, Pike LM (2005b) Development of a co-dominant, PCR-based marker for allelic selection of the pink trait in onions (*Allium cepa*), based on the insertion mutation in the promoter of the anthocyanidin synthase gene. Theor Appl Genet 110:573–578

Kim S, Yoo KS, Pike LM (2005c) Development of a PCR-based marker utilizing a deletion mutation in the dihydroflavonol 4-reductase (DFR) gene responsible for the lack of anthocyanin production in yellow onions (*Allium cepa*). Theor Appl Genet 110:588–595

Kim S, Bang H, Yoo KS, Park S, Pike LM (2006) Identification of the fourth allele of the ANS (anthocyanidin synthase) gene and its effect on red color intensity in onions (*Allium cepa*). Euphytica 149:45–51

Kim S, Baek D, Cho DY, Lee E, Yoon M (2009a) Identification of two novel inactive DFR-A alleles responsible for failure to produce anthocyanin and development of a simple PCR-based molecular marker for bulb color selection in onion (*Allium cepa* L.). Theor Appl Genet 118:1391–1399

Kim S, Lee E, Cho DY, Han T, Bang H, Patil BS, Ahn YK, Yoon M (2009b) Identification of a novel chimeric gene, *orf725*, and its use in development of a molecular marker for distinguishing among three cytoplasm types in onion (*Allium cepa* L.). Theor Appl Genet 118:433–441

Kim S, Kim CW, Park M, Choi D (2015a) Identification of candidate genes associated with fertility restoration of cytoplasmic male-sterility in onion (*Allium cepa* L.) using a combination of bulked segregant analysis and RNA-seq. Theor Appl Genet 128:2289–2299

Kim S, Kim MS, Kim YM, Yeom SI, Cheong K (2015b) Integrative structural annotation of de novo RNA-Seq provides an accurate reference gene set of the enormous genome of the onion (*Allium cepa* L.). DNA Res 22:19–27

Kim S, Park JY, Yang T (2015c) Comparative analysis of the complete chloroplast genome sequences of a normal male-fertile cytoplasm and two different cytoplasms conferring cytoplasmic male sterility in onion (*Allium cepa* L.). J Hort Sci Biotech 90:459–468

Kim S, Park JY, Yang TJ (2015d) Characterization of three active transposable elements recently inserted in three independent *DFR-A* alleles and one high-copy DNA transposon isolated from the *Pink* allele of the *ANS* gene in onion (*Allium cepa* L.). Mol Genet Genomics 290:1027–1037

Kim B, Kim K, Yang TJ, Kim S (2016a) Completion of the mitochondrial genome sequence of onion (*Allium cepa* L.) containing the CMS-S male-sterile cytoplasm and identification of an independent event of the ccmFN gene split. Curr Genet 62:873–885

Kim EY, Kim CW, Kim S (2016b) Identification of two novel mutant ANS alleles responsible for inactivation of anthocyanidin synthase and failure of anthocyanin production in onion (*Allium cepa* L.). Euphytica 212:427–437

King JJ, Bradeen JM, Bark OH, McCallum JA, Havey MJ (1998) A low-density genetic map of onion reveals a role for tandem duplication in the evolution of an extremely large diploid genome. Theor Appl Genet 96:52–62

Kirk JTO, Rees H, Evans G (1970) Base composition of nuclear DNA within the genus *Allium*. Heredity 11:507–512

Kirov I, Kiseleva AV, Van Laere K, Van Roy N, Khrustaleva LI (2017) Tandem repeats of *Allium fistulosum* associated with major chromosomal landmarks. Mol Genet Genomics 292:453–464

Klaas M (1998) Applications and impact of molecular markers on evolutionary and diversity studies in the genus *Allium*. Z Pflanzenzüchtg (Plant Breed) 117:297–308

Kofoet A, Zinkernagel V (1990) Resistance to downy mildew (*Peronospora destructor*) in *Allium* species. J Plant Dis Protect 97:13–23
Koul AK, Gohil RN (1970) Cytology of the tetraploid *Allium ampeloprasum* with chiasma localization. Chromosoma 29:12–19
Koul AK, Gohil RN (1971) Further studies on natural triploidy in viviparous onion. Cytologia 36:253–261
Kuhl JC, Cheung F, Yuan Q, Martin W, Zewdie Y, McCallum J, Catanach A, Rutherford P, Sink KC, Jenderek M, Prince JP, Town CD, Havey MJ (2004) A unique set of 11,008 onion (*Allium cepa*) ESTs reveals expressed sequence and genomic differences between monocot orders Asparagales and Poales. Plant Cell 16:114–125
Labani R, Elkington T (1987) Nuclear DNA variation in the genus *Allium* L. (Liliaceae). Heredity 59:119–128
Li Q, Zhou S, He X, Yu Y, Zhang Y, Wei X (2010) Phylogeny and biogeography of *Allium* based on nuclear ribosomal internal transcribed spacer and chloroplast *rps16* sequences, focusing on the inclusion of species endemic to China. Ann Bot 106:709–733
Li QQ, Zhou SD, Huang DQ, He XJ, Wei XQ (2016) Molecular phylogeny, divergence time estimates and historical biogeography within one of the world's largest monocot genera. AoB Plants 8: plw041. https://doi.org/10.1093/aobpla/plw041
Liu Q, Wen C, Zhao H, Zhang L, Wang J, Wang Y (2014) RNA-seq reveals leaf cuticular wax-related genes in Welsh onion. PLoS ONE 9:e113290
Maaβ HI (1997) Studies on triploid viviparous onions and their origin. Genet Res Crop Evol 44:95–99
Maeda T (1937) Chiasma studies in *Allium fistulosum* L., *Allium cepa* L. and their F_1, F_2 and backcross hybrids. Jap J Genet 13:146–169
Maggini F, Garbari F (1977) Amounts of ribosomal DNA in *Allium* (Liliaceae). Plant Syst Evol 128:201–208
Maggini F, Barsanti P, Marazia T (1978) Individual variation of the nucleolus organizer regions in *Allium cepa* and *Allium sativum*. Chromosoma 66:173–183
Martin W, McCallum J, Shigyo M, Jakse J, Kuhl JC, Yamane N, Joyce MP, Gokce AF, Sink KC, Town CD, Havey MJ (2005) Genetic mapping of expressed sequences in onion and in silico comparisons show scant colinearity with rice. Mol Genet Genomics 274:197–204
Masamura N, McCallum J, Khrustaleva L, Kenel F, Pither-Joyce M, Shono J, Suzuki G, Mukai Y, Yamauchi N, Shigyo M (2012) Chromosomal organization and sequence diversity of genes encoding lachrymatory factor synthase in *Allium cepa* L. G3: Genes, Genomes, Genetics 2:643–651
Masuzaki S, Araki N, Yamauchi N, Yamane N, Wako T, Kojima A, Shigyo M (2006) Chromosomal locations of microsatellites in onion. HortScience 41:315–318
McCallum J, Leite D, Pither-Joyce M, Havey MJ (2001) Expressed sequence markers for genetic analysis of bulb onion (*Allium cepa*). Theor Appl Genet 103: 979–991
McCallum J, Thomson S, Pither-Joyce M, Kenel F, Clarke A, Havey MJ (2008) Genetic diversity analysis and single-nucleotide-polymorphism marker development in cultivated bulb onion based on expressed sequence tag–simple sequence repeat markers. J Amer Soc Hort Sci 133:810–818
Moue T, Uehara T (1985) Inheritance of cytoplasmic male sterility in *Allium fistulosum* L. (Welsh onion). J Jpn Soc Hortic Sci 53:432–437
Narayan RKJ (1988) Constraints upon the organisation and evolution of chromosomes in *Allium*. Theor Appl Genet 75:319–329
Ohara T, Song YS, Tsukazaki H, Wako T, Nunome T, Kojima A (2005) Genetic mapping of AFLP markers in Japanese bunching onion (*Allium fistulosum*). Euphytica 144:255–263
Ohri D, Pistrick K (2001) Phenology and genome size variation in *Allium* L.—a tight correlation? Plant Biol 3:654–660
Ohri D, Fritsch RM, Hanelt P (1998) Evolution of genome size in *Allium* (Alliaceae). Plant Syst Evol 210:57–86
Ohsumi C, Kojima A, Kinata K, Etoh T, Hayashi T (1993) Interspecific hybrid between *Allium cepa* and *Allium sativum*. Theor Appl Genet 85:969–975
Park J, Bang H, Cho DY, Yoon M, Patil BS, Kim S (2013) Construction of high-resolution linkage map of the *Ms* locus, a restorer-of-fertility gene in onion (*Allium cepa* L.). Euphytica 192:267–278
Pearce SR, Pich U, Harrison G, Flavell AJ, Heslop-Harrison J, Schubert I, Kumar A (1996) The Ty1-copia group retrotransposons of *Allium cepa* are distributed throughout the chromosomes but are enriched in the terminal heterochromatin. Chrom Res 4:357–364
Peffley EB (1986) Evidence for chromosomal differentiation of *Allium fistulosum* and *Allium cepa*. J Amer Soc Hort Sci 111:126–129
Peffley EB, Hou A (2000) Bulb type onion introgressants possessing *Allium fistulosum* L. genes recovered from interspecific hybrid backcrosses between *A. cepa* L. and *A. fistulosum* L. Theor Appl Genet 100:528–538
Peffley EB, Mangum PD (1990) Introgression of *Allium fistulosum* L. into *Allium cepa* L.: cytogenetic evidence. Theor Appl Genet 79:113–118
Perkins D, Kehr A, Brown R, Tims E, Miller J (1958) Delta Giant: A new long season shallot. Louisiana Exp, Station Circular No, p 52
Pich U, Schubert I (1998) Terminal heterochromatin and alternative telomeric sequences in Allium cepa. Chrom Res 6:315–321
Pich U, Fuchs J, Schubert I (1996) How do alliaceae stabilize their chromosome ends in the absence of TTTAGGG sequence? Chrom Res 4:207–213
Potz H, Tatlioglu T (1993) Molecular analysis of cytoplasmic male sterility in chives (*Allium schoenoprasum* L.). Theor Appl Genet 87:439–445
Puizina J, Papes D (1996) Cytogenetical evidence for hybrid structure and origin of diploid and triploid

shallots (*Allium cepa* var *viviparum*, Lilliaceae) from Dalmatia (Croatia). Plant Syst Evol 203:203–215

Puizina J, Javornik B, Bohanec B, Schweizer D, Maluszynska J, Papes D (1999) Random amplified polymorphic DNA analysis, genome size, and genomic in situ hybridization of triploid viviparous onions. Genome 42:1208–1216

Ranjekar PK, Pallotta D, Lafontaine JG (1978) Analysis of plant genomes. V. Comparative study of molecular properties of DNAs of seven *Allium* species. Biochem Genet 16:957–970

Ricroch A, Peffley EB, Baker RJ (1992) Chromosomal location of rDNA in *Allium*: in situ hybridization using biotin- and fluorescein-labelled probe. Theor Appl Genet 83:413–418

Ricroch A, Yockteng R, Brown SC, Nadot S (2005) Evolution of genome size across some cultivated *Allium* species. Genome 48:511–520

Rieman GH (1931) Genetic factors for pigmentation in the onion and their relation to disease resistance. J Agr Res 42:251–278

Romanov D, Divashuk M, Havey MJ, Khrustaleva L (2015) Tyramide-FISH mapping of single genes for development of an integrated recombination and cytogenetic map of chromosome 5 of *Allium cepa* L. Genome 58:111–119

Sato Y (1998) PCR amplification of CMS-specific mitochondrial nucleotide sequences to identify cytoplasmic genotypes of onion (*Allium cepa* L.). Theor Appl Genet 96:367–370

Satoh Y, Nagai M, Mikami T, Kinoshita T (1993) The use of mitochondrial DNA polymorphism in the classification of individual onion plants by cytoplasmic genotypes. Theor Appl Genet 86:345–348

Scholten OE, Van Heusden AW, Khrustaleva LI, Burger-Meijer K, Mank RA, Antonise GC, Harrewijn JL, Van Haecke W, Oost EH, Peters RJ, Kik C (2007) The long and winding road leading to the successful introgression of downy mildew resistance into onion. Euphytica 156:345–353

Schubert I, Wobus U (1985) In situ hybridization confirms jumping nucleolus organizing regions in *Allium*. Chromosoma 92:143–148

Schubert I, Ohle H, Hanelt P (1983) Phyogenetic conclusions from Giemsa banding and NOR staining in top onions (Liliaceae). Plant Syst Evol 143:245–256

Schweisguth B (1973) Etude d'un nouveau type de sterilite male chez l'oignon, *Allium cepa* L. Ann Amelior Plantes 23:221–233

Schwinn KE, Ngo H, Kenel F, Brummell DA, Albert NW, McCallum JA, Pither-Joyce M, Crowhurst RN, Eady C, Davies KM (2016) The onion (*Allium cepa* L.) *R2R3-MYB* gene *MYB1* regulates anthocyanin biosynthesis. Front Plant Sci 7:1865

Shibata F, Hizume M (2002a) Evolution of 5S rDNA units and their chromosomal localization in *Allium cepa* and *Allium schoenoprasum* revealed by microdissection and FISH. Theor Appl Genet 105:167–172

Shibata F, Hizume M (2002b) The identification and analysis of the sequences that allow the detection of *Allium cepa* chromosomes by GISH in the allodiploid *A. wakegi*. Chromosoma 111:184–191

Shigyo M, Tashiro Y, Isshiki S, Miyazaki S (1996) Establishment of a series of alien monosomic addition lines of Japanese bunching onion (*Allium fistulosum* L.) with extra chromosomes from shallot (*A. cepa* L. Aggregatum group). Genes Genet Syst 71:363–371

Shigyo M, Imamura K, Iino M, Yamashita K, Tashiro Y (1998) Identification of alien chromosomes in a series of *Allium fistulosum*—*A. cepa* monosomic addition lines by means of genomic in situ hybridization. Genes Genet Syst 73:311–315

Sohn S, Ahn YK, Lee TH, Lee JE, Jeong MH, Seo CH, Chandra R, Kwon YS, Kim CW, Kim DS, Won SY, Kim JS, Choi D (2016) Construction of a draft reference transcripts of onion (*Allium cepa*) using long-read sequencing. Plant Biotech Rep 10:383

Song Y, Suwabe K, Wako T, Ohara T, Nunome T, Kojima A (2004) Development of microsatellite markers in bunching onion (*Allium fistulosum* L.). Breeding Sci 54:361–365

Song S, Kim CW, Moon JS, Kim S (2014) At least nine independent natural mutations of the DFR-A gene are responsible for appearance of yellow onions (*Allium cepa* L.) from red progenitors. Mol Breed 33:173–186

Stack SM, Comings DE (1979) The chromosomes and DNA of *Allium cepa*. Chromosoma 70:161–181

Stack SM, Roelofs D (1996) Localized chiasmata and meiotic nodules in the tetraploid onion *Allium porrum*. Genome 39:770–783

Stevenson M, Armstrong SJ, Ford Lloyd BV, Jones GH (1998) Comparative analysis of crossover exchanges and chiasmata in *Allium cepa* × *fistulosum* after genomic in situ hybridization (GISH). Chrom Res 6:567–574

Suzuki G, Ura A, Saito N, Do G, Seo B, Yamamoto M, Mukai Y (2001) BAC FISH analysis of *Allium cepa*. Genes Genet Syst 76:251–255

Suzuki G, Ogaki Y, Hokimoto N, Xiao L, Kikuchi-Taura A, Harada C, Okayama R, Tsuru A, Onishi M, Saito N, Do GS, Lee SH, Ito T, Kanno A, Yamamoto M, Mukai Y (2012) Random BAC FISH of monocot plants reveals differential distribution of repetitive DNA elements in small and large chromosome species. Plant Cell Rep 31:621–628

Tatlioglu T (1982) Cytoplasmic male sterility in chives (*Allium schoenoprasum* L.). Z Pflanzenzüchtg 89:251–262

Tatlioglu T (1987) Genetic control of temperature-sensitivity of cytoplasmic male sterility (cms) in chives (*Allium schoenoprasum* L.). Plant Breed 99: 65–76

Tatlioglu T, Wricke G (1988) Genetic control of tetracycline sensitivity of cytoplasmic male sterility (cms) in chives (*Allium schoenoprasum* L.). Plant Breed 100:34–40

Tsukazaki H, Nunome T, Fukuoka H, Kanamori H, Kono I, Yamashita K, Wako T, Kojima A (2007) Isolation of 1,796 SSR clones from SSR-enriched DNA libraries of bunching onion (*Allium fistulosum*). Euphytica 157:83–94

Tsukazaki H, Yamashita K, Yaguchi S, Masuzaki S, Fukuoka H, Yonemaru J, Kanamori H, Kono I, Hang TTM, Shigyo M, Kojima A, Wako T (2008) Construction of SSR-based chromosome map in bunching onion (*Allium fistulosum*). Theor Appl Genet 117:1213–1223

Tsukazaki H, Yaguchi S, Sato S, Hirakawa H, Katayose Y, Kanamori H, Kurita K, Itoh T, Kumagai M, Mizuno S, Hamada M, Fukuoka H, Yamashita K, McCallum JA, Shigyo M, Wako T (2015) Development of transcriptome shotgun assembly-derived markers in bunching onion (*Allium fistulosum*). Mol Breeding 35:55

Umehara M, Sueyoshi T, Shimomura K, Iwai M, Shigyo M, Hirashima K, Nakahara T (2006) Interspecific hybrids between *Allium fistulosum* and *Allium schoenoprasum* reveal carotene-rich phenotype. Euphytica 148:295–301

Van Heusden AW, Ooijen JW, Vrielink-Van, Ginkel R, Verbeek WHJ, Wietsma WA, Kik C (2000) A genetic map of an interspecific cross in *Allium* based on amplified fragment length polymorphism (AFLP) markers. Theor Appl Genet 100:118–126

van Raamsdonk LWD, Wietsma WA, Vries JN (1992) Crossing experiments in *Allium* L. section *Cepa*. Bot J Linnean Soc 109:293–303

Ved Brat S (1965) Genetic systems in *Allium*. Chromosoma 16:486–499

Vitte C, Estep MC, Leebens-Mack J, Bennetzen JL (2013) Young, intact and nested retrotransposons are abundant in the onion and asparagus genomes. Ann Botany 112:881–889

Volk GM, Henk AD, Richards CM (2004) Genetic diversity among U.S. garlic clones as detected using AFLP methods. J Amer Soc Hort Sci 129:559–569

Von Kohn C, Kiełkowska A, Havey MJ (2013) Sequencing and annotation of the chloroplast DNAs of normal (N) male-fertile and male-sterile (S) cytoplasms of onion and single nucleotide polymorphisms distinguishing these cytoplasms. Genome 56:737–742

Vu HQ, Iwata M, Yamauchi N, Shigyo M (2011) Production of novel alloplasmic male sterile lines in *Allium cepa* harbouring the cytoplasm from *Allium roylei*. Plant Breed 130:469–475

Vu HQ, Yoshimatsu Y, Khrustaleva LI, Yamauchi N, Shigyo M (2012) Alien genes introgression and development of alien monosomic addition lines from a threatened species, *Allium roylei* Stearn, to *Allium cepa* L. Theor Appl Genet 124:1241–1257

Wilkie SE, Isaac PG, Slater RJ (1993) Random amplified polymorphic DNA (RAPD) markers for genetic analysis in *Allium*. Theor Appl Genet 86:497–504

Yamashita K, Tashiro Y (1999) Possibility of developing a male sterile line of shallot (*Allium cepa* L. Aggregatum group) with cytoplasm from *A. galanthum* Kar. et Kir. J Jap Soc Hort Sci 68:256–262

Yamashita K, Tashiro Y (2001) RFLP analysis of mitochondrial DNA in Wakegi onion. J Jap Soc Hort Sci 70:232–234

Yamashita K, Tashiro Y (2004) Seed productivity test of CMS lines of Japanese bunching onion (*Allium fistulosum* L.) possessing the cytoplasm of a wild species, *A. galanthum* Kar. et Kir. Euphytica 136:327–331

Yamashita K, Arita H, Tashiro Y (1999) Cytoplasm of a wild species, *Allium galanthum* Kar. et Kir., is useful for developing the male sterile line of *A. fistulosum* L. J Jap Soc Hort Sci 68:788–797

Yamashita K, Iino M, Shigyo M, Tashiro Y (2000) Visualization of nucleus substitution between *Allium galanthum* and shallot (*A. cepa*) by genomic in situ hybridization. J Jap Soc Hort Sci 69:189–191

Yamashita K, Oyama T, Noda R, Miyazaki T, Isshiki S, Tashiro Y (2001) Phylogenetic relationships among cultivated and wild species in section *Cepa* of *Allium* based on RFLPs of mitochondrial and chloroplast DNAs. J Jap Soc Hort Sci 70:195–201

Yamashita K, Takatori Y, Tashiro Y (2005) Chromosomal location of a pollen fertility-restoring gene, Rf, for CMS in Japanese bunching onion (*Allium fistulosum* L.) possessing the cytoplasm of *A. galanthum* Kar. et Kir. revealed by genomic *in situ* hybridization. Theor Appl Genet 111:15–22

Yamashita K, Tsukazaki H, Kojima A, Ohara T, Wako T (2010) Inheritance mode of male sterility in bunching onion (*Allium fistulosum* L.) accessions. Euphytica 173:357–367

Yang Y, Huo Y, Miao J, Liu B, Kong S, Gao L, Liu C, Wang Z, Tahara Y, Kitano H, Wu X (2012) Identification of two SCAR markers co-segregated with the dominant Ms and recessive ms alleles in onion (*Allium cepa* L.). Euphytica 190:267–277

Zewdie Y, Havey MJ, Prince JP, Jenderek MM (2005) The first genetic linkages among expressed regions of the garlic (*Allium sativum* L.) genome. J Amer Soc Hort Sci 130:569–574

Zhao W, Chung J, Lee G, Ma K, Kim H, Kim K, Chung M, Lee J, Kim N, Kim S, Park Y (2011) Molecular genetic diversity and population structure of a selected core set in garlic and its relatives using novel SSR markers. Plant Breed 130:46–54

Future Perspective

15

Mostafa Abdelrahman, Anil Khar and Masayoshi Shigyo

Abstract

Allium, an enormous genus, contains hundreds of species with significant economics, culinary, and medicinal value. Several *Allium* species, including onion, shallot, garlic, leek, and Japanese bunching onion are extensively consumed, but genetically remain poorly understood. *Allium* species are famous for their large genomes, mostly in the range of 10–20 Gbps, which have complicated the genomic studies and genetic map development. Recent growth in NGS technology is making the whole genome sequencing easier than before and the ongoing progress for completing *Allium cepa* genome sequencing is being expected through Sequon-Onion Genome Sequencing Project (http://www.onion-genome.net/). In addition, transcriptome analyses have been widely used in various *Allium* species and an open-access *Allium* transcriptome database (http://alliumtbd.kazusa.or.jp) has recently been released, which contains valuable transcriptome and sequence information of onion, shallot, Japanese bunching onion, and *Allium roylei*. Progress in *Allium* metabolomics is still slow in comparison with other vegetable crops. Therefore, extensive efforts are needed in this area, which can bridge the gap between phenotype and genotype. In this future perspective chapter, we have given a brief introduction to the current understandings of *Allium* omics as well as genetic resources, and provide perspective on future directions for research in this field.

M. Abdelrahman
Laboratory of Molecular Breeding, Arid Land Research Center, Tottori University, 1390, Hamasaka, Tottori-shi, Tottori 680-0001, Japan
e-mail: meettoo2000@tottori-u.ac.jp

A. Khar
Division of Vegetable Science, ICAR-Indian Agricultural Research Institute, Pusa 110012, New Delhi, India
e-mail: anilkhar@iari.res.in

M. Shigyo (✉)
Laboratory of Vegetable Crop Science, College of Agriculture, Graduate School of Sciences and Technology for Innovation, Yamaguchi University, Yamaguchi-City Yamaguchi 753-8515, Japan
e-mail: shigyo@yamaguchi-u.ac.jp

Since the "Green Revolution" occurred between the 1930s and 1960s, research into fundamental aspects of plant growth and development, and stress physiology have revolutionized global food production. However, as the global climate becomes increasingly unpredictable and the global population continues to rise, it is clear that the need for sustainable food production will push current crop cultivars and modern agriculture production methods to the edge in the near future. Advances in understanding plant biology and

M. Shigyo et al. (eds.), *The Allium Genomes*, Compendium of Plant Genomes,
https://doi.org/10.1007/978-3-319-95825-5_15

omics technologies generate new solutions for food security under changing environmental conditions. New germplasm and gene candidates that are projected to improve crop yields and other agronomic traits under stress have to pass long development phases based on large-scale field evaluation. Therefore, the combination of novel molecular tools, automated screening methods, and economic assessment combined with decision-making algorithms should be the key goal for crop genetic improvement in the future.

Although significant progress has been made in *Allium* genomics, there is still a considerable gap between the phenotype and the genotype; a gap which must be linked in order to breed elite cultivars suitable for sustainable agriculture under unpredicted and dynamic global climate changes. The genus *Allium* is one of the most economically important monocot families which include several important crop plants such as bulb onion, garlic, shallot, and Japanese bunching onion. Bulb onion and garlic are consumed daily all over the world because of their cuisinal and medicinal properties. Bulb onion is characterized by a large diploid genome (haploid genome size is about 15,000 Mb) and unique telomeric sequences with lower GC content (~36.8% of onion DNA) in the coding region. Onion genome is five times greater than the human genome, and 120 times greater than *Arabidopsis*, which makes complete genome sequencing of onion a significant challenge. Enormous genome-size differences exist among closely related *Allium* species. For example, bulb onion has 28% more nuclear DNA than *A. fistulosum*, whereas garlic has 7% smaller nuclear DNA than bulb onion. Recent sequence studies of bulb onion BACs demonstrated that the majority of the sequences were similar to retroelement and transposons, indicating that onion genome exhibited low gene densities and high frequencies of repetitive DNA as major components of large chromosomes. The BAC clones of onion developed as a partial library can be used not only for molecular cloning of novel onion genes but also can be used as FISH probes in molecular cytogenetic studies.

Although the availability of the structural genomic resources in *Allium* species has been restricted to date, the recent rapid progress of NGS technology is making the whole genome sequencing remarkably easier than before. Actually, genome sequencing of *A. cepa* is now ongoing through SEQUON project (http://www.oniongenome.net) by using Illumina HiSeq 2500 sequencer. In addition, transcriptome sequencing is an efficient approach to generate sequence information from expressed regions of the genome and has been widely used in the *Alliums*. For example, transcriptome sequencing has been completed for garlic, Japanese bunching onion, and bulb onion, and open access *Allium* transcriptome database (http://alliumtdb.kazusa.or.jp) was recently released which contained a valued transcriptome and sequence information of *A. roylei*, shallot, and bulb onion. These efforts have produced reference transcriptomes for these *Alliums*. In the future, a coordinated effort is necessary to sequence cDNAs from a variety of tissues or treatments so that a more global view of gene expression can be developed. Future sequencing efforts should concentrate on doubled haploid lines to ensure complete genetic uniformity of the lines, and for mining of the sequences for SSRs and SNPs. An international *Allium* research community must work together to ensure that independent sequencing efforts are coordinated and results available to the community will enable researchers to identify causal mutations for major phenotypes. The output information will enable breeders to select desirable phenotypes to improve selection efficiency, toward the production of new, value-added cultivars for growers, processors, and consumers.

The introduction and further development of web-based genebanking information systems (e.g., GENESYS, PLANT SEARCH) has facilitated the exchange of data to a large extent between *Allium* collection holders worldwide and provided an overview of the *Allium* genetic resources managed worldwide. However, a significant gap still exists in *Allium* collections worldwide, especially with respect to the wild crossable relatives of the various *Allium* crops, and thus, a more contribution from collection holders is needed in the coming future. In addition, significant future progress in *Allium* genetic resources maintenance is required. The development of new methods for the preservation of *Allium* germplasm through the introduction of cryopreservation will enable the maintenance of *Allium* accessions in a cheap and efficient way. The cryopreservation method is important for the conservation of vegetatively maintained germplasm. Fundamental changes could also be observed with respect to acquisition and exchange of plant genetic resources everywhere in the world due to many and complex new regulations on the legal and organizational levels. Therefore, we need to underline that the threats of genetic erosion and extinction due to the disappearance of traditional cultivation methods, the devastation of our environment and climatic change, the conservation of genetic resources remains of prime importance for agriculture. Together with a number of the brightest young generations all over the world, a highly integrated initiative that brings together scientists and natural genetic resources worldwide is a necessary step and perhaps the only practical and efficient way to identify the functional diversity of alleles for agriculturally useful genes from the wild/landrace progenitors, and to apply the findings to *Allium* crop genetic improvement.

MIX
Papier aus verantwortungsvollen Quellen
Paper from responsible sources
FSC® C105338

If you have any concerns about our products,
you can contact us on
ProductSafety@springernature.com

In case Publisher is established outside the EU,
the EU authorized representative is:
Springer Nature Customer Service Center GmbH
Europaplatz 3, 69115 Heidelberg, Germany

Printed by Libri Plureos GmbH
in Hamburg, Germany